Mining and Communities in Northern Canada

Canadian History and Environment Series
Alan MacEachern, Series Editor
ISSN 1925-3702 (Print) ISSN 1925-3710 (Online)

The Canadian History & Environment series of edited collections brings together scholars from across the academy and beyond to explore the relationships between people and nature in Canada's past. Published simultaneously in print and open-access form, the series then communicates that scholarship to the world.

ALAN MACEACHERN, FOUNDING DIRECTOR
NiCHE: Network in Canadian History & Environment
Nouvelle initiative canadienne en histoire de l'environnement
http://niche-canada.org

No. 1 · **A Century of Parks Canada, 1911–2011**
Edited by Claire Elizabeth Campbell

No. 2 · **Historical GIS Research in Canada**
Edited by Jennifer Bonnell and Marcel Fortin

No. 3 · **Mining and Communities in Northern Canada: History, Politics, and Memory**
Edited by Arn Keeling and John Sandlos

UNIVERSITY OF CALGARY
Press

Mining and Communities in Northern Canada

History, Politics, and Memory

EDITED BY | Arn Keeling and John Sandlos

Canadian History and Environment Series
ISSN 1925-3702 (Print) ISSN 1925-3710 (Online)

© 2015 Arn Keeling and John Sandlos

University of Calgary Press
2500 University Drive NW
Calgary, Alberta
Canada T2N 1N4
www.uofcpress.com

This book is available as an ebook which is licensed under a Creative Commons license. The publisher should be contacted for any commercial use which falls outside the terms of that license.

LIBRARY AND ARCHIVES CANADA CATALOGUING IN PUBLICATION

    Mining and communities in Northern Canada : history, politics, and memory / edited by Arn Keeling and John Sandlos.

(Canadian history and environment series, ISSN 1925-3702 ; no. 3)
Includes bibliographical references and index.
Issued in print and electronic formats.
ISBN 978-1-55238-804-4 (paperback).—ISBN 978-1-55238-806-8 (pdf).—
ISBN 978-1-55238-807-5 (epub).—ISBN 978-1-55238-808-2 (mobi)

    1. Mineral industries—Canada, Northern—History. 2. Mineral industries—Political aspects—Canada, Northern. 3. Mineral industries—Social aspects—Canada, Northern. 4. Mineral industries—Environmental aspects—Canada, Northern. 5. Mineral industries—Economic aspects—Canada, Northern. 6. Native peoples—Canada, Northern—History. 7. Oral history—Canada, Northern. 8. Collective memory—Canada, Northern. I. Keeling, Arn, author, editor II. Sandlos, John, 1970-, author, editor III. Series: Canadian history and environment series ; 3

HD9506.C32M55 2015   338.20971   C2015-905147-9
                                                              C2015-905148-7

The University of Calgary Press acknowledges the support of the Government of Alberta through the Alberta Media Fund for our publications. We acknowledge the financial support of the Government of Canada through the Canada Book Fund for our publishing activities. We acknowledge the financial support of the Canada Council for the Arts for our publishing program.

This book has been published with the help of a grant from the Canadian Federation for the Humanities and Social Sciences, through the Awards to Scholarly Publications Program, using funds provided by the Social Sciences and Humanities Research Council of Canada.

  Canada Council for the Arts Conseil des Arts du Canada ArcticNet

Cover design, page design, and typesetting by Melina Cusano

Table of Contents

Acknowledgments	ix
Glossary of Key Mining Terms	xi
A Note on Terminology	xiii
Summary of Key Locations and Characteristics of Case Study Mine Sites	xiv
Introduction (Arn Keeling and John Sandlos) The Complex Legacy of Mining in Northern Canada	1

Section 1: Mining and Memory — 33

1| **Arn Keeling and Patricia Boulter** — 35
From Igloo to Mine Shaft: Inuit Labour and Memory at the Rankin Inlet Nickel Mine

2| **Sarah M. Gordon** — 59
Narratives Unearthed, or, How an Abandoned Mine Doesn't Really Abandon You

3| **Alexandra Winton and Joella Hogan** — 87
"It's Just Natural": First Nation Family History and the Keno Hill Silver Mine

4| **Jane Hammond** 117
Gender, Labour, and Community in a Remote
Mining Town

5| **John Sandlos** 137
"A Mix of the Good and the Bad": Community
Memory and the Pine Point Mine

Section 2: History, Politics, and Mining Policy — 167

6| **Jean-Sébastien Boutet** 169
The Revival of Québec's Iron Ore Industry:
Perspectives on Mining, Development, and History

7| **Hereward Longley** 207
Indigenous Battles for Environmental Protection
and Economic Benefits during the Commercialization
of the Alberta Oil Sands, 1967–1986

8| **Andrea Procter** 233
Uranium, Inuit Rights, and Emergent
Neoliberalism in Labrador, 1956–2012

9| **Tyler Levitan and Emilie Cameron** 259
Privatizing Consent? Impact and Benefit Agreements
and the Neoliberalization of Mineral Development
in the Canadian North

Section 3: Navigating Mine Closure — 291

10| Scott Midgley — 293
Contesting Closure: Science, Politics, and Community Responses to Closing the Nanisivik Mine, Nunavut

11| Heather Green — 315
"There Is No Memory of It Here": Closure and Memory of the Polaris Mine in Resolute Bay, 1973–2012

12| Kevin O'Reilly — 341
Liability, Legacy, and Perpetual Care: Government Ownership and Management of the Giant Mine, 1999–2015

Conclusion — 377

Notes on Contributors — 383

Bibliography — 387

Index — 425

Acknowledgments

A project of this magnitude would not have been remotely possible without the generous support of many people and organizations. The editors want to express their thanks to our funders: the Social Sciences and Humanities Research Council of Canada, ArcticNet, and the Social Economy Research Network. Research grants from these organizations allowed the editors and their students to engage in very expensive but important travel to remote northern communities to speak with the people whose lives have been so deeply affected, for good or ill, by mining. Our funding also enabled our team to present our results at various academic meetings and, in many cases, to the communities themselves. The Faculty of Arts at Memorial University supported this work through teaching releases for the two editors; the School of Graduate Studies provided matching fellowship support for students at Memorial. The Rachel Carson Center for Environment and Society provided John Sandlos with a generous one-year writing fellowship and residency that supported the bulk of his editorial and writing work on this project. Many thanks to office staff Fran Warren and Pam Murphy who processed the endless travel claims from the mining history group. Research assistant Emma LeClerc provided fantastic support work as we brought the different strands of the book together. Thanks also to Charlie Conway for cartographic support, and to the editorial team at University of Calgary Press for their efforts throughout the publication process. Funding support for open access publication of this volume was generously provided by the Network in Canadian History and Environment (NiCHE).

While many authors have included acknowledgments for their individual chapters, the editors would like to thank all the communities and individuals who supported this work by sharing their thoughts in interviews, or helping us with research and local logistics. We hope this volume will help preserve your stories and provide a way to share them with others.

Glossary of Key Mining Terms

Abandoned Mine: A site where advanced exploration (diggings, pits, trenches), or mineral extraction has ceased, without effective remediation or reclamation. This term is often used to refer to orphaned mines (see below).

Acid Mine Drainage: A pollution issue where mine wastes (tailings, waste rock, etc.) from sulphide rock formations react with air and water to produce sulphuric acid. The resulting acidic water has the potential to oxidize heavy metals (lead, cadmium, copper, etc.), exacerbating the water pollution problem.

Base Metals: Metals that are not considered precious (iron, copper, lead, etc.).

Cyanide: A chemical used to dissolve gold or silver in order to facilitate separation from ore.

Open Pit (or Open Cast) Mining: A mining method that removes ore deposits through the mechanized digging of large holes directly from the surface (usually after the removal of overburden such as vegetation and soils).

Ore Body: The entire body of rock and other material that is extracted to process and produce one or more valuable minerals.

Ore Concentrates: Produced through a milling process (often crushing and chemical separation) that results in a fine powder with a high percentage of the target metal. Ore concentrates are not a finished product, but are often produced in situ for more efficient transport to a smelter.

Ore Reserves: An assessment of the total amount of ore that can be extracted to produce minerals, usually categorized as possible, probable, or proven.

Orphaned Mine: An abandoned mine for which no private owner can be identified in order to establish liability. Such sites typically revert to public ownership and responsibility.

Placer Mining: Recovery of surface or stream-bed deposits of a target mineral (often gold), typically by washing, dredging, or hydraulic mining.

Prospecting: The earliest stage of the development process involving the active search for possible mineral claims.

Reclamation: A process of converting abandoned (or soon to be abandoned) mining lands to a usable state, as opposed to allowing them to become derelict.

Rehabilitation: In mining landscapes where full restoration (see below) is impossible, a partial repair of the structure and function of the previous ecosystem.

Remediation: Environmental cleanup at operating or abandoned mines, usually focused on lands and waters contaminated with heavy metals, radiation, and other toxic substances.

Restoration: An attempt to address the ecological impacts of mining through a return (as nearly as possible) to the ecological conditions that existed prior to mining.

Strip Mining: Mining near the surface through the removal of overburden and scraping of the ore over large areas. Strip mining is common with coal and sometimes bitumen deposits.

Tailings: Waste material (often a fine dust or slurry) emitted from an ore-processing mill after separating valuable minerals from the surrounding ore.

Tailings Pond: An artificially constructed body of water meant to confine tailings and prevent associated toxic material from spreading to local bodies of water or escaping as airborne dust. Leakage from tailings ponds has historically proved a major problem at mining operations.

Underground Mining: Removal of valuable minerals through the digging of mining shafts, tunnels, and chambers.

Waste Rock: Larger chunks of rock (and sometimes coarse gravel) produced through the mining process but containing no valuable minerals. Waste rock is often left in large piles at abandoned mines, but can also be used as construction material for roads or as fill during reclamation activities.

A Note on Terminology

Throughout this volume, we generally use the terms *indigenous* to connote first peoples in a global context, *Aboriginal* and *Native* to indicate first peoples in a Canadian context, and *First Nations* or *Inuit* to describe distinct cultural/linguistic groups or legally recognized bands in Canada.

TABLE 1: Summary of key locations and characteristics of case study mine sites

Mine	Relevant Chapter	Operational Period	Mineral Type	Extraction Methods	Province or Territory	Nearby Towns	Aboriginal Groups
Rankin Inlet Nickel Mine	1	1957-62	Nickel	Underground	Nunavut	Rankin Inlet (Kangiqiniq)	Inuit
Port Radium	2	1933-40; 1942-60; 1975-82	Radium-uranium-silver	Underground	Northwest Territories	Cameron Bay, Déline	Sahtúot'ı̨nę (Sahtu Dene) of Déline
Keno Hill	3	1913-17 1919-89	Silver-lead-zinc	Underground	Yukon	Keno City, Elsa, Mayo	Na-cho Nyäk Dun
Carol Lake Mine	4	1962-present	Iron	Open pit	Newfoundland and Labrador	Labrador City, Wabush	Innu
Pine Point	5	1964-1989	Lead-zinc	Open pit and underground	Northwest Territories	Pine Point, Fort Resolution, Hay River	Deninu Kue (Dene) First Nation, K'at'odeeche (Dene) First Nation
Schefferville Mining District	6	1954-1982	Iron ore	Open pit	Northern Quebec	Schefferville	Naskapi Nation of Kawawachichamak; Innu Nation of Matimekush-Lac John
Athabasca oil sands (several projects)	7	1967-present	Bitumen	Surface mining	Northern Alberta	Fort McMurray, Fort McKay	First Nations of Athabasca Tribal Council (Cree, Dene)
Nanisivik	10	1974-2002	Lead-zinc	Underground	Nunavut	Arctic Bay (Ikpiarjuk)	Inuit
Polaris	11	1982-2002	Lead-zinc	Underground	Nunavut	Resolute (Qausuittuq)	Inuit
Giant Mine	12	1948-2004	Gold	Underground and open pit	Northwest Territories	Yellowknife	Yellowknives Dene First Nation

Introduction: The Complex Legacy of Mining in Northern Canada

Arn Keeling and John Sandlos

In the midst of his annual northern tour in August 2011, Prime Minister Stephen Harper visited the Meadowbank Gold Mine, one hundred kilometres or so north of Baker Lake, Nunavut. Speaking in front of assembled workers and massive ore-hauling trucks, Harper hailed the mine—Nunavut's only producing site at the time—as a beacon of future prosperity for the people of the Arctic, and promised his government would help industry "unlock development possibilities in the North."[1] The PM reiterated these grand pronouncements to party supporters in Yukon the following year, when, echoing former Conservative Prime Minister John Diefenbaker, he declared that "the North's time has come" and that mineral development would help fulfill a "great national dream."[2]

Fast-forward to early 2013: as gold prices continued their long slide in the wake of the global recession, Agnico-Eagle, the company that owns Meadowbank, announced that it would close the mine three years earlier than planned (in 2017) and shift operations to another northern site. Mineral exploration funds began drying up, and new development proposals were scaled back across the North. Most dramatically, the massive Baffinland Mary River iron mine project in Nunavut radically curtailed its investment and development plans, cancelling a proposed railway and port development on Baffin Island and scaling back its project investment to $740 million from an initial projection of $4 billion.[3]

These events illustrated, in a very short time frame, the extreme volatility of the mining sector and the uncertainties that stalk mining-dependent communities and regions. Since the early 2000s, this industry has seen strong growth, as global demand and concerns over mineral scarcity began driving a "commodity supercycle" that spurred strong investment.[4] High mineral prices meant companies began to seek opportunities in regions previously considered too remote or expensive to operate profitably, including Northern Canada. Exploration and development expenditures skyrocketed, attracting renewed interest in historic mining areas—whether former mines or previously explored but unexploited deposits—that had been idled by the industry's prolonged slump in the 1980s and 1990s. After the "bust" that closed mines and destroyed communities in the 1980s, mining seemed ready to "boom" again in the heady days of the pre-recession 2000s. Most observers remain cautiously optimistic about the medium to long-term trends in commodity prices, and the prospects for northern development, but there's little doubt recent bumps in the road to northern riches have caused some to reflect on the industry's risks, as well as benefits.[5]

As a historian and a historical geographer, our perspective on these developments is informed by the long view of mining's place in Canada's northern and national history. Hardrock mining was the most important activity that brought industrial development to Northern Canada in the twentieth century. Large-scale industrial mining projects in the Canadian North began to appear even as placer gold mining in the Klondike began to decline in the early part of the century. Before 1945, industrial mining was concentrated at three main sites across the North: the Keno Hill silver camp near Mayo in the Yukon, the radium/uranium mines around Port Radium, NWT, and the gold mines at Yellowknife, NWT (Map 1). Surging postwar demand for industrial and strategic minerals, fuelled by both Cold War military needs and an expanding consumer society in North America, stimulated widespread exploration and development in the Canadian North. The postwar boom saw several major new developments, including the lead-zinc deposits at Pine Point, NWT, lead deposits at Faro in the Yukon (the Cyprus-Anvil Mine), and nickel at Rankin Inlet, NWT (now Nunavut), as well as the expansion of gold mining in the Yellowknife district.

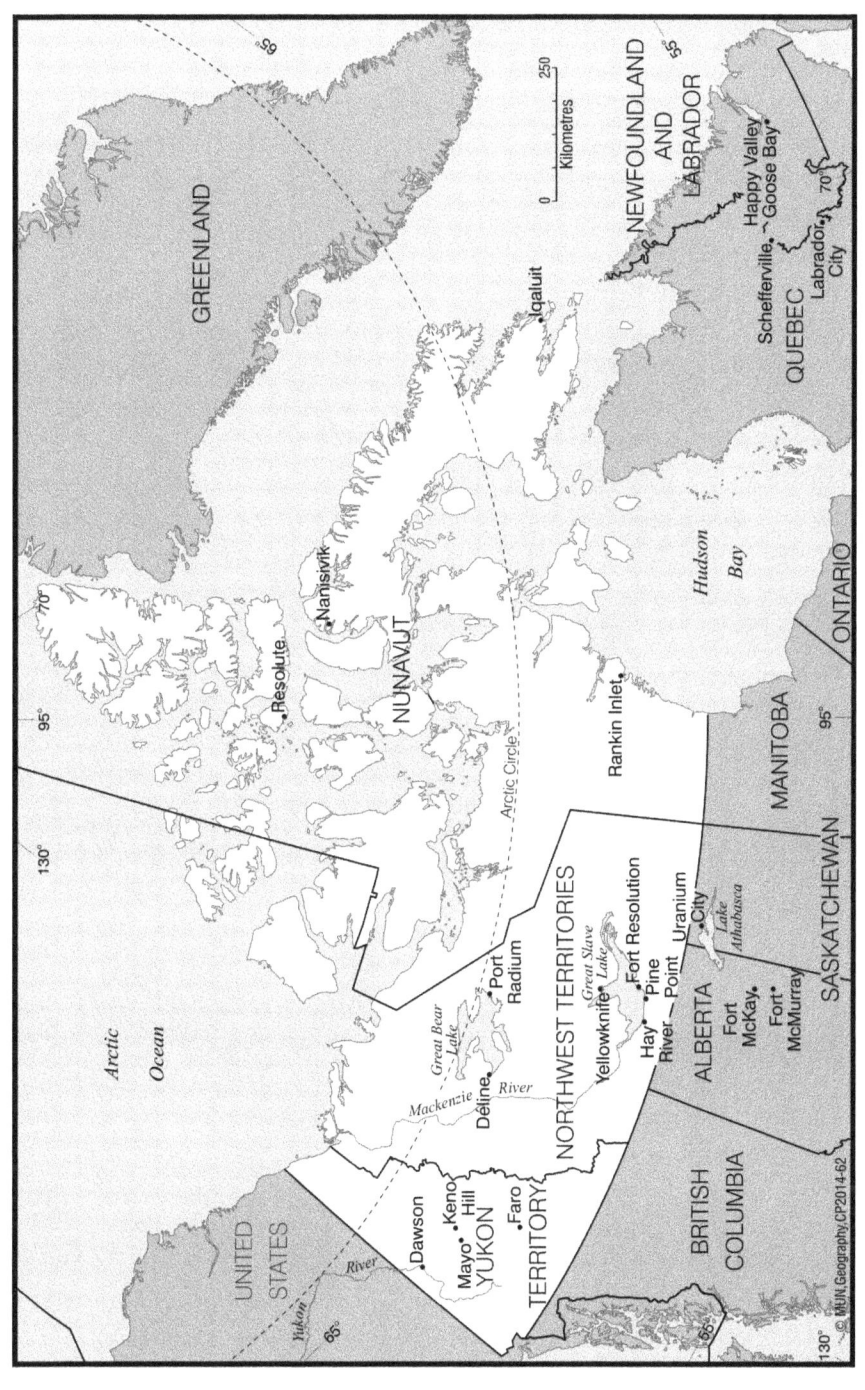

FIGURE 1: Locations of case study mines and communities in Northern Canada. Map by Charlie Conway.

Despite the relatively small number and wide geographic dispersal of these mines, industrial mining had a transformative impact on the North. By the 1950s, metal mining and fuel production (the latter almost entirely for local consumption in industrial developments and settlements) accounted for over 80 per cent of territorial economic output.[6] Mining in the provincial norths also expanded, including major iron mines in Quebec and Labrador (Schefferville, Labrador City), uranium developments in Saskatchewan and Ontario, and several base metal mines in northern Manitoba. Mineral development in this period stimulated non-Native exploration and settlement, investment in infrastructure from roads to railways to power developments, and the increasing integration of these once-remote territories into the national economy. Yet the costs of such development have become increasingly evident in recent decades. Former mine sites have left in their wake not only a toxic legacy of tailings ponds and waste rock dumps, but also a history of social and economic dislocation that continues to disproportionately impact northern Native communities. The promise of development and prosperity for northern regions—the "northern dream" of prime ministers since John Diefenbaker—has frequently delivered only ephemeral benefits, while leaving behind lingering social and environmental problems.

Mining and Communities in Northern Canada traces the history and legacies of the region's encounter with industrial mining in the twentieth century. With chapters spanning Canada's territorial north (and two provincial norths), this book aims to place the contemporary mineral boom (and accompanying hyperbolic rhetoric) into a critical historical context, as well as documenting the tremendous environmental, economic, and socio-cultural changes wrought by this transformative industry. Certainly the northern mining industry imported many significant historical tensions and contradictions worthy of their own book-length analyses, not least the ongoing conflict between capital and a labour force composed largely of outsiders. However, the studies in this volume focus largely on the often-neglected historical experiences of northern Native communities and their encounters with mineral development. Addressing the paucity of detailed historical studies on mining in the region, this volume represents an important collective contribution to our understanding of northern history, industrial development, and environmental change in the North, even as the region stands on the brink of another transformative period.

Most of the chapters emerged from a research project based at Memorial University called "Abandoned Mines in Northern Canada," which sought to illuminate the complex historical geography of mineral development, as well as its impacts on local communities and environments in the North. Generous funding from the Social Sciences and Humanities Research Council, ArcticNet, the Northern Scientific Training Program and other sources enabled the recruitment of talented students who, along with the project leaders and other collaborators, undertook extensive archival and field-based research in northern communities. Using a series of abandoned mines—sites of mineral exploration, development, and production that had ceased operation—as case studies, project researchers explored the impacts of industrial modernization on northern communities and environments, as well as the contemporary problems related to closure, abandonment, and remediation of these sites. In addition to producing a series of fine theses related to their case studies, many of the students involved in the project contributed chapters to this book. In addition, project partners from the communities themselves feature not only as informants or research partners in the field, but also in some instances as chapter authors.[7] The result (summarized at the end of this introduction) is a unique collaborative volume documenting the complex and contentious historical engagement with mining in the twentieth-century North.

FRAMING CANADA'S NORTHERN MINING HISTORY

If our encounters with northern mining people and landscapes provided the most immediate and powerful influence on the chapters contained herein, our collective attempts to contend with the vast and complex historical literature on Canadian and international mining have also shaped our work in profound ways. From the outset of the project, Harold Innis's foundational work on the rise and fall of hinterland staples economies provided a key pathway into our study. Although certainly not a new perspective, Innis's analysis of economic vulnerability among communities dependent on single resources provided a template for understanding the patterns of boom and bust in northern mining developments whose fate was often determined by distant economic forces. Ironically, Innis's most important work on mining, *Settlement and the Mining Frontier*, was quite positive about the economic impact of the Klondike gold rush and optimistic about the potential for mining to spread modern energy and transportation infrastructure throughout Northern Canada.[8]

Innis's boosterism reflected a broader optimism about northern development in the 1920s, but his staples thesis nevertheless influenced a newer generation of scholars who analyzed more critically the cyclical nature of northern mining development. In a series of articles, geographer John Bradbury examined the massive post–World War II development of the Quebec–Labrador iron ore range, highlighting the rise and near-collapse of communities such as Schefferville.[9] In a more theoretical vein, Trevor Barnes and others have elaborated on Innis's concept of cyclonics, the idea the hinterland resource developments proceed in storm-like fashion, with a sudden flood of capital, labour, materials, and knowledge into remote areas that dissipates just as suddenly when conditions change.[10] Even the few American environmental historians who have turned their gaze northward to Alaska or the Yukon emphasize the deep connection of mining to distant metropolitan centres where materials, animals, people, and capital are mobilized for the single-minded exploitation of a single resource, and where overdependence on these "paths out of town" can lead to economic shock and collapse in mining's instant towns.[11] Mining is a risky business in remote regions such as the Canadian North: high transportation and operating costs combined with vulnerability to volatile markets have often made community breakdown and economic collapse the inevitable endgame for northern mining towns.

But what of the Native communities that predate the development of mining in the Canadian North? Too often the staples-inflected literature has focused only on the rise and fall of largely non-Native mining towns, stories that ignored First Nations and Inuit communities who experienced sudden and rapid social, economic, and environmental change due to mining, and who persisted in spite of the abandonment of adjacent mining activities. As we began our research, we found surprisingly little scholarly work on this theme, particularly from a longer-term historical perspective. Much of what has been done stems from the work of NGOs such as MiningWatch and the Canadian Arctic Resources Committee, or emanates from the formal assessment processes and reporting requirements of mines with significant environmental liabilities.[12] Aware of the burgeoning literature on mining and indigenous conflicts in the US Southwest, Australia, Africa, and Latin America, we immediately sought to frame mining in Northern Canada within the global literature on environment and justice.[13] We ultimately adopted historical political ecology as a major theoretical frame for our work, a geographical subdiscipline that highlights the unequal distribution of environmental harms and benefits due to colonial environmental management and resource production schemes in Third World regions. We argued that the territorial north in the twentieth century, much like these other resource extraction zones in the developing world, was a thinly populated but still largely indigenous space where the long reach of mining severely impacted pre-existing subsistence economies, provided few local employment or investment opportunities, and often left severe environmental problems with which Aboriginal communities have to contend.[14]

The historical political ecology framework thus helps connect mineral development's local impacts and conflicts to larger questions surrounding the links between political economy, state-led or promoted development projects, and the settler colonial dispossession of indigenous people.[15] In Canada, past northern development visions and policies have been strongly linked to the exploitation of the region's natural resources; particularly after the Second World War, politicians and bureaucrats promoted resource development, especially mining, as the key to assimilating northern people and territories into the national economy.[16] Throughout much of Arctic and Subarctic Canada, this agenda advocated the transition of Aboriginal economies away from land-based subsistence and trade economies, and toward wage economies and settlement life—often with dire consequences for Aboriginal people themselves.[17] Our research has found, similar to cases in the developing world, that the southern Canadian perception of the North as both an underdeveloped territory and as a potentially rich resource frontier underwrote industry-friendly government policies and subsidies for industrial infrastructure, including dams, townsites, roads, and railways, all of which further affected Native communities and territories.

These impacts were most stark at major mines where Native communities faced acute toxic threats or other environmental clean-up issues over the long term: sites such as the Port Radium mine, the Cyprus-Anvil lead-zinc mine, the Pine Point Mine, or the Giant Mine. Indeed, many chapters in this volume reveal that Native people maintain strong feelings of historical injustice about the social and environmental legacies of mining in their regions. Some regard mining as a key agent of colonialism in their region, bringing sudden influxes of outside workers, instant communities, state agencies, and environmental degradation, all of which combined to compromise subsistence economies based on hunting and trapping.[18] For instance, particularly in early to midcentury developments, mines drove significant local deforestation through their voracious demand for wood for fuel and construction.[19] Many large northern mineral developments were accompanied by hydroelectric dams (including at Faro, Pine Point, Yellowknife, Uranium City, Lynn Lake, and Schefferville) in order to power towns and industrial facilities, dams that altered local rivers and displaced land users.[20] In places such as Port Radium or Yellowknife, Native people were exposed to acute and low-level toxic contaminants (radiation at the former site, arsenic at the latter) that caused death and illness in some cases. The precise health and ecological impacts of past mine pollution is difficult to assess due to the inherent limits of historical epidemiology (and poor contemporary monitoring practices), but the persistence of pollutants long after the closure of these mines has contributed to a sense that the land is sick and dangerous rather than a source of sustenance.[21]

Very often the impacts of mining and its related developments occurred alongside the other profound environmental, social, and health challenges that northern Native people faced throughout the twentieth century. New diseases, acute hunger, poor nutrition associated with store-bought food, declining fur trade economies, community relocations, military activities, and poor housing often coincided with mining development or arose as a direct result of conditions within mining communities, as many of the case studies make clear.[22] In spite of these wide-ranging changes, for northern Native communities on the front lines of toxic or radiological exposure due to mining, the use of their traditional lands as repositories for pollution and waste is often remembered as the most significant of the changes that outsiders brought with them.

As with other historical works highlighting the dire ecological impacts of mining, it was difficult to conclude that the environmental issues associated with northern mining could lead anywhere but to stories of decline and dislocation.[23] However, a major challenge to this negative perspective has come from what might loosely be described as the "resilience school" of mining history. Inspired by the pioneering work of geographer Richard Francaviglia (and perhaps influenced by the general revolt against "declensionist" narratives among environmental historians and historical geographers), several scholars and popular writers have argued against a "mining imaginary" in which the death of mining communities and environmental catastrophe inevitably follow the closure of a mine. Mining communities, these writers suggest, often survive after the end of mining, through any combination of economic diversification, mining-related tourism, redevelopment (as sites of mineral production or as brownfield redevelopments for other industries), or large-scale ecological restoration projects. Central to these studies is the idea that mining communities identify very strongly with their own history; meaningful commemoration of mine work, community life, and mining landscapes remains a key concern in many mining towns facing closure and remediation.[24]

In Northern Canada, factors such as distance and limited infrastructure present very real challenges to the survival of mining communities, but we did find ample evidence of resilience and a strong mining identity among former mining communities in the region. Since the final closure of the Giant and Con gold mines adjacent to Yellowknife in the early 2000s, the city has thrived as a government town (it is the territorial capital) and as a supply and labour centre for the major diamond mines that commenced operation on the tundra to the northeast in the late 1990s. The city still retains a strong identification with its origins as a mining town: local history societies have published several oral histories of life in the early gold mining years, while the NWT Mining Heritage

Society hopes to develop a museum preserving some of the industrial heritage (including at least one headframe) at the Giant Mine site.[25] Former Inuit miners in Rankin Inlet also maintain a strong identification with the community's mining history, while the community has survived as a relatively large (by northern standards) government and service centre, and regional transportation hub.[26] Other sites such as Port Radium, the Keno Hill silver mines, and the Quebec-Labrador iron region have received almost hagiographic treatment in popular histories produced by former mine officials or in histories commissioned by mining companies.[27] Some of our other case study towns, such as Schefferville, Quebec; Uranium City, Saskatchewan; and Keno City, Yukon, clung to life despite being depopulated when the mines closed due to plunging mineral prices in the 1980s, and now anticipate a rebirth if commodity prices remain high and redevelopment of abandoned mineral deposits moves forward. Even in cases where the physical community did not survive, former town residents keep the memory of their community alive. One example is Pine Point's very active web-based memorial, a grassroots commemoration that was recently highlighted in a brilliant online interactive documentary by the National Film Board.[28] Mining towns in Northern Canada sometimes *do* survive the collapse of their main raison d'etre, whether they are sustained by new economic activity or kept alive in the memories and virtual worlds of former residents.

How to reconcile the theme of resilience with the historical political ecology framework we had adopted became a key question for our collective work on abandoned mines. Ultimately we found that many Native communities exhibited their own forms of resilience both during and after mining boom periods, taking advantage of wage labour opportunities when presented but also returning to hunting and trapping activities when prices were favourable. From very early in our research, we realized it was too simple to suggest that Native northerners were simply the passive victims of mining. Native people took advantage of opportunities in the mining industry when they could, adapting to mineral development through strategies that ranged from engaging in ad hoc labour to eventually applying political pressure for the establishment of Inuit and Dene mineral rights and/or royalty regimes through impact and benefit agreements and the comprehensive land claims process.[29]

And yet, abandoned mines in Northern Canada remain places that are too deeply contested to fit neatly within the resilience framework. Indeed, the resilience school too often ignores the multiple ways in which mines impact communities, particularly in light of the colonial and political ecology context that we highlight above. For many researchers on this project, one of the most remarkable findings was the extent to which many Native northerners embraced the complexity of their mining histories, critiquing the colonialism and environmental degradation that were invariably tied to mining on the one hand, but minutes later expressing the same pride in their work and nostalgia for the good old mining days as their non-Native former co-workers. As so many of the people we spoke with stated in many different ways, mining often brought with it a complicated and mixed legacy.

Nowhere are these deep contradictions surrounding mining history more acute than the contemporary debates surrounding the remediation and redevelopment of many of our case study mines. Again, contrary to the arguments of the resilience school, the "afterlife" of mines can have extremely negative implications for northern Native communities. Many northern mines live on as sources of local controversy because of severe long-term environmental degradation, where legacies including massive landscape changes, waste rock piles, abandoned industrial facilities, and toxic contaminants such as heavy metals, acid mine drainage, or radioactivity have forced the federal government to establish expensive remediation programs. In addition, prompted by recent spikes in mineral prices, many abandoned northern mines are being resurrected through redevelopment of remaining ore deposits—in some cases simultaneously with remediation of the original mining development. We have come to think of these examples as "zombie" mines, because in their afterlife they continue to haunt communities with many of the same issues—environmental risks, unequal wealth distribution, decision making by outsiders—that emerged with the original development.[30]

Even where projects are undergoing only remediation—often assumed to be a form of healing the land—the mobilization and containment of toxic material and the perpetual care and monitoring required at some sites raise profoundly complex issues associated with community risks and intergenerational equity. In 2002 a federal Auditor General's

FIGURE 2: Yellowknife, NWT's abandoned Giant Mine in 2008, before remediation activities began. Remediation costs at Giant are expected to exceed $1 billion. Photo by Arn Keeling.

report prioritized thirty abandoned mines in the territorial north as requiring remediation, with an estimated cost of $555 million (a very low figure given the fact that remediation for the Giant Mine alone is forecasted to cost over $1 billion).[31] As of this writing, Aboriginal Affairs and Northern Development Canada is conducting remediation activity, planning, and/or monitoring programs at seventeen of these sites, with four simultaneously undergoing redevelopment.[32] During the course of our research, we have been able to document community memories and contemporary reactions to six mines in the territorial or provincial norths that exhibit zombie-like characteristics—Giant, Keno Hill, Pine Point, Nanisivik, Rankin Inlet, and Port Radium—as well as the proposed redevelopment of iron deposits near Schefferville. These examples represent only a small fraction of abandoned mines in the Canadian North or the thousands of sites globally, but they reveal the complex and multifaceted ways that the fractious histories and community memories of cyclonic development projects in Northern Canada have profoundly shaped local responses to contemporary mine remediation efforts.

UNDERSTANDING MINES THROUGH ORAL HISTORY

Local memories of abandoned mines have persisted in much the same way as the after-effects of the sites themselves. In most of our case study communities, we found that undertaking oral history research was essential to capture the varied local memories of mining and to document the complex and layered community stories that rarely emerged in archival documents or government reports. A lively scholarly and popular literature on mining and communities makes extensive use of oral history in order to document the experiences of workers and families in these uniquely hardscrabble settings.[33] Oral histories have also explored the environmental degradation associated with mining, while highlighting the various ways these "wasted" landscapes are perceived by area residents themselves, who may identify positively with industrial ruins or mine waste sites.[34] In this way, oral history permits researchers to explore the multiple meanings and experiences of mining places.[35] Ranging from collections of edited transcripts to the use of oral interviews in conjunction with ethnographic, archaeological, and archival sources, oral history is regarded as a potent source for the documentation of otherwise "hidden voices" of mining history.

Typically absent from this body of oral history work is the experience of Native people in many mining regions. Few studies in Canada have used oral history to capture indigenous people's parallel historical development of mining identities and their experience of mine closure, either as workers or as broader participants in local economic and settlement life.[36] As people with lifeways and knowledge systems intimately tied to local environments, northern Native communities often have borne the brunt of the environmental changes associated with mining, whether deforestation, local resource depletion caused by habitat change and harvesting by newcomers, or pollution and toxicity stemming from mining and mineral processing wastes. Indigenous oral histories can also provide important insight into these processes and experiences of mining-driven environmental change.

Our initial motivation, then, for using oral history was one common to many practitioners: to address the gaps in the archival record and to explore the hidden histories and untold experiences of industrial development in the North. Most of the chapters in this volume include detailed archival research in federal and territorial archives, which document government policy and, to a lesser extent, corporate and individual mining-related activities in the North.[37] But Native voices are conspicuously absent from the vast government archival record on northern mining, despite the fact that responsibility for Aboriginal affairs and northern development rested within a single federal department—the Department of Northern Affairs and National Resources, in its various incarnations—after the Second World War. Where Native people are present, they appear as objects of state policies around employment, settlement, or interactions with non-Native newcomers; rarely, before the 1970s at least, were Native people's opinions, reactions, or experiences of development recorded or sought. Our interest in "recovering" Native experiences with industrial development resonates with one of the core motivations of oral historians: to record (literally and figuratively) these ignored stories and thereby to challenge dominant historical narratives.[38]

In the process of engaging with communities and individuals over the course of this project, however, some of the paradoxes of this approach became apparent, prompting a deeper reflection on questions of positionality and knowledge of the sort urged by advocates of indigenous methodologies.[39] Although the communities and individuals we worked with expressed enthusiasm for documenting their stories, the notion of "restoring hidden voices" to the public transcript in many ways reflected our own outsider status and interests (as southern, non-Native academics and students).[40] It quickly became apparent that stories of industrial development and environmental damage are far from "hidden" in the communities themselves; rather, they continually circulate and are mobilized from time to time in communities' ongoing engagements with land-use planning, land claims, and new development proposals. This question also arose when we returned to communities to report on the results of the oral history studies, which sometimes placed us in the awkward position of repeating to community members their own stories. It has become important for us to recognize that the absences in the archival record and ignorance of southern Canadians (including scholars) regarding Native experiences do not necessarily extend to the communities themselves.

Similar questions arose surrounding cross-cultural communication, translation, and representation in this project. As outsider researchers, we typically worked with community researchers, who participated in the research process by connecting us with informants, asking questions themselves, or translating interviews with elders who preferred to speak their language. These community researchers represented a vital bridge between ourselves as non-Native visitors (and, to informants, unknown quantities) and these knowledge-holders. Nevertheless, questions of translation—both linguistic and epistemological—remain. With respect to language, the two-way communication of an interview became mediated through a third party and the back-and-forth translation of ideas and terms, raising ample potential for misunderstanding and miscommunication. Where possible, we have sought full translations of interviews after the fact, but the potential gulf in meaning and understanding remains, however tentatively bridged by our able translators. As outsiders and non-Native people, we do not expect to share or completely understand Native knowledge and experience; we also recognize that, because they were speaking to outsiders, interviewees may have chosen to share certain versions of stories with us, based on the questions we asked. In that sense, we follow Linda Shopes and others in regarding oral histories as products of this interaction, rather than as objectively collected "knowledge."[41]

In this dynamic of the co-production of knowledge, the outcomes of oral history research are best regarded in terms of shared authority and negotiated meanings. Shopes suggests this shared authority is a check on all participants in the oral history process: "Scholars do not get to exercise critical judgment quite so forcefully or conform to current historiographic thinking quite so deftly; laypeople do not get to romanticize the past quite so easily."[42] Nevertheless, our ongoing interpretation and re-presentation of these interviews becomes a further layer of mediation between the informant and various audiences (including the readers of this volume). As we began to undertake transcription (itself a problematic textual rendering of oral expression) and analysis of our interviews, we confronted questions of respectful yet critical representation of this knowledge in both public and scholarly contexts. Concerns about reliability and validity stalk oral history: in analyzing stories about industrial development and environmental change, we remain alert for the effects of nostalgia or "social memory" in shaping individual recollections of particular events and issues.[43] Nevertheless, we have approached the oral history component of the project not with the desire to document concrete events or specific environmental impacts (i.e., to answer questions of causality or proof), but rather with the goal of capturing personal and collective experiences and perceptions of mining-induced social and environmental change. As archaeologist Karen Metheny asserts, "Oral history, then, is as much an exercise in verification as it is learning how people create their own versions of the past and determining the meaning of those constructions."[44] Thus, we regard oral stories not as hard "evidence" about environmental and social change (although respondents have often suggested environmental impacts and problems not well documented in the archival record), but rather as personal insights enriching and deepening narratives of industrial development with grounded, individual experiences and observations.

In seeking these stories, it was critical to acknowledge community motivations (collectively represented by First Nation, Métis, and Inuit governments or other organizations) and the interests of individual participants themselves in sharing their experiences, and to undertake the research through processes and agreements approved by them. Each of the case study communities brought varying motivations and levels of participation and oversight to the research. In some cases, our proposals to study the mining past were greeted with more or less instant enthusiasm. In other cases, relationship building took repeated visits and discussions over years. In all cases, the communities' engagement with the research (quite properly) reflected their interest in documenting the past, whether for the purposes of land claims, resource management, and regulatory processes, or simply recording for posterity elders' experiences with mining and the coming of non-Native settlement. Rather than limiting or biasing the research, these interests validated our presence in the community and the goals of our study, while allowing us more or less free rein to conduct the interviews as we saw fit. Guided by the Tri-Council Policy on Ethical Conduct for Research Involving Humans

(TCPS2),[45] project researchers also followed the relevant community, territorial, and institutional protocols for gaining informed consent from participants. While particular individuals and organizations are acknowledged in each of the chapters, it is appropriate here to acknowledge and thank the many northerners who worked with us to help make this research possible.

Rather than undertaking "life history" or traditional knowledge studies, we employed a semi-structured, "directed interview" technique that sought to highlight individual experiences and knowledge of mining and the local environment. Because we aimed to document lived experience of events that were (mostly) within living memory of the subjects, we did not seek folkloric or other stories of a "traditional" nature (though ironically, most of the research agreements we established with First Nations are governed by traditional knowledge protocols). This approach had benefits and costs. While it enabled a focused and purposeful interview (allowing respondents to direct the discussion as they saw fit), it perhaps divorced the topic somewhat from the wider contexts of interviewees' lives: family, community, land, and culture.[46] Whether or not interviewees worked there or had strong opinions about it, the mine was likely not the central feature of their lives—except perhaps during the hour or so spent in conversation with researchers. These contexts nevertheless occasionally forced their way into the transcript, through references to family origins, histories of movement, or important personal milestones, and provided important glimpses of interviewees' lives.

The result is a rich tapestry of personal and community memories and stories from small mining towns ranging from the High Arctic to the provincial northlands; from Nunatsiavut, Labrador, in the East to Mayo, Yukon, in the West. The vastness of the subject matter and the geographic territory, not to mention the usual constraints of time and budgets, prevented us from covering all major mining developments in Northern Canada. Indeed, some major historical mines such as Con Mine near Yellowknife, the Cyprus-Anvil Mine in the Yukon, and the uranium mines of northern Saskatchewan receive only brief mention in this book. Innumerable smaller mineral developments are ignored. We chose our existing case studies in part because of their size, longevity, and severity of local impacts, but also based on the extent to which individual mines represented the leading edge of mineral-led colonialism in particular regions of Northern Canada. Keno Hill silver mine, for instance, was the first industrial mine in the territorial north; other sites were the first to push mining into remote regions such as the Eastern Arctic (Rankin Inlet), the Arctic Islands (Polaris and Nanisivik), isolated areas along the Quebec–Labrador border (Schefferville and Labrador City), and the south side of Great Slave Lake (Pine Point). Certainly we wanted a collection that represented diverse regions of Northern Canada (including examples from the provincial norths) and a variety of time periods, but individual research interests and life experiences (some authors live close to their case studies) also shaped the choices of cases and subject matter in this book.

The first section of the book focuses most intensely on our oral history research projects and the complex community memories of mining in Northern Canada. Arn Keeling and Patricia Boulter's chapter highlights Inuit memories of the North Rankin Nickel Mine, the first large-scale Arctic mining project, opened in 1957, and the first to employ Inuit labour as a deliberate social and economic development strategy. The chapter traces the challenges faced by the Inuit, but also their strategic adaptation, during the rapid transition "from igloo to mine shaft," arguing that Inuit developed a strong identity as miners even as they critiqued shortcomings of the project such as poor housing and the short five-year life of the mine. In the second chapter, Sarah Gordon analyzes one of Northern Canada's most controversial mining sites, the Port Radium mines that produced radium and then uranium from the eastern shores of Sahtú (Great Bear Lake) from 1931 to 1960 (and intermittently produced silver until 1982). The mine is notorious, not only for its contribution of uranium to the Manhattan Project, but also due to the controversy surrounding radiological exposure among Dene workers from Déline. Gordon highlights how community memory surrounding the discovery of pitchblende (which they contend was passed on by elder Old Beyonnie rather than "discovered" by legendary prospector Gilbert Labine) intersects with more recent attempts to reconcile the colonial past of the mine with community healing in the present. In Chapter 3, Alexandra Winton and Joella Hogan trace the long history of interaction between the United Keno Hill Mine and the Na-Cho Nyäk Dun First

Nation in the central Yukon. Through a close reading of the experience of one elder, Henry Melancon, the authors argue (similar to Keeling and Boulter) that the mines brought a mixed legacy of opportunities and negative impacts, historical experiences that inform contemporary reactions to Alexco Resource Corporation's ongoing simultaneous remediation and redevelopment of the old mining area. Jane Hammond's work in Chapter 4 moves away from northern Native communities, instead providing an oral history that accounts for gender relations in the iron-mining town of Labrador City. Hammond's work reveals that, despite women's advances in the workforce elsewhere in Canada in the 1970s, social pressure, company policies, and masculine workplace culture meant that women entered mine work only very slowly in Labrador City, often finding themselves caught uneasily between the twin pressures of the domestic and wage labour realms. John Sandlos's final chapter in this section surveys the intensive oral history research conducted in Fort Resolution, NWT, near the massive Pine Point lead-zinc mining complex that operated on the south shores of Great Slave Lake between 1964 and 1988. The many collected memories of individuals suggest a mixed historical legacy for mining in the region, with interviewees recalling their great fondness for mine work and town life, while at the same time lamenting the negative social, economic, and environmental changes that accompanied the introduction of mining to their region. As with the previous studies, oral history research throws into bold relief the manner in which northern Native communities remember mining, in the words of one interviewee, as a mix of the good and the bad.

However remote they may be geographically, northern communities are never isolated from the political and policy frameworks that simultaneously promote and regulate mining development in Canada. The second section of the book thus examines the many ways that northern communities have attempted to insert their voices into provincial, territorial, and federal regulatory processes or to directly engage mining companies. In Chapter 6, Jean-Sébastien Boutet evaluates the Quebec government's *Plan Nord*, a comprehensive northern development program centred on major hydro and mineral projects, in light of the mixed legacy of the abandoned iron mines near Schefferville. Boutet suggests that previous mine closures and the sudden loss of a wage economy at Schefferville continue to haunt contemporary discussions of development among the Innu communities of Matimekush–Lac John (Schefferville) and Uashat mak Mani-utenam (Sept-Îles), despite current promises of long-term jobs and prosperity. Hereward Longley, in Chapter 7, turns toward the initial period of oil sands growth in the 1970s, tracing the attempts of the Fort McKay First Nation to assert some control over the economic and environmental impacts of development through the courts and regulatory hearings. Andrea Procter's chapter examines the growth of an Inuit rights discourse in response to uranium exploration in northern Labrador, an assertion of resource claims that was expressed formally through the emergence of the Labrador Inuit Association in the 1970s and the land claims process that produced Nunatsiavut in 2005. And yet, even though resource rights and land claims may represent a step forward from total exclusion of Inuit from development, Procter questions whether granting Inuit a share of the dominant development paradigm may also represent an entrenchment of neoliberal ideas of self-sufficiency and economic autonomy from state "handouts." In a similar vein, the final chapter of this section, by Tyler Levitan and Emilie Cameron, provides a pointed critique of the recent move toward impact and benefit agreements (IBAs), the private deals that are now typically struck between mining companies and northern Native communities in order to delineate jobs, economic benefits, and environmental liabilities associated with nearby mining developments. As with Procter's work, Levitan and Cameron suggest that the seemingly inclusionary step of inviting northern Native communities into negotiations with companies may actually represent the affirmation of a neoliberal regime whereby northern social development programs become at least partially privatized within the (disturbingly) boom and bust capital flows associated with the mining industry. While mining companies and governments point to new industry–Native community partnerships as a marker of major change, the chapters in this section suggest that unequal power relations and colonial legacies still play a major role in shaping northern development projects.

The final section presents three chapters that examine the zombie-like afterlife of many mines, and the manner in which the history of these places is reflected in the contemporary reality of nearby communities. The first of these is Scott Midgley's discussion of the abandoned

Nanisivik lead-zinc mine on Baffin Island, which operated from 1976 to 2002. Midgley considers the post-closure debates over mine remediation at Nanisivik, arguing that, far from being valueless, the abandoned mine became a site of contested valuation (of land and environment), as the government and company insisted on scientific and cost-effective approaches to mine remediation while the Inuit of Arctic Bay expressed deep skepticism about the long-term efficacy of the proposed remediation plan. By contrast, Heather Green suggests in Chapter 11 that the fly-in, fly-out Polaris lead-zinc mine that operated on Little Cornwallis Island figures hardly at all in the historical memory of the service town of Resolute, located about ninety kilometres to the south on Cornwallis Island. Through oral history interviews, however, Green discovered that the relative amnesia surrounding Polaris is in part a product of Resolute's own experience of marginalization from the social and economic benefits of the development, and the mine persists mainly as a symbol of relative Inuit exclusion from previous development projects. The book's final chapter belongs to Yellowknife-based social and environmental activist Kevin O'Reilly, who provides a detailed yet passionate account of attempts to insert local voices into the remediation of the abandoned Giant Mine. Here, the federal department of Aboriginal Affairs and Northern Development Canada (AANDC) is proposing to contain through freezing 237,000 tons of arsenic trioxide stored underground at the mine, a staggering amount of highly toxic material that will require care and maintenance in perpetuity. O'Reilly tells a deeply disturbing story of how AANDC has resisted public involvement, environmental assessment, and independent oversight over the remediation project, reducing the deeply conflicted issues surrounding the historical mine site to a mere technical engineering problem. Perhaps more than any other site, the immense scale and complexity of remediating Giant illustrate the costly legacies associated with northern mining, and the deep conflicts these sites continue to provoke in the present.

It has been more than one hundred years since mining began in the Keno Hill Silver District, the site of the territorial north's first industrial-scale hardrock mines. What has a century of mining brought the Canadian North? Did the mines bring civilization or advance the agenda of settler colonialism? Did they bring untold riches or siphon wealth from the region for other people living in other places? Did the mines bring economic development or dislocation, all the while bequeathing long-term and costly environmental problems that may not be solved for generations? The chapters in this book suggest that mining in Northern Canada brought all of these things. There is no doubt, for instance, that mining was one of the major stimulants of northward expansion in Canada, a colonial incursion into Native territory that has only recently been redressed, however inadequately, through land claims and IBA processes. The materials and much of the wealth from these mines did indeed follow many pathways out of town—gold to the vaults of central banks, lead and zinc to massive smelting facilities in Southern Canada or overseas, and uranium to the research laboratories of the Los Alamos atomic bomb project—but often provided only marginal employment benefits to the Native communities adjacent to the mines. At the same time, Native communities did often embrace mining development and mining labour as a hedge against tough times in the fur trade and hunting economies. Mining brought new communities and settlers to Northern Canada, but the collapse of many major developments left a legacy of nearly abandoned communities or ghost towns that stand as testaments to the ephemeral nature of the mineral economy. In some cases, the abandoned mines of Northern Canada have also produced environmental problems that afflicted the past and the present, and potentially could persist far into the future.

Almost all of our case studies illustrate the deeply conflicted historical experience of northern mining development that inevitably hangs over current debates about mine remediation, redevelopment, or new mining projects close to Native communities. If the history of northern mining is indeed a tangled legacy, the chapters in this book allow us a close-up view of these contradictory stories, and the ongoing attempts to reconcile the complex past with the opportunities and challenges that mining may present northern communities in the future. The renewed northern mining boom of the past decade has brought territorial governments, Native leaders (particularly the newly empowered land-claims organizations), and mining companies into new relationships, under circumstances very different than those surrounding twentieth-century mining. Yet as Virginia Gibson suggested in her study of diamond mining in the

Northwest Territories, while modern miners "may seek to enter the political geography of the north without acknowledging the past, [a] relational view of history reveals they will arrive with the shadows of ghost-mines behind them."[47] Our research into abandoned mines confirms that reckoning with the history, geography, and ongoing legacies of past rounds of extractive development is critical if large-scale mining has any chance of generating enduring prosperity and opportunities for sustainable economies in Northern Canada.

NOTES

1   Gloria Galloway, "Prime Minister Stephen Harper Pushes Mining Exploration in Arctic," *Globe and Mail*, August 24, 2011.

2   Stephanie Levitz, "Moiling for Gold: Harper Spends Second Day in the North at Area Mine," *Canadian Press*, August 21, 2012, accessed February 19, 2014, http://www.huffingtonpost.ca/2012/08/21/harper-arctic-trip-gold_n_1815787.html.

3   Pav Jordan, "Baffinland Iron Mines Sharply Scales Back Mary River Project," *Globe and Mail*, January 11, 2013; "Mining Slowdown Hurts North's Economy," *Canadian Press*, October 16, 2013, accessed February 19, 2014, http://www.cbc.ca/news/canada/north/mining-slowdown-hurts-north-s-economy-1.2075144. Meadowbank's travails are documented in Alistair MacDonald and John W. Miller, "Mining at Minus 45 Celsius Is No Picnic," *Wall Street Journal*, February 23, 2014, accessed February 25, 2014, http://online.wsj.com/news/articles/SB10001424052702303636404579392912016780616.
Excellent sources for tracking the ups and downs of mining investment and activity in the North include the annual mining issues of the NWT and Nunavut *News/North* papers, as well as the Natural Resources Canada quarterly and annual mining statistics reports, available at http://www.nrcan.gc.ca/mining-materials/statistics/8848.

4   Brenda Bouw and David Ebner, "The Global Commodity Cycle Speeds Up," *Globe and Mail*, January 24, 2011; Daina Lawrence, "Canadian Government Fostering Arctic Exploration," *Resource World* 11 (December 2011/January 2012): 82–83.

5   Philip Cross, "Canada's North Finally Opens Up," *Financial Post*, March 22, 2012. Notably, mining's potential for northern development has been highlighted in a recent series of reports from the Conference Board of Canada's new "Centre for the North": see Conference Board of Canada, "Mapping the Economic Potential of Canada's North" (Ottawa: December 2010);

Conference Board of Canada, "Toward Thriving Northern Communities" (Ottawa: December 2010); Conference Board of Canada, "The Future of Mining in Northern Canada" (Ottawa: January 2013).

6   Kenneth J. Rea, *The Political Economy of the Canadian North* (Toronto: University of Toronto Press, 1968), 439.

7   Indeed, not all the students or community researchers involved in the project are represented in this volume, for various reasons. Nevertheless, their contributions to this collective research effort certainly shaped the overall project in important ways through writing, conversations, and interactions in communities.

8   For Innis's important writings on mining and/or staples production, see Daniel Drache, ed., *Staples, Markets, and Cultural Change: Selected Essays*, Innis Centenary Series (Montreal: McGill-Queen's University Press, 1995); Harold Adams Innis, *The Fur Trade in Canada: An Introduction to Canadian Economic History*, rev. ed. (Toronto: University of Toronto Press, 1956); Harold Adams Innis, *Settlement and the Mining Frontier* (Toronto: Macmillan, 1936); Harold Adams Innis, *The Problems of Staple Production in Canada* (Toronto: Ryerson Press, 1933). For discussion, see Matthew Evenden, "The Northern Vision of Harold Innis," *Journal of Canadian Studies* 34 (1999): 162–86. For a discussion of Innis's linking of mining, industrialism, and settlement in Northern Canada, see Liza Piper, "Innis, Biss, and Industrial Circuitry in the Canadian North, 1921–1965," in *Harold Innis and the North: Appraisals and Contestations*, ed. William J. Buxton (Montreal and Kingston: McGill-Queen's University Press, 2013), 127–48.

9   John H. Bradbury, "Towards an Alternative Theory of Resource-Based Town Development in Canada," *Economic Geography* 55, no. 2 (1979): 147–66; John H. Bradbury, "Declining Single-Industry Communities in Quebec-Labrador, 1979–1983," *Journal of Canadian Studies* 19, no. 3 (1984): 125–39; John H. Bradbury, "The Impact of Industrial Cycles in the Mining Sector: The Case of the Quebec-Labrador Region in Canada," *International Journal of Urban and Regional Research* 8 (1984): 311–31; John H. Bradbury and Isabelle St.-Martin, "Winding Down in a Quebec Mining Town: A Case Study of Schefferville," *Canadian Geographer* 27, no. 2 (1983): 128–44. See also Cecily Neil, Markku Tykkyläinen, and John Bradbury, eds., *Coping with Closure: An International Comparison of Mine Town Experiences* (New York: Routledge, 1992).

10 Trevor Barnes, "Borderline Communities: Canadian Single Industry Towns, Staples, and Harold Innis," in *B/ordering Space*, eds. Henk Van Houtum, Olivier Kramsch, and Wolfgang Zierhofer (Burlington, VT: Ashgate Publishing, 2005), 109–22. For other scholars of mining influenced by Innis, see Arn Keeling, "'Born in an Atomic Test Tube': Landscapes of Cyclonic Development at Uranium City, Saskatchewan," *Canadian Geographer* 54, no. 2 (2010):

228–52; Mary Louise McAllister, "Shifting Foundations in a Mature Staples Industry: A Political Economic History of Canadian Mineral Policy," *Canadian Political Science Review* 1 (June 2007): 73–90. See also Jody Berland, "Space at the Margins: Critical Theory and Colonial Space after Innis," in *Harold Innis in the New Century*, eds. Charles R. Acland and William J. Buxton (Montreal and Kingston: McGill-Queen's University Press, 1999).

11   William Cronon, "Kennecott Journey: The Paths out of Town," in *Under an Open Sky: Rethinking America's Western Past*, eds. William Cronon, George Miles, and Jay Gitlin (New York: W. W. Norton, 1992); Kathryn Morse, *The Nature of Gold: An Environmental History of the Klondike Gold Rush* (Seattle: University of Washington Press, 2003). Important recent mining histories from south of the border that make similar connections include Thomas G. Andrews, *Killing for Coal: America's Deadliest Labor War* (Cambridge, MA: Harvard University Press, 2008); Gray Brechin, *Imperial San Francisco: Urban Power, Earthly Ruin* (Berkeley: University of California Press, 1999); Kathleen A. Brosnan, *Uniting Mountain and Plain: Cities, Law, and Environment along the Front Range* (Albuquerque: University of New Mexico Press, 2002); Andrew C. Isenberg, *Mining California: An Ecological History* (New York: Hill and Wang, 2005); Eugene Moehring, *Urbanism and Empire in the Far West, 1840–1890* (Reno: University of Nevada Press, 2004). Sociologists have made useful contributions in this regard as well: see for instance Scott Frickel and William R. Freudenburg, "Mining the Past: Historical Context and the Changing Implications of Natural Resource Extraction," *Social Problems* 43, no. 4 (1996): 444–66. For a solid contemporary overview focused on the Lower 48, see Lisa J. Wilson, "Riding the Resource Roller Coaster: Understanding Socioeconomic Differences between Mining Communities," *Rural Sociology* 69, no. 2 (2009): 261–81.

12   The first major critique of the Pine Point Mine and its impact on nearby Dene communities appeared in Thomas Berger's report on the Mackenzie Valley Pipeline Inquiry in the 1970s. This section merges staples theory with the Dene critiques that emerged during the hearings. See Thomas Berger, *Northern Frontier, Northern Homeland: The Report of the Mackenzie Valley Pipeline Inquiry, Vol. 1* (Toronto: James Lorimer and Co., 1977). One scholarly work that flowed out of the pipeline debate was Mel Watkins, ed., *Dene Nation: The Colony Within* (Toronto: University of Toronto Press, 1977). See also Lisa Sumi and Sandra Thomsen, *Mining in Remote Areas: Issues and Impacts* (Ottawa: MiningWatch Canada, 2001); Délįnę Uranium Team, *If Only We Had Known: The History of Port Radium as Told by the Sahtúot'įnę* (Délįnę, NWT: Délįnę Uranium Team, 2005); Heather Myers, *Uranium Mining in Port Radium, N.W.T: Old Wastes, New Concerns* (Ottawa: Canadian Arctic Resources Committee, August 1982); Janet E. Macpherson, "The Pine Point Mine" and "The Cyprus Anvil Mine," in *Northern Transitions, Volume I: Northern*

Resource Use and Land Use Policy Study, eds. Everett B. Peterson and Janet B. Wright (Ottawa: Canadian Arctic Resources Committee, 1978), 65–148. For brief scholarly overviews, see Claudia Notzke, *Aboriginal Peoples and Natural Resources in Canada* (North York, ON: Captus University Publications, 1994), 216–17; Saleem H. Ali, *Mining, the Environment, and Indigenous Development Conflicts* (Tucson: University of Arizona Press, 2003); William Hipwell, Katy Mamen, Viviane Weitzner, and Gail Whiteman, *Aboriginal People and Mining in Canada: Consultation, Participation, and Prospects for Change: Working Discussion Paper* (Ottawa: North-South Institute, 2002). For an excellent popular publication on the issue, see ethnographer Ellen Bielawski's *Rogue Diamonds: Northern Riches on Dene Lands* (Vancouver: Douglas and McIntyre, 2003). There is a large literature on the more recent negotiation of impact and benefit agreements between mining companies and northern Native communities. For an overview, see Levitan and Cameron, this volume.

13   Arn Keeling and John Sandlos, "Environmental Justice Goes Underground? Historical Notes from Canada's Mining Frontier," *Environmental Justice* 2 (2009): 117–25; links between mining and environmental justice are explored in: Subhabrata Bobby Banerjee, "Whose Land Is It Anyway? National Interest, Indigenous Stakeholders, and Colonial Discourses," *Organization & Environment* 13 (March 2000): 3–38; Al Gedicks, *Resource Rebels: Native Challenges to Mining and Oil Companies* (Cambridge, MA: South End Press, 2001); Robert Wesley Heber, "Indigenous Knowledge, Resources Use, and the Dene of Northern Saskatchewan," *Canadian Journal of Development Studies* 26 (2005): 247–56; Richard Howitt, *Rethinking Resource Management: Justice, Sustainability, and Indigenous Peoples* (London: Routledge, 2001); Marcus B. Lane and E. Rickson Roy, "Resource Development and Resource Dependency of Indigenous Communities: Australia's Jawoyn Aborigines and Mining at Coronation Hill," *Society and Natural Resources* 10 (1997): 121–42; Nicholas Low and Brendan Gleeson, "Situating Justice in the Environment: The Case of BHP at the Ok Tedi Copper Mine," *Antipode* 30 (1998): 201–26; Joan Martinez-Alier, "Mining Conflicts, Environmental Justice, and Valuation," *Journal of Hazardous Materials* 86 (2001): 153–70. For uranium mining and indigenous struggles in the US Southwest, see Doug Brugge, Timothy Benally, and Esther Yazzie-Lewis, *The Navajo People and Uranium Mining* (Albuquerque: University of New Mexico Press, 2006); Doug Brugge and R. Goble, "The History of Uranium Mining and the Navajo People," *American Journal of Public Health* 92 (2002): 1410–19.

14   Christian Brannstrom, "What Kind of History for What Kind of Political Ecology?" *Historical Geography* 32 (2004): 71–88; Karl Offen, "Historical Political Ecology: An Introduction," *Historical Geography* 32 (2004): 7–18. For introductions to the general field of political ecology, see Paul Robbins, *Political Ecology: A Critical Introduction* (Oxford: Blackwell, 2004); Karl S.

Zimmerer and Thomas J. Bassett, "Approaching Political Ecology: Society, Nature, and Scale in Human-Environment Studies," in *Political Ecology: An Integrative Approach to Geography and Environment-Development Studies*, eds. Karl S. Zimmerer and Thomas J. Bassett (New York: Guilford Press, 2003); Peter Walker, "Politics of Nature: An Overview of Political Ecology," *Capitalism, Nature, Socialism* 9 (March 1998): 131–44; Nicholas Low and Brendan Gleeson, *Justice, Society, and Nature: An Exploration of Political Ecology* (London: Routledge, 1998). For the application of "Third World" political ecology in a "First World" setting such as North America (and the problem of this dualism), see James McCarthy, "First World Political Ecology: Lessons from the Wise Use Movement," *Environment and Planning A* 34 (2002): 1281–1302; Richard A. Schroeder, Kevin St. Martin, and E. Albert Katherine, "Political Ecology in North America: Discovering the Third World Within?" *Geoforum* 37 (2006): 163–68; Peter A. Walker, "Reconsidering 'Regional' Political Ecologies: Toward a Political Ecology of the Rural American West," *Progress in Human Geography* 27 (2003): 7–24.

15 For an overview of mining and the political geography of resource extractive zones, see Gavin Bridge, "Resource Triumphalism: Postindustrial Narratives of Primary Commodity Production," *Environment and Planning A* 33 (2001): 2149–73, and Gavin Bridge, "Contested Terrain: Mining and the Environment," *Annual Review of Environment and Resources* 29 (2004): 205–59. The connections between the state, industry, and indigenous dispossession are explored by many authors, including Anthony Bebbington et al., "Contention and Ambiguity: Mining and the Possibilities of Development," *Development and Change* 39, no. 6 (2008): 965–92; Richard Howitt, *Rethinking Resource Management: Justice, Sustainability, and Indigenous Peoples* (London: Routledge, 2001). A Marxist literature examining these processes focuses on mining's role in what is termed "accumulation by dispossession" in various settings: see Rebecca Hall, "Diamond Mining in Canada's Northwest Territories: A Colonial Continuity," *Antipode* 45, no. 2 (2013): 376–93; William Holden, Kathleen Nadeau, and R. Daniel Jacobson, "Exemplifying Accumulation by Dispossession: Mining and Indigenous Peoples in the Philippines," *Geografiska Annaler: Series B, Human Geography* 93 (2): 141–61; Todd Gordon and Jeffery R. Webber, "Imperialism and Resistance: Canadian Mining Companies in Latin America," *Third World Quarterly* 29, no. 1 (2008): 63–87.

16 For studies that highlight northern resource development policy, see Rea, *The Political Economy of the Canadian North*; Frances Abele, "Canadian Contradictions: Forty Years of Northern Political Development," *Arctic* 40 (1987): 310–20; Morris Zaslow, *The Northward Expansion of Canada, 1914–1967* (Toronto: McClelland and Stewart, 1988); Shelagh Grant, *Sovereignty Or Security?: Government Policy in the Canadian North, 1936–1950* (Vancouver: UBC Press, 1988); Kerry Abel, *Drum Songs: Glimpses of Dene History*

(Montreal and Kingston: McGill-Queen's University Press, 1993); Liza Piper, *The Industrial Transformation of Subarctic Canada* (Vancouver: UBC Press, 2009); John Sandlos and Arn Keeling, "Claiming the New North: Mining and Colonialism at the Pine Point Mine, Northwest Territories, Canada," *Environment and History* 18, no. 1 (2012): 5–34.

17 Frank Tester and Peter Kulchyski, *Tammarniit (Mistakes): Inuit Relocation in the Eastern Arctic, 1939–63* (Vancouver: UBC Press, 1994); Jean-Sébastien Boutet, "Développement ferrifère et mondes autochtones au Québec subarctique, 1954–1983," *Recherches amérindiennes au Québec* 40, no. 3 (2010): 35–52; Caroline Desbiens, "Producing North and South: A Political Geography of Hydro Development in Quebec," *Canadian Geographer* 48, no. 2 (2004): 101–18; Mark O. Dickerson, *Whose North?: Political Change, Political Development, and Self-Government in the Northwest Territories* (Vancouver: UBC Press, 1992); R. Quinn Duffy, *The Road to Nunavut: The Progress of the Eastern Arctic Inuit since the Second World War* (Montreal and Kingston: McGill-Queen's University Press, 1988); Matthew Farish and P. Whitney Lackenbauer, "High Modernism in the Arctic: Planning Frobisher Bay and Inuvik," *Journal of Historical Geography* 35 (2009): 517–44; Frank Tough, *'As Their Natural Resources Fail': Native Peoples and the Economic History of Northern Manitoba, 1870–1930* (Vancouver: UBC Press, 1996); David Quiring, *CCF Colonialism in Northern Saskatchewan: Battling Parish Priests, Bootleggers, and Fur Sharks* (Vancouver: UBC Press, 2004).

18 See Heather Goodall, "Indigenous Peoples, Colonialism, and Memories of Environmental Injustice," in *Echoes from the Poisoned Well: Global Memories of Environmental Injustice*, eds. Sylvia Hood Washington, Paul C. Rosier, and Heather Goodall (Oxford: Lexington Books, 2006), 74–75.

19 Piper, *The Industrial Transformation of Subarctic Canada*, 95–96; Keeling, "Born in an Atomic Test Tube," 236.

20 The example of the Taltson Dam, built to power the Pine Point Mine, is discussed in Sandlos and Keeling, "Claiming the New North," 31, and Ellen Bielawski (in collaboration with the community of Lutsel k'e), The Desecration of Nanula Kué: Impact of the Talston Hydroelectric Development on Dene Soline, Unpublished report for the Royal Commission on Aboriginal Peoples (December 1993).

21 See chapters by O'Reilly and Gordon, this volume. See also Yellowknives Dene First Nation, *Weledeh Yellowknives Dene: A History* (Dettah, NWT: Yellowknives Dene First Nation Council, 1997), 53–54; Canada-Délı̨nę Uranium Table, *Final Report Concerning Health and Environmental Issues Related to the Port Radium Mine* (Ottawa: Indian and Northern Affairs Canada, 2005).

22 For more on Inuit relocations and hunger in the Eastern Arctic, see Tester and Kulchyski, *Tammarniit (Mistakes)*. For nutrition and disease, see Liza Piper

and John Sandlos, "A Broken Frontier: Ecological Imperialism in the Canadian North," *Environmental History* 12, no. 4 (2007): 759–95, and Liza Piper, "Nutritional Science, Health, and Changing Northern Environments," in *Big Country, Big Issues: Canada's Environment, Culture, and History*, eds. Nadine Klopfer and Christof Mauch (Munich: Rachel Carson Center Perspectives Series, 2011), 60–81; P. Whitney Lackenbauer and Matthew Farish, "The Cold War on Canadian Soil: Militarizing a Northern Environment," *Environmental History* 12, no. 4 (2007): 920–50.

23 Timothy J. LeCain, *Mass Destruction: The Men and Giant Mines That Wired America and Scarred the Planet* (New Brunswick, NJ: Rutgers University Press, 2009); Martin Lynch, *Mining in World History* (London: Reaktion, 2002); LeCain, *Mass Destruction*; Timothy LeCain, "The Limits of 'Eco-Efficiency': Arsenic Pollution and the Cottrell Electric Precipitator in the U.S. Copper Smelter Industry," *Environmental History* 5 (2000): 336–51; Chad Montrie, *To Save the Land and People: A History of Opposition to Surface Coal Mining in Appalachia* (Chapel Hill: University of North Carolina Press, 2002); Duane A. Smith, *Mining America: The Industry and the Environment, 1800–1980* (Lawrence: University Press of Kansas, 1987); David Stiller, *Wounding the West: Montana, Mining, and the Environment* (Lincoln: University of Nebraska Press, 2000). For the environmental impact and cross-border policy implications of secondary production facilities, specifically smelters, see John D. Wirth, *Smelter Smoke in North America: The Politics of Transborder Pollution* (Lawrence: University of Press of Kansas, 2000).

24 Richard V. Francaviglia, *Hard Places: Reading the Landscape of America's Historic Mining Districts* (Iowa City: University of Iowa Press, 1991); Ben Marsh, "Continuity and Decline in the Anthracite Towns of Pennsylvania," *Annals of the Association of American Geographers* 77, no. 3 (1987): 337–52; William Wyckoff, "Postindustrial Butte," *The Geographical Review* 85, no. 4 (1995): 478–96; David Robertson, *Hard as the Rock Itself: Place and Identity in the American Mining Town* (Boulder: University Press of Colorado, 2006). See also Peter Goin and C. Elizabeth Raymond, *Changing Mines in America* (Santa Fe: Center for American Places, 2004); Peter Goin and Elizabeth Raymond, "Living in Anthracite: Mining Landscape and Sense of Place in Wyoming Valley, Pennsylvania," *The Public Historian* 23, no. 2 (2001): 29–45; John Harner, "Place Identity and Copper Mining in Sonora, Mexico," *Annals of the Association of American Geographers* 91, no. 4 (2001): 660–80. For a nuanced application of these ideas to Northern Canada, see Liza Piper, "Subterranean Bodies: Mining the Large Lakes of North-West Canada, 1921–1960," *Environment and History* 13, no. 2 (2007): 155–86.

25 See Susan Jackson, ed., *Yellowknife, NWT: An Illustrated History* (Yellowknife, NWT: Nor'West Publishing, 1990); Terry Foster and Ronne Heming,

eds., *Yellowknife Tales: Sixty Years of Stories from Yellowknife* (Yellowknife, NWT: Outcrop, 2003); Northwest Territories Mining Heritage Society, "The Gold Mines Built Yellowknife," accessed February 28, 2012, http://www.nwt-miningheritage.com/files/frontend-static mininghistory/The%20Gold%20 Mines%20Built%20Yellowknife.pdf.

26   Tara Cater and Arn Keeling, "'That's Where Our Future Came From': Mining, Landscape, and Memory in Rankin Inlet, Nunavut," *Études/Inuit/Studies* 37, no. 2 (2013): 59–82.

27   Aaro E. Aho, *Hills of Silver: The Yukon's Mighty Keno Hill Mine* (Madeira Park, BC: Harbour Pub, 2006); Richard Geren and Blake McCullogh, *Cain's Legacy: The Building of Iron Ore Company of Canada* (Sept-Îles, QC: Iron Ore Company of Canada, 1990); Robert Bothwell, *Eldorado: Canada's National Uranium Company* (Toronto: University of Toronto Press, 1984).

28   For the web memorial, see "Pine Point Revisited," accessed May 7, 2012, http://pinepointrevisited.homestead.com/Pine_Point.html. For the multimedia documentary, see Paul Shoebridge and Michael Simons, *Welcome to Pine Point* (Montreal: National Film Board, 2011).

29   See, for example, Robert McPherson, *New Owners in Their Own Land: Minerals and Inuit Land Claims* (Calgary: University of Calgary Press, 2003). For a discussion of impact and benefit agreements, see Levitan and Cameron, this volume.

30   John Sandlos and Arn Keeling, "Zombie Mines and the (Over)Burden of History," *Solutions Journal* 4, no. 3 (2013): 80–83.

31   Office of the Auditor General of Canada, "Report of the Commissioner of the Environment and Sustainable Development, 2002" (Ottawa: Minister of Public Works and Services, 2002), chapters 2 and 3. A 2012 report from the same office reviewed the growing financial liabilities for federal contaminated sites remediation, highlighting Giant and Faro mines as two of the four highest reported liabilities. "Report of the Commissioner of the Environment and Sustainable Development – Spring 2012" (Ottawa: Minister of Public Works and Services, 2012), chapter 3.

32   Mines in the territorial north underdoing remediation and/or monitoring under the Northern Contaminants Program include Cantung, Colomac, Con, Contact Lake, Discovery, El Bonanza, Faro, Giant, Indor/Beaverlodge, Keno Hill, Mt. Nansen, Nanisivik, North Inca, Port Radium, Rayrock, Sawmill Bay, and Silver Bay Properties. Of these sites, Cantung, Contact Lake, Discovery, and Keno Hill are simultaneously being redeveloped for new mining activity. Pine Point and Lupin also have proposals for mine redevelopment pending. See AANDC, Contaminants and Remediation Directorate, "Contaminated Sites Remediation: What's Happening in the Sahtu?," March 2009, accessed February 19, 2014, http://publications.gc.ca/collections/

collection_2010/ainc-inac/R1-27-2009-eng.pdf; AANDC, Major Mineral Projects and Deposits North of 60 Degrees, accessed March 3, 2012, http://www.aadnc-aandc.gc.ca/DAM/DAM-INTER-HQ/STAGING/texte-text/mm_mmpd-ld_1333034932925_eng.html; AANDC, Northern Contaminants Program, last updated August 15, 2012, http://www.aadnc-aandc.gc.ca/eng/1100100035611.

33   For examples of publications using oral history in analyses of mining community formation and history, see Thomas Dublin, *When the Mines Closed: Stories of Struggle in Hard Times* (Ithaca, NY: Cornell University Press, 1998); Janet Finn, *Tracing the Veins: Of Copper, Culture, and Community from Butte to Chuquicamata* (Berkeley: University of California Press, 1998); Mary Murphy, *Mining Cultures: Men, Women, and Leisure in Butte, 1914–41* (Urbana: University of Illinois Press, 1997); Lynne Bowen, *Boss Whistle: The Coal Miners of Vancouver Island Remember*, rev. ed. (Nanaimo, BC: Rocky Point Books, 2002). Community-driven and public oral history projects related to mining are popular in mining regions and are too numerous to list!

34   Robertson, *Hard Places*; Goin and Raymond, "Living in Anthracite."

35   Karen Bescherer Metheny, *From the Miner's Doublehouse: Archaeology and Landscape in a Pennsylvania Coal Company Town* (Knoxville: University of Tennessee Press, 2007); Ashley Ward, "Reclaiming Place through Remembrance: Using Oral Histories in Geographic Research," *Historical Geography* 40 (2012): 133–45.

36   Déline Uranium Team, *If Only We Had Known*; Lianne Leddy, "Cold War Colonialism: The Serpent River First Nation and Uranium Mining, 1953–1988" (PhD thesis, Wilfrid Laurier University, 2011); Lianne Leddy, "Interviewing *Nookomis* and Other Reflections: The Promise of Community Collaboration," *Oral History Forum d'histoire orale* 30 (2010): 1–18; Brugge, Benally, and Yazzie-Lewis, *The Navajo People and Uranium Mining*.

37   Project researchers collected thousands of documents related to mining activity and mining policy from Library and Archives Canada in Ottawa, the Prince of Wales Northern Heritage Centre in Yellowknife, the Yukon Archives in Whitehorse, and other repositories. These records tend to be scattered and deal with individual mines and projects, but federal Indian Affairs and Northern Affairs records (RG22 and RG85 in the national archives) from this period were particularly significant for understanding government policy respecting mining and northern development. On our practice of digitizing archival documents for collaborative research, see Arn Keeling and John Sandlos, "Shooting the Archives: Document Digitization for Historical-Geographical Collaboration," *History Compass* 9, no. 5 (May 2011): 423–32.

38   Donald A. Ritchie, *Doing Oral History*, 2nd ed. (Oxford: Oxford University Press, 2003), chapter 1; Alistair Thomson, "Four Paradigm Transformations

in Oral History," *Oral History Review* 34, no. 1: 49–70; David Neufeld, "Parks Canada, The Commemoration of Canada, and Northern Aboriginal Oral History," in *Oral History and Public Memories*, ed. Paula Hamilton (Philadelphia: Temple University Press, 2008), 26–48.

39 Margaret Kovach, "Emerging from the Margins: Indigenous Methodologies," in *Research as Resistance: Critical, Indigenous, and Anti-oppressive Approaches*, ed. Leslie A. Brown and Susan Strega (Toronto: Canadian Scholars' Press, 2005), 19–36; Kathy Absolon and Cam Willett, "Putting Ourselves Forward: Location in Aboriginal Research," in *Research as Resistance*, 97–126; Shawn Wilson, *Research Is Ceremony: Indigenous Research Methods* (Winnipeg: Fernwood Publishing, 2008).

40 On the importance of contexts of production of oral histories, see Laura Cameron, "Listening for Pleasure," *Native Studies Review* 11, no. 1 (1996): 109–29; Julie Cruikshank, *The Social Life of Stories: Narrative and Knowledge in the Yukon Territory* (Vancouver: UBC Press, 1998).

41 Linda Shopes, "Oral History and the Study of Communities: Problems, Paradoxes, and Possibilities," *Journal of American History* 89, no. 2 (Sept. 2002): 588–98.

42 Ibid., 597.

43 Paula Hamilton and Linda Shopes, "Introduction: Building Partnerships between Oral History and Memory Studies," in *Oral History and Public Memories*, vii–xvii; David Lowenthal, *The Past Is a Foreign Country* (Cambridge: Cambridge University Press, 1985), chapter 5; Dydia DeLyser, "Authenticity on the Ground: Engaging the Past in a California Ghost Town," *Annals of the Association of American Geographers* 89, no. 4 (1999): 602–32. For a useful introduction to the idea of collective "social" memory, see Bernd Steinbock, *Social Memory in Athenian Public Discourse: Uses and Meanings of the Past* (Ann Arbor: University of Michigan Press, 2013), 7–19.

44 Metheny, *From the Miner's Doublehouse*, 252.

45 These guidelines are outlined at Panel on Research Ethics, *TCPS 2—2nd edition of Tri-Council Policy Statement: Ethical Conduct for Research Involving Humans*, accessed February 19, 2014, http://www.pre.ethics.gc.ca/eng/policy-politique/initiatives/tcps2-eptc2/Default/.

46 A methodological and political risk highlighted in the thoughtful introduction in Jean-Sébastien Boutet, "Opening Ungava to Industry: A Decentring Approach to Indigenous History in Subarctic Québec, 1937–54," *Cultural Geographies* 21, no. 1 (January 2014): 79–97.

47 Virginia Valerie Gibson, "Negotiated Spaces: Work, Home, and Relationships in the Dene Diamond Economy" (PhD diss., University of British Columbia, 2008), 81–82.

Section 1
Mining and Memory

| CHAPTER 1

From Igloo to Mine Shaft: Inuit Labour and Memory at the Rankin Inlet Nickel Mine

Arn Keeling and Patricia Boulter

The North Rankin Nickel Mine in Kangiqiniq (Rankin Inlet)[1] has been closed now for over fifty years, and its iconic headframe lost to fire in the late 1970s. Yet the sense of connection and identification with the mine remains strong in the town. At a workshop held in 2011, nearly seventy people gathered to share stories, examine historical photographs, and watch a screening of the National Film Board documentary "People of the Rock."[2] The film, made in 1961 and depicting a somewhat sanitized version of the Rankin Inlet mine story, nevertheless elicited poignant memories from elders in the room, who shared stories of lost loved ones and memories of the mining days and their youth. Community members gathered around archival photos of Inuit miners at work, pointing out friends and relations, and helping in some cases with identification (Figs. 1, 2). The mood of the gathering, while reflective, was also celebratory, a

FIGURE 1: Workers joking with a supervisor at the nickel mine [Harry Liberal, Titi Kudlu, Noah Kumakjuaq and Andy Easton]. Photo by Kryn Taconis. Library and Archives Canada photo PA-175565.

chance for the community to once again honour its founding members and partake in stories and images of the mining days.

In many ways, a commemorative gathering such as this workshop might seem unremarkable. Indeed, a wide-ranging literature on mining history and heritage explores the persistence of local identities as "mining communities" long after the end of mining. Historically, mining communities have been known for the strong sense of worker solidarity forged in the often-extreme "workscapes" of mining, a solidarity that may extend to community responses to the crisis of deindustrialization.[3] Through rituals, commemorative activities, and a typically strong identification with the mining landscape, many mining communities demonstrate their resilience in the face of the often-devastating economic and social changes associated with mine closure. Describing this tenacious sense of place within marginal communities and landscapes, Ben Marsh notes that, for many people, "land retains its meaning long after the means are exhausted."[4] While perhaps easily dismissed as exercises in nostalgia, or critiqued for their masking of the social inequities and environmental degradation associated with mineral development and deindustrialization, these perspectives nevertheless capture something of the intense identity formation connected with the experience of mining labour and the role of these identities in fostering community spirit and resilience in the face of economic decline.[5]

FIGURE 2: Two men stand and talk in front of Rankin Inlet Nickel Mines Ltd., July 1961. Douglas Wilkinson fonds/Nunavut Archives/N-1979-051: 2316

But the Kangiqiniq workshop also highlights the absence from these accounts of the experience of indigenous people in many mining regions. Although indigenous people have long worked in and for mines in various capacities, their mining history is usually explored in terms of their dispossession, exclusion, marginalization, and experience of landscape degradation associated with (neo)colonial mineral development. As Bridge and Frederiksen note (with reference to Nigeria), mining was "part and parcel of the process of socio-ecological modernization" of indigenous territories globally, and "the principal means by which [these territories] became incorporated into a world economy under conditions of colonial rule."[6] Yet indigenous people did not merely suffer through or resist mineral development; they also participated (willingly or otherwise) and, in so doing, became miners. For instance, classic anthropological studies in Bolivia explored the development of an indigenous mining identity there, associated with the long history of colonial and modern mining in the Andean region.[7] Nevertheless, few studies (beyond those in this volume) have sought to capture indigenous people's parallel historical development of mining identities and their experience of mine closure, either as workers or as broader participants in local economic and settlement life.[8]

The history of the North Rankin Nickel Mine provides insights into the complex indigenous experience of mining as an agent of socio-economic change. Founded on a rich nickel deposit located on the western shore of Hudson Bay in present-day Nunavut, the mine formed the basis for the settlement at Kangiqiniq in the late 1950s. Regarded as an experiment in Inuit modernization and a solution to a perceived crisis affecting traditional resources, the mine's short operational life (1957–1962) belied its importance as Canada's first Arctic mine and the first to actively promote the employment of indigenous workers. At its peak, Inuit employees, virtually all of whom moved to Rankin Inlet with no experience of wage work, comprised about 70 per cent of the mine's workforce, as both underground and surface workers. The mine's sudden closure in 1962 devastated the local economy, threatening the community's very survival and forcing many Inuit to leave the community to seek alternate employment or to return to traditional harvesting activities.

While the story of the Rankin Inlet mine has been explored by various authors, few have incorporated first-hand perspectives of Inuit miners themselves.[9] Based on oral histories conducted in Kangiqiniq in 2011 with the assistance of Inuit researchers, this essay explores the history of mining at Kangiqiniq and the emergence of a mining identity among Inuit miners.[10] These perspectives are supplemented by archival research on the history of the mine and its relation to government and company policy in the region. In spite of its short, tumultuous life and sudden collapse, the mine remains central to the identity of the community and its Inuit and non-Inuit residents alike. The memories shared by Inuit miners about Kangiqiniq provide important insights into the experience of indigenous workers with mineral development, the transition to industrial modernity, and the impacts of mine closure. The continued close identification with the mine by these elders, and by the broader Kangiqiniq community, reveals the unique and persistent sense of heritage and place connected to the mining experience.

****

The North Rankin Nickel Mine, which commenced construction in the early 1950s and produced its first ore in 1957, operated in the context of rapid socio-economic change in the postwar Eastern Arctic. Before the 1950s, northern agencies, including the RCMP and federal government officials, strove to prohibit Inuit from congregating around permanent settlements by enforcing a strict "policy of dispersal." This policy aimed to protect traditional Inuit land-based culture, but more importantly, to ensure Inuit would not become a financial burden to the nation. At the same time, contradictory educational and settlement policies sought both to elevate Inuit from their "primitive" position through modernizing their social and cultural practices while at the same time preserving their "independent" and "traditional" lifeways.[11] Early in the 1950s, with the collapse of Arctic fox fur prices and a perceived crisis in caribou populations, the Department of Northern Affairs adopted an increasingly interventionist policy in the Eastern Arctic.[12] These developments, and the negative publicity surrounding desperate Inuit living conditions (including episodes of starvation among inland Inuit groups) largely ended the laissez faire attitude of the government, which shifted to the promotion of wage labour as a solution to the "Eskimo problem."

Along with work in construction at Distant Early Warning (DEW) Line stations, the mine at Rankin Inlet appeared to offer an opportunity to shift Inuit away from their seemingly precarious land-based economy and toward industrial wage labour and settlement life. As a Canadian Press reporter noted in 1958, "[Government] officials here call the Rankin experiment a 'bright shining light' against the general background of the Eskimo Problem. Sustained success would mean a lot in the program to integrate the Eskimo from his stone-age past into the 'time clock'

FIGURE 3: One of the supervisors at the nickel mine [Singiituq]. Photo by Kryn Taconis/ Library and Archives Canada photo PA-175593.

world."[13] Writing to the mine company's secretary for information about the operation, Deputy Minister of Northern Affairs R. Gordon Robertson suggested that "because of the steadily increasing inroads on the wildlife resources of the North, it is going to be necessary to have more Eskimos adapted to wage employment as their means of livelihood."[14] To this end, the Department of Northern Affairs and National Resources and other government agencies in the region, such as the RCMP, assisted the North Rankin Nickel Mine in identifying suitable Inuit candidates for employment.[15] As the mine's first general manager Ken Whatmough recalled, the company also drew on the connections of Singiituq, a boat pilot from Chesterfield Inlet, to locate local labour when needed.[16]

For its part, the company initially embraced Inuit labour as a seasonal workforce, and Inuit were engaged in construction and stevedore work (as well as trade) at Rankin Inlet as early as 1953. By 1956–57, as the mine shifted to production, Inuit were recruited to Rankin Inlet in increasing numbers. North Rankin Nickel Mine (NRNM) president W. W. Weber told Northern Affairs officials he was "strongly in favor of employing as many Eskimos as possible" and integrating them permanently into the life of the mining camp.[17] That year, Inuit employment increased from fourteen to eighty workers and, under mine manager J. Andrew Easton, Inuit workers became integrated into nearly all aspects of the operation, including (eventually) underground work. For NRNM, in spite of language and cultural barriers, Inuit workers provided a ready and "cheerful" labour force that helped the company deal with the challenge and expense of attracting and retaining southern mine workers in this remote location. With the mine's financial viability in question from the outset, Inuit labour presented an important means of reducing costs.[18]

Though the bulk of the recruits were from the Chesterfield Inlet area of northwestern Hudson Bay, Inuit migrated to the new settlement from across the Kivalliq (then known as Keewatin) region. Interviewees recalled travelling by dog sled, airplane, and Peterhead boat to Rankin Inlet in the mid- to late 1950s. Ollie Ittinuar, who was an RCMP special constable at Chesterfield Inlet in the early 1950s, recalled a community meeting at which people were asked if they would like to work at the mine. Seeking better wages, Ittinuar travelled by dog team to Rankin Inlet with another family. "As soon as we got over there the mine people came over and asked us to work for them right away, so we got to work right away upon arrival. So that's how the job started with the mine."[19] Others, like Joachim Kavik and Francis Kapuk, lived with their families in the area (Kavik at Meliadine Lake, Kapuk at Baker Foreland), so they were well aware of the developments at Rankin Inlet. Kapuk recalled:

> When we came here from caribou hunting we would come back here and stay with people here . . . They were living in igloos at that time when we would come here . . . There was white people who came in at that time as well who came to tell us there were employment opportunities at the mine, who said if you hear of anyone who wants to work for the mine, pass it around.

Others, like Thomas Tudlik (from north of Chesterfield Inlet) and John Towtoongie (from Coral Harbour), had previously left their communities to seek opportunities elsewhere before coming to Rankin Inlet. Veronica Manilak flew to Rankin Inlet from Repulse Bay with her husband in 1961, where they joined her father, who was already working at the mine.

In the *Keewatin Journal* (a recollection of life in the region from the 1950s–70s published in 1979), several people recounted how between 1958 and 1959 they were no longer able to make a living or survive off the land due to the shortage in caribou, and therefore had no alternative other than to work in the mine.[20] Similarly, several miners we interviewed recalled how their moves brought a dramatic, almost immediate change from a predominantly land-based lifestyle of hunting and trapping to industrial work and settlement life. "We lived as a true natured Inuit when I was growing up," noted Thomas Tudlik. "For example, we had no family allowances. In those days we had no governments, no established governments during the time I was growing up. We had the Hudson Bay Company, where we traded furs and things like that." Joachim Kavik said that his father, while happy living on the land, was encouraged to come work at the mine and eventually did so. "My father was a very traditional Inuk he did not speak one word of English when he started working for the mine." His family wanted to keep their dogs for hunting, so they lived about a mile from the Rankin settlement.

The transition to settlement life posed challenges for Inuit workers and their families. Both settlement and work life for Rankin Inuit was highly structured by the paternalistic yet often contradictory policies of the company and Northern Affairs officials. Initially, Inuit workers lived in sod houses and tents at the fringes of the nascent settlement (see Fig. 4), while southern workers occupied bunkhouses near the mine. Francis Kapuk recalled difficult living conditions: "In the early stages when we came here we were living in a tent, we didn't really stay in an igloo but we stayed in a tent. We fixed it up by putting together whatever cardboard we could find as insulation inside the tent." Veronica Manilak remembers the challenges of starting over in a new community: "I found it rather boring, extremely lonely [and] strange, because when we left we left everything at home including our husky dogs including our belongings. We left home with nothing, we left our tent, we left everything in there, the only thing we took were our children and our rifle and that's pretty much it."

Migration to Rankin Inlet brought Inuit from different regional dialect and kinship groups together with white southerners in the "cosmopolitan" setting of Rankin Inlet, and the settlement was initially segregated by race and kinship. As it developed, the townsite was divided into

FIGURE 4: Reinforced tent house in Rankin Inlet, early 1950s. Photo by Kenneth Whatmough, Consulting Engineer and General Manager, North Rankin Nickel Mines Limited/Nunavut Archives.

three distinct sections: two containing Inuit residents (the New and Old Eskimo Settlements) and the other meant for non-Inuit, Euro-Canadian personnel of the mine, government, or other institutions.[21] The "White Settlement" comprised two neat rows of buildings: houses for the federal northern service officer, the HBC store, the school, and the Roman Catholic and Anglican missions. In this section, houses were supplied with heat, water, and sewage lines from the mine. "The New Eskimo Settlement" was separated from the white settlement and located further from the mine buildings. It was closely monitored by the NRNM and considered off limits to non-Inuit mine workers. The houses here were prefabricated, three-room structures. Although these houses all had electrical lighting and regular garbage collection provided for by the mine, they did not have centralized heating or running water. Instead, Inuit heated their houses with cooking stoves and had to walk to get water. Makeshift outhouses were often shared by several families. Located a quarter mile north of the mine's headframe, the "Old Eskimo Settlement" consisted of several clusters of shacks and tents, many built by Inuit themselves. In this section, the mine provided no essential services and "debris [was] . . . scattered everywhere."[22] This area was typically occupied by those who had lower-paying positions within the mine. However, many Inuit families preferred living in the Old Settlement, for they could live near kin and practise their traditional pursuits easily in the less monitored environment. The settlement's segregated nature in this period was reinforced by government officials when interior Padlirmiut suffering from starvation in the winter of 1957–58 were relocated and settled in a nearby Keewatin Rehabilitation Project camp called Itivia.[23]

Work for wages was, of course, the principal attraction of Rankin Inlet. Initially, Inuit workers, particularly those doing construction work, were paid less than non-Inuit (as little as 60 cents per hour).[24] The company and federal officials reacted strongly to accusations from the local Oblate missionary, Father Fafard, and others that they perpetuated second-class status and wages among Inuit, noting that Inuit workers also received meals in the mine's mess hall and, by the end of the 1950s, free housing and stove oil.[25] By the time Inuit workers began working underground, wages were on par with those paid non-Inuit workers for similar jobs. Although the hours were long and the underground environment unfamiliar, Inuit miners appreciated the security of regular wages and, as one miner's wife recalled, the opportunity to purchase goods at the mine commissary and Hudson Bay store.[26] Nevertheless, Inuit miners' pay packets were controlled at first by the northern service officer, "in order to ensure that at the outset at least their earnings are used to purchase only essential goods and not frittered away on non-essentials" like luxury goods.[27]

As in most aspects of settlement life, workplace segregation (Inuit and non-Inuit initially ate separately, for instance) gave way to gradual, if partial, integration of Inuit and non-Inuit. Although few spoke any English at all (and no white miners spoke Inuktitut), several Inuit miners recalled how they learned by example how to drill, blast, and operate underground: according to Peter Ipkarnerk, "The white people that we worked [with], they were there to show us how to do things, so we learned by example, by looking, by observing." In this way, Inuit workers learned to operate machinery above and below ground, trained as mechanics (such as plumbers), and joined non-Inuit workers on the mine's emergency rescue team. Others, like Jack Kabvitok, worked in construction and performed other services around the settlement before working in the mine itself. The former miners suggested work relations with qallunaat (non-Inuit) workers were good, overall, even jokingly remembering learning how to swear in English from them. Although some tensions and conflicts with supervisors were recalled, the "big boss," mine manager Andy Easton, was fondly remembered as a fair man who cared for the Inuit workers.

For the mine and Northern Affairs officials, one of the main challenges in employing Inuit was the inculcation of the norms and values of wage work, including time discipline. Southern employees, working on seasonal rotation and living in bunkhouses away from families, worked seven days a week, but Inuit workers balked at this schedule (although they worked these hours during construction). As early as 1956, a northern service officer reported that "the Eskimos stated that they were satisfied with the pay and working conditions but that they would like to have time off for hunting meat for their wives and children."[28] From time to time, Inuit employees would fail to show up for shifts in order to go hunting, a source of friction with the mine. Working with Northern Affairs officials, mine management devised a system whereby Inuit could request leave to hunt, so long as they helped to find a shift replacement in advance. As former miner Peter Ipkarnerk recounted, "We . . . made a request to our supervisors, to the authorities, to go out and hunt . . . as long as they agreed then we could go out in the middle of the week." This option was important to the miners, not only materially in helping feed their families, but also culturally. According to Thomas Tudlik: "the work was very important, the fact that we had to work all the time. But, at the same time we are meat eaters, so we used to go out hunting." In 1959, when the mine reinstituted seven-day weeks, Inuit miners complained and the northern service officer wrote to manager Andrew Easton, noting that Inuit needed time off to practise "their traditional occupations," with a view to a "return to life on the land" after mining.[29]

Invoking ideas of acculturation and citizenship, many federal officials regarded employment at Rankin Inlet as furthering the Inuit "adjustment and integration into our industrial society" with "the same opportunities to develop their talents as other Canadians."[30] Inuit from the coastal regions north of Rankin Inlet (Chesterfield Inlet and Repulse Bay) were considered more acculturated to living in settled communities and participating in wage labour ventures, due to their interaction with whalers and history of Distant Early Warning Line construction employment. For its part, the mining company (which dominated community life) remained mainly concerned with the performance of Inuit as workers, not their status as citizens.[31] As the anthropologist Robert Dailey and his wife Lois noted in their 1958 report on the community, "Those Eskimo [sic] that are frugal, hard working, punctual, and cooperative, are in the eyes of the mine 'desirable.' Those who do not readily adjust or who do not pay close enough attention to orders, or who malinger, are rejected and forced to leave the community."[32] Similarly, the anthropologist, social worker, and long-time Rankin Inlet resident Bob Williamson concluded that in spite of the ready adaptation by many to mine work, "the Eskimo did not completely identify with the mine and management objectives." Indeed, some Inuit who worked at the mine chose to leave, returning to former communities or moving to more "traditional" communities like Whale Cove.[33] Clearly some, like John Towtoongie, who told us of splitting his time between living in Whale Cove and working in Rankin Inlet, preferred to retain their connections to hunting life over a full-time commitment to mining.

The emerging social life of the community reflected a similar pattern of segregation, then partial integration. Early efforts to limit and monitor contact between the Inuit and non-Inuit of the community extended to social functions.[34] In spite of a desire to acculturate and "modernize" Inuit, the many entertainments organized within the community were initially racially segregated, and a separate movie night and dance were held each week for Inuit and non-Inuit. Non-Inuit men were strictly prohibited from entering the "Eskimo Village," or they would be fired.[35] Similarly, Inuit women were not allowed access to the male bunkhouses and could not enter the commissary except on Saturdays and only in the company of their husbands. Although later relaxed, these policies aimed to prevent drinking, gambling, and, most of all, liaisons between non-Inuit men and Inuit women, to avoid sexually transmitted diseases, adultery, prostitution, and unwanted pregnancies. In spite of these restrictions, as Ollie

Ittinuar's son Peter, who grew up in Rankin Inlet, commented, "in a way it was more integration than had ever existed before, even though it was segregated. People took part in the same activities, there was community dances, movies . . . white people and Inuit people went to these things at the same time, went to church at the same time." Elders fondly recalled fiddle dances and cowboy films attended by both Inuit and qallunaat. They noted that beer drinking, too, became a feature of community life, a source of bonding between Inuit and non-Inuit workers, but also a source of problems for some workers.

By the early 1960s, Rankin Inlet was a thriving community of about 600 Inuit and non-Inuit, with government offices (housed by the mine), an RCMP detachment, three religious missions, and a Hudson's Bay store. But the North Rankin Nickel Mine, from the outset a financially precarious operation, began to seriously founder in 1960 and on April 3, 1962, the *Globe and Mail* reported that the mine had closed, throwing seventy "Eskimos" out of work and threatening the future of the town. Deputy Minister of Northern Affairs R. G. Robertson ruefully reflected, "The problem with mines is that they run out," while expressing hopes that displaced Inuit workers might eventually find work elsewhere and promising government support in the interim.[36] In another story, business reporter Stanley Twardy, playing on the popular image in the southern press of the modernized Inuit miner, was less sanguine: "Hoping to bring civilization to the Rankin Eskimos the Government sold them on installment-plan-buying of wooden huts and the Hudson's Bay store accepted their credit on other long-term purchases. Some Eskimos learned to live in style and purchased refrigerators, which are operated from the mine's electric supply."[37] A month later, newspapers were reporting the government's emergency response to the emerging "Keewatin crisis" and forecasting "an end to the unique inland caribou people" occasioned by declining caribou herds, relocation to the coast and, now, economic displacement.[38]

Northern Affairs officials, who had so eagerly encouraged Inuit to migrate to Rankin Inlet for work, veered between paralysis and panic at the prospect of the mine's closure and mass unemployment in the region. As early as 1959, the area administrator for Rankin Inlet had urged government planning for closure, noting "wage employment is the basic need of the people here" but concluding, grimly, that "some of the people might have been better off in the long run had they never entered the field of wage employment."[39] Through a series of reports, correspondence, and conferences on the future of the region in the wake of mine closure, suggestions for the community ranged from the creation of alternative industries to a return to traditional semi-subsistence activities to the complete depopulation of Rankin Inlet and relocation of its residents (either to other Arctic communities, or to Southern Canada).[40] In the end, the unofficial "plan" for Rankin Inlet's post-mining future consisted of a chaotic series of initiatives that involved elements of voluntary relocation, migration (back to previous settlements), economic diversification, and (eventually) the move of Northern Affairs' Keewatin regional headquarters from Churchill, Manitoba, to Rankin Inlet as an economic stimulus. For its part, the mining company and its non-Native employees simply walked away from the settlement, after selling much of the town's infrastructure to the federal government.[41]

Inuit workers and their families recalled closure as a time of hardship and adjustment. As two workers interviewed by filmmaker Peter Ittinuar in the early 1970s recalled, with few work opportunities in Rankin Inlet, many Inuit returned to their former communities at their own expense. "I thought there was going to be permanent mining activity, and I thought there was going to be a lot of employment," David Iglukak told Ittinuar.[42] While Northern Affairs officials encouraged Inuit to leave the community and to return to land-based activities for survival, many who had left this life struggled to re-adjust. Veronica Manilak recalled:

> We became extremely poor after the mine closed. We were as a matter of fact very hungry at times. . . . There was a man named Batiste who had dogs, and with his dogs my husband went out caribou hunting one day and he got lots of caribou and we got lots of meat at that point. We had no more snowmobiles and things like that because the mine had closed and we became extremely poor.

As Shingituk [Singiituq], the Inuk foreman, pointed out in a meeting with Rankin Inlet's Eskimo Affairs Council in February 1962, many people who had moved to Rankin Inlet "no longer had the type of equipment they would need to return to the land"—especially dogs.[43] Belying their supposed status as working Canadian citizens, Inuit workers were excluded from unemployment insurance benefits, and welfare payments in the community skyrocketed in 1963. Reflecting the efforts to push Inuit back onto the land, welfare rates were adjusted to account for an individual's ability (though not necessarily success) to obtain "country food."[44] For its part, the mining company, Veronica Manilak suggested, "just left us behind," and the interviewees in Peter Ittinuar's film lamented the hardship and reliance on government assistance that resulted.

Contemporary interviews illustrate the diverse strategies for survival pursued by the unemployed miners. Many families (indeed the majority in Rankin Inlet) did not want to leave; as Jack Kabvitok noted, "Rankin Inlet had become my home." He sold carvings to stake hunting activities ("it was a bit of a struggle at that point because we had practically nothing, even though at that time I had five dogs"), and eventually found work with the town government. John Towtoongie recalled that people who returned to Arviat and Whale Cove also found it difficult to hunt for a living. "I was actually about the only one with a team of dogs, along with [another] man. When we would go out caribou hunting for example, there were a few caribou around at that time, when we would catch caribou we would distribute the meat among the other people." Francis Kapuk briefly returned to Chesterfield Inlet to hunt seal, then was recruited to return to Rankin Inlet to work in the government-established fish canning enterprise. This and other government-sponsored arts and crafts initiatives, including a sewing centre and ceramics studio, provided some income for the families of displaced miners, and opened up wage-earning opportunities for women.[45]

As many noted at the time, however, Rankin Inuit no longer desired to live fully off the land, but instead wished to be given the choice of practising their traditional lifeways, participating in wage labour opportunities, or balancing the two. Many of the miners, in fact, wanted to continue mine work, and Northern Affairs officials sought to match Inuit mine workers with other industrial opportunities elsewhere in the North, with some success. With government assistance, Inuit from Rankin Inlet relocated to work in mines in Quebec (at the request of former NRNM manager Andy Easton), Manitoba, the Yukon, and the

Northwest Territories; as well, they went to work on the construction of the Great Slave Lake Railway line from Alberta to the Pine Point Mine in the NWT.[46] Peter Ipkarnerk and Francis Kapuk, along with two other miners, were flown to Yellowknife in late 1963 to work at Con Mine. Ollie Ittinuar, after a short stint at the Asbestos Hill Mine in Quebec, joined several other Rankin families at the Sherritt Gordon mine in Lynn Lake, Manitoba, where he worked for nine years, eventually becoming a shift boss. Some relocated miners brought their families (like the Ittinuars); others, like Joachim Kavik, lived in bunkhouses or in rented accommodations with other miners.

These moves (as with the relocations to Rankin Inlet) were treated by Northern Affairs officials as an "experiment," so they were closely monitored and reported on, including in the press.[47] As with Rankin Inlet, the redeployment of Inuit miners at other northern sites was intended to reduce labour turnover costs at northern mines and to continue the process of Inuit social development. Government reports and correspondence document the miners' struggle to adjust to new surroundings and their separation from families, but also the successful transplant of many workers.[48] Several workers moved repeatedly between Rankin Inlet and different locations or jobs as they sought personal stability and opportunity. Separation from family was, oftentimes, an obstacle to long-term employment at mines and communities far from Rankin Inlet. Others, particularly those like the Ittinuars whose family remained together, thrived in their new locations. Many miners gained reputations as valued employees and went on to work at mines across the North, including Cullaton Lake, Nanisivik, and others.

Nevertheless, for many miners, Rankin Inlet continued to draw them back, in spite of the community's challenges. Upon returning from a few years working in Yellowknife, Peter Ipkarnerk found that "it was lonely when we came back here to Rankin Inlet. Looking at the mine . . . seeing the mine closed it was very lonely because that was the only place where we were making money, where we were able to work and make money." He continued working on long rotations away from the settlement before retiring as a miner in 1969. After a long period as a miner in Lynn Lake, eventually Ollie Ittinuar also brought his family back to Rankin Inlet, where he opened a coffee shop. As Williamson and Foster noted in their

1975 report on the relocations, most non-Inuit observers regarded the return of the miners to Rankin Inlet as a failure of the relocation and re-establishment program.[49] But it appears many workers simply seized the opportunity to return to the community they and their families now called home, particularly once the establishment of regional administration in Rankin Inlet somewhat stabilized the community in the 1970s. Many of the original Rankin mining families became leading families in the town; their names adorn the lists of past town councillors and mayors posted on the wall of the hamlet chamber, and they are regularly honoured as the founding generation of the community. As several miners we interviewed told us, mining is still regarded as the community's reason for being and as a shaper of its character.[50]

****

Asked about the hard work performed by the miners of Rankin Inlet, Peter Ipkarnerk offered this analogy:

> Inuit are no strangers to hard work . . . many years ago Inuit lived a very difficult life with the contact with the white man, you know. We would receive matches for example, in order to save one stick of match, for example, we used to split in half so that we'd have another match, another a bit more match . . . to light the Inuit oil lamp. My parents, for example, had maybe two bags of tea and those bags of tea would last for a very long time. So we were always aware of the hard work that we did, that we used to do years and years ago.

Ipkarnerk's story, like those of some others we interviewed, reflects the connection of his Inuit values of industriousness and adaptability with his life experience of mine work. Although wage employment, as Pamela Stern points out, is often contrasted with subsistence work,[51] Ipkarnerk's comment suggests a kind of continuity or connection between these activities and the struggles associated with them.

In the interviews we conducted, the miners' articulation of a strong sense of identity *as* miners does not supplant, but is folded into their sense of Inuitness and seen as contiguous with it. While clearly influenced and perhaps to some degree controlled by federal officials and mine managers in the settlement, Inuit miners also pursued goals and practices commensurate with their own "life projects," whether hunting for country food or leaving jobs altogether to return to their families and homes.[52] In some cases (perhaps among the miners we were unable to interview), former workers might not have identified so closely or positively with their mining experience. Nevertheless, the miners we interviewed talked not only with evident pride about their achievements working underground in the Rankin Inlet mine and elsewhere, but also about the importance of hunting and language, their work ethic, and their sense of connection to Kangiqiniq as an Arctic place. In the crucible of their struggles to adapt and survive in difficult circumstances, whether environmental or economic, identities were forged and transformed: the miners' and that of the Kangiqiniq community.[53]

The Rankin Inlet story, as told in the archives, in film, and in the oral histories of miners and their spouses, adds a significant indigenous dimension to the stories of work, community life, and survival in mining and post-mining communities. As Katharine Rollwagen observed in relation to the Britannia Beach mine in British Columbia, "employees' experiences during these crises . . . remind us that resource-town closures cannot be characterized as inevitable or tragic; these are dynamic periods of intense change, shaped by both material realities, such as income and commodity prices, and discursive factors, such as loyalty and community"—to which we would add, identity.[54] In some cases, elders' memories of the mining experience at Rankin Inlet seem to have been filtered, to some extent, through the lens of personal and collective nostalgia, and the sharp edges of social struggle and economic hardship dulled somewhat by the passage of time. But as Piita Irniq, former commissioner of Nunavut and Kangiqiniq resident in the 1980s and 1990s, noted in an interview, the experience of moving "from igloo to mine shaft" in a single generation also resonates with the larger story of Inuit resilience in the face of colonialism and rapid socio-economic and cultural change. While federal officials at the time promoted mineral development as an instrument of Inuit acculturation and assimilation into modern Canadian society, what the mine's history and its aftermath show is that while Inuit successfully, in many cases, became miners, they definitively remained

Inuit. Similar to the "entanglements of industry and indigeneity" documented by Jean-Sébastien Boutet in his account of Naskapi and Innu communities near Schefferville, Quebec, Inuit in Rankin Inlet embraced a variety of strategies as they pursued their life projects in the context of rapidly changing historical-geographical circumstances, including environmental change, the growing influence of colonial forces in their lives, and the opportunities and challenges presented by industrial development and decline.[55]

NOTES

1. Since the conclusion of the Nunavut Land Claims Agreement and the creation of the Nunavut Territory in 1998, many of the English-language place names in Nunavut have reverted to their Inuktitut names. Although the Hamlet of Rankin Inlet continues to use its English name, in this chapter, generic reference to the contemporary place will be the Inuktitut "Kangiqiniq," meaning "deep bay." Historical references to Rankin Inlet will remain in English.

2. *People of the Rock*, directed by Clarke Daprato (National Film Board of Canada, 1961). The workshop was hosted by the authors, with the help of the research assistants noted below (see note 10) and the vital translation assistance of Piita Irniq. The authors wish to thank these contributors, the workshop participants, and the Hamlet of Rankin Inlet, and to acknowledge funding support from ArcticNet for this research.

3. Thomas G. Andrews, *Killing for Coal: America's Deadliest Labor War* (Cambridge, MA: Harvard University Press, 2008); Janet L. Finn, *Tracing the Veins: Of Copper, Culture, and Community from Butte to Chuquicamata* (Berkeley: University of California Press, 1998); Katharine Rollwagen, "When Ghosts Hovered: Community and Crisis in a Company Town, Britannia Beach, British Columbia, 1957–1965," *Urban History Review/Revue d'histoire urbaine* 35, no. 2 (2007): 25–36; William Wyckoff, "Postindustrial Butte," *Geographical Review* 85, no. 4 (1995): 478–96.

4. Ben Marsh, "Continuity and Decline in the Anthracite Towns of Pennsylvania," *Annals of the Association of American Geographers* 77, no. 3 (1987): 351.

5. Thomas Dublin, *When the Mines Closed: Stories of Struggle in Hard Times* (Ithaca, NY: Cornell University Press, 1998); Peter Goin and Elizabeth Raymond, "Living in Anthracite: Mining Landscape and Sense of Place in Wyoming Valley, Pennsylvania," *The Public Historian* 23, no. 2 (2001): 29–45; John Harner, "Place Identity and Copper Mining in Sonora, Mexico," *Annals of*

the Association of American Geographers 91, no. 4 (2001): 660–80; Rosemary Power, "'After the Black Gold': A View of Mining Heritage from Coalfield Areas of Britain," *Folklore*, no. 119 (2008): 160–81; David Robertson, *Hard as the Rock Itself: Place and Identity in the American Mining Town* (Boulder: University Press of Colorado, 2006); Robert Summerby-Murray, "Interpreting Personalized Industrial Heritage in the Mining Towns of Cumberland County, Nova Scotia: Landscape Examples From Springhill and River Hebert," *Urban History Review* 35, no. 3 (2007): 51–59.

6   Gavin Bridge and Tomas Frederiksen, "'Order Out of Chaos': Resources, Hazards, and the Production of a Tin-Mining Economy in Northern Nigeria in the Early Twentieth Century," *Environment and History* 18, no. 3 (2012): 371. See also John Sandlos and Arn Keeling, "Claiming the New North: Development and Colonialism at the Pine Point Mine, Northwest Territories, Canada," *Environment and History* 18, no. 1 (2012): 5–34; William Holden, Kathleen Nadeau, and R. Daniel Jacobson, "Exemplifying Accumulation by Dispossession: Mining and Indigenous Peoples in the Philippines," *Geografiska Annaler: Series B, Human Geography* 93, no. 2 (2011): 141–61.

7   June Nash, *We Eat the Mines and the Mines Eat Us: Dependency and Exploitation in Bolivian Tin Mines*, 2nd ed. (New York: Columbia University Press, 1993); Michael T. Taussig, *The Devil and Commodity Fetishism in South America* (Chapel Hill: University of North Carolina Press, 1980).

8   One notable exception is the Navajo Uranium Miner Oral History Project, although the focus of many of the interviews was on the health legacies of uranium mining in Navajo Country in the US Southwest. See Doug Brugge, Timothy Benally, and Esther Yazzie-Lewis, *The Navajo People and Uranium Mining* (Albuquerque: University of New Mexico Press, 2006).

9   Most notably, Robert G. Williamson, *Eskimo Underground: Socio-Cultural Change in the Canadian Central Arctic* (Uppsala, Sweden: Institutionen för Allmän och Jämförande Etnografi, 1974). Williamson, a social worker and anthropologist living in Rankin Inlet in the 1960s, based his research on intensive local interaction with Rankin Inlet residents, but tended to be preoccupied with questions of collective cultural adjustment and dislocation. Robert McPherson, in *New Owners in Their Own Land: Minerals and Inuit Land Claims* (Calgary: University of Calgary Press, 2003), based much of his analysis on Williamson. For histories of the mine based on archival research, see Mary Josephine Taylor, "The Development of Mineral Policy for the Eastern Arctic, 1953–1985" (MA thesis, Carleton University, 1985) and Patricia J. Boulter, "The Survival of an Arctic Boom Town: Socio-economic and Cultural Diversity in Rankin Inlet, 1956–63" (MA major paper, Memorial University, 2011), the latter of which formed the basis for portions of this article.

10   Interviews were conducted with nine Inuit elders, including seven former miners and two miners' wives. Interviews took place in English and Inuktitut, with Piita Irniq, former commissioner of Nunavut and an expert translator, simultaneously translating and facilitating the interviews. Although the questions were developed and posed by the authors, other participants included research assistants Pallulaaq Kusugak Friesen of Rankin Inlet and Jordan Konek, a young Inuit filmmaker from Arviat who video recorded the interviews. Keeling also interviewed Irniq, a former Rankin Inlet resident but not a miner, about his memories of this period, and (later) Peter Ittinuar, son of former miner Ollie Ittinuar, now living in Ontario.

11   David Damas, *Arctic Migrants, Arctic Villagers: The Transformation of Inuit Settlement in the Central Arctic* (Montreal: McGill-Queen's University Press, 2002); "Notes Respecting the Administration of Eskimo Affairs," 10 March 1948, RG 85 vol. 2081 file 1012-4 pt. 3, Library and Archives Canada (hereafter LAC).

12   Peter K. Kulchyski and Frank J. Tester, *Kiumajut (Talking Back): Game Management and Inuit Rights, 1900–70* (Vancouver: UBC Press, 2007); John Sandlos, *Hunters at the Margin: Native People and Wildlife Conservation in the Northwest Territories* (Vancouver: UBC Press, 2007).

13   Arch MacKenzie, "Eskimos Said May Be Big Factor in Northern Mining," *Saskatoon Star-Phoenix*, September 22, 1958, 13.

14   R. G. Robertson, Letter to Secretary, North Rankin Nickel Mine, 18 November 1955, RG22 R216 vol. 210 file 40-3-22 pt. 1, LAC.

15   Indeed, it appears that the mine approached the RCMP detachment in Chesterfield Inlet for help in recruiting workers in early 1956. Report by Cpl. C. E. Boone, RCMP Chesterfield Inlet Detachment, 14 March 1956, RG 85 vol. 1268 file 1000-184 pt. 1, LAC.

16   File 1, Ken Whatmough Fonds, Nunavut Archives. Electronic copies of two files from these fonds, as well as numerous photographs, were provided by Nunavut territorial archivist, Edward Atkinson. Singiituq went on to play an important role in the community as an intermediary between the mine and its Inuit workers.

17   Cited in C. H. Herbert, Memorandum to the Deputy Minister, 25 April 1957, RG22 R216 vol. 832 file 40-3-22 pt. 2, LAC.

18   "Employment for Northern People," no date (likely 1950s), RG 22 vol. 1339-180 file 40-8-23 pt. 2, LAC.

19   Ollie Ittinuar interview, August 2011. Peter Ittinuar also related the story of his father's move in an interview with the author in May 2012, as well as in Thierry Rodon, ed., *Teach an Eskimo How to Read...: Conversations with Peter Freuchen Ittinuar* (Iqaluit: Nunavut Arctic College, 2008).

20 Dale Smith, ed., *Keewatin Journal* (Rankin Inlet, NU: Dale Smith, 1979), 17. Held in National Library of Canada.

21 The following description comes from Robert C. Dailey and Lois Dailey, "The Eskimo of Rankin Inlet: A Preliminary Report," Ottawa: Northern Co-ordination and Research Centre, June 1961.

22 Dailey and Dailey, "Eskimo of Rankin Inlet," 17.

23 On the relationship between the Keewatin Relocation Project and wider developments in the region, including Rankin Inlet, see Frank James Tester and Peter Kulchyski, *Tammarniit (Mistakes): Inuit Relocation in the Eastern Arctic, 1939–63* (Vancouver: UBC Press, 1994), chapters 7 and 8, and "The Keewatin Project" (November 1958), RG 85 vol. 1071 file 251-6 pt. 2, LAC.

24 Ipkarnerk interview with Arn Keeling, August 2011. A government memorandum from 1956 lists the rate as 75 cents an hour, and suggests the rate be increased: Letter from Sivertz to Kerr, "Eskimos of Rankin Inlet, N.W.T.," 10 December 1956, RG 85 vol. 1268 file 1000-184 pt. 1, LAC. These low wages were justified due to the "primitive" nature of Inuit lifeways at this time. Letter to Maurice Marrinan from Jean Lesage (March 18, 1957), RG 85 vol. 1268 file 1000-184 pt. 1, LAC.

25 These accusations, published in southern newspapers, and the responses, including from the minister of Northern Affairs, are detailed in RG 85 R216 vol. 1268 file 1000-184 pt. 2, LAC.

26 Irkootee interview with Patricia Boulter, August 2011.

27 Letter, B. G. Sivertz to H. Larsen, RCMP G. Division Chief, 12 April 1956, and B. G. Sivertz, Memorandum for W. G. Kerr re: Eskimos, Rankin Inlet N.W.T., 10 December 1956, RG 85 vol. 1268 file 1000-184 pt. 1, LAC.

28 Memorandum, F. G. Cunningham to Deputy Minister of Northern Affairs, 10 December 1956, RG 85 vol. 1268 vol. 1000-184 pt. 1, LAC.

29 Letter, D. W. Grant to Andrew Easton, May 1959, RG 85 vol. 1512 vol. 1000-184 pt. 3, LAC.

30 "Employment of Eskimos," 26 December 1956, Alexander Stevenson fonds, N1992-023 box 31 file 1, Prince of Wales Northern Heritage Centre (hereafter PWNHC), Northwest Territories Archives.

31 McPherson, *New Owners*, 12.

32 Dailey and Dailey, "Eskimo of Rankin Inlet," 94. For instance, in 1958, eleven men were returned to Baker Lake after failing to report for work at the mine. Monthly Report by W. G. Kerr (July 31, 1958), RG 85 vol. 623 file A205-184, LAC.

33 F. G. Vallee, *Kabloona and the Eskimo in the Central Keewatin* (Ottawa: Northern Co-ordination and Research Centre, 1962), 56; Letter from Easton to Grant (Aug. 26, 1959), RG 85 vol. 1512 file 1000-184 pt. 3, LAC.

34 Letter by B. G. Sivertz to Kerr, "Eskimos, Rankin Inlet, N.W.T." (Dec. 10, 1956), RG 85 vol. 1268 vol. 1000-184 pt. 1, LAC.

35 In spite of these restrictions, a Northern Affairs welfare officer reported in 1957 two instances of liaisons between non-Inuit men and Inuit women resulting in pregnancies. F. W. Thompson, "Report on Visit to Rankin Inlet, November 11–15, 1957," RG 85 vol. 1268 vol. 1000-184 pt. 2, LAC.

36 "Eskimo Job Problem Posed by Mine Closure," *Globe and Mail*, April 3, 1962, 25.

37 Stanley Twardy, "Ebbing Ore Blow to Eskimos," *Globe and Mail*, April 3, 1962, cited in RG 22 vol. 832 file 40-3-22 pt. 2, LAC.

38 "Displaced by Caribou Shortage, Eskimos to Get Aid From Ottawa," Canadian Press story clipping, 24 May 1962, cited in RG 22 vol. 832 file 40-3-22 pt. 2, LAC.

39 Report by D. W. Grant, Area Administrator, N1992-023 file 35-8, PWNHC.

40 See the extensive correspondence and reports in RG85 R216 vol. 1448 file 1000-184 vols. 7 and 8, LAC, as well as D. M. Brack and D. McIntosh, "Keewatin Mainland Area Economic Survey and Regional Appraisal," report for Industrial Division, Department of Northern Affairs and National Resources, March 1963, held in National Library of Canada.

41 See documents in RG85 D-3-a A251-3-500 file 1933 pt. 3, LAC.

42 Interviews in film, "Rankin Inlet Mine," Peter Ittinuar, film held in Rankin Inlet Community Resource Centre. This remarkable film (shot on black-and-white video) consists of a series of interviews in Inuktitut between Ittinuar and former miners, documenting the challenges the community faced after the sudden closure of the mine. This quote is based on a recorded simultaneous translation of the film by Piita Irniq. In an interview with Keeling, Ittinuar himself noted the film's "poignancy" and the miners' eloquence in discussing their post-mining struggles.

43 "Proceedings of a Meeting with the Eskimo Affairs Council and People of Rankin Inlet on February 26, 1962," RG85 Series D-1-A R216 vol. 1448 file 1000-184 pt. 7, LAC.

44 T. D. Stewart, Memorandum, Social assistance payments—Rankin Inlet, 6 March 1964, RG85 Series D-1-A R216 vol. 1962 file 1009-10 pt. 1, LAC. These welfare policies are also discussed in Williamson, *Eskimo Underground*.

45 Several interviewees mentioned these activities; they are discussed in detail in Stacy Neale, "The Rankin Inlet Ceramics Project: A Study in Development and Influence" (MA thesis, Concordia University, 1997).

46   These relocations are discussed in detail in Robert G. Williamson and Terrence W. Foster, "Eskimo Relocation in Canada," Ottawa: Department of Indian and Northern Affairs, 1975.

47   See, for instance, Bob Hill, "Transplanted Eskimos Doing Well," *Edmonton Journal*, November 28, 1963, 8; "Great Slave Lake Railway Begins Training Program," *North* magazine, March–April 1966, 44–45.

48   In addition to Williamson and Foster, several files in boxes 31, 51, and 57 of the R. G. Williamson fonds (MG 216) at the University of Saskatchewan Archives in Saskatoon also document the relocations and their challenges. Another report dealing with worker relocation is D. S. Stevenson, "Problems of Eskimo Relocation for Industrial Employment: A Preliminary Study," Northern Science Research Centre report, Department of Indian Affairs and Northern Development, May 1968.

49   Williamson and Foster, "Eskimo Relocation in Canada," 108.

50   The crest of the Hamlet of Rankin Inlet, for instance, depicts the mine headframe (ironically, now gone from the landscape) fronted by an inukshuk, with a crossed miner's pick and Inuit harpoon. On the identification of contemporary residents with the mining past and landscape, see Tara Cater and Arn Keeling, "That's where our future came from": Mining, Landscape, and Memory in Rankin Inlet, Nunavut," *Études/Inuit/Studies* 37, no. 2 (2013): 59–82.

51   Pamela Stern, "Upside-Down and Backwards: Time Discipline in a Canadian Inuit Town," *Anthropologica* 45, no. 1 (2003): 155.

52   For a discussion of indigenous "life projects" and development, see Mario Blaser, Harvey Feit, and Glenn McRae, "Indigenous Peoples and Development Processes: New Terrains of Struggle," in *In the Way of Development: Indigenous Peoples, Life Projects, and Globalization*, eds. Mario Blaser, Harvey Feit, and Glenn McRae (London: Zed Books, 2004), 1–25.

53   This comment reflects the discussion of the politics of Inuit cultural identity in Edmund (Ned) Searles, "Anthropology in an Era of Inuit Empowerment," in *Critical Inuit Studies: An Anthology of Contemporary Arctic Ethnography*, eds. Pamela Stern and Lisa Stevenson (Lincoln: University of Nebraska Press, 2006), 89–101.

54   Rollwagen, "When Ghosts Hovered," 33.

55   Jean-Sébastien Boutet, "Opening Ungava to Industry: A Decentring Approach to Indigenous History in Subarctic Québec, 1937–1954," *Cultural Geographies* 21, no. 1 (2014): 79–97.

| CHAPTER 2

Narratives Unearthed, or, How an Abandoned Mine Doesn't Really Abandon You

Sarah M. Gordon

In Denendeh, the traditional territory of the Dene Nation, there are two places called Sǫbak'e, "the money place." One is Yellowknife, capital of the Northwest Territories and administrative hub for most industry of the region, which centred on gold mining beginning in 1935 and diamond mining beginning in 1998.[1] The other is on the eastern shore of Great Bear Lake, where the Port Radium mine, and its associated village, used to stand. Arguably the most striking cultural collision between the Dene of the Sahtú (Great Bear Lake) and the forces of urban Canada took place at Port Radium. For sixty years, the only settlements on the lake were the Dene town of Délįnę (formerly Fort Franklin) on the western shore, the Port Radium mining town on the eastern shore, and a small Dene settlement at Sawmill Bay near Port Radium. Délįnę is the only one remaining. In the 1940s and 1950s, companies associated with the Port Radium mine hired Dene workers to load and transport uranium ore across the lake and downriver and to supply wood for fuel and construction at the mine site. The long-term impacts of the mine have been devastating and controversial to the Sahtúot'ı̨nę, or Great Bear Lake Dene. The story of Port Radium has become, in Délı̨nę, a cautionary tale about what happens when trust is given to the wrong people, local interests are not given equal weight to outside interests, and when outside influence is allowed to progress unchecked by local knowledge on Dene land.

Port Radium has been the subject of numerous histories that have foregrounded different perspectives on its impacts and importance: the community of Délı̨nę itself has published a book of personal histories;[2] historians and academics have produced texts that have sought to give the mine broader historical context;[3] it has been the subject of at least two documentary films;[4] and countless pages in magazines, newspapers, and other periodicals have been devoted to its story.[5] Collectively, these texts tell conflicting stories about the origin of the mine and its relationship with the Aboriginal people who lived and worked there. This chapter does not seek to evaluate any of those narratives, nor does it seek to add yet another voice to the cacophony. Rather, its goal is to assess some of these conflicting narratives *as narratives*. These are stories that people tell and that they believe, and as such they reflect larger epistemic paradigms at work in the context of their circulation. In the words of Julie Cruikshank: "More interesting than the question of which versions more accurately account for 'what really happened' is what differing versions tell us about the values they commemorate."[6] A story never exists in isolation. In all cases, there are people who tell the story, people who listen, and people who remember; all of these people do the work of contextualizing that story within the framework of the relationships surrounding it, the history that precedes it, and the future that flows forward from it. J. L. Austin introduced to linguistic circles the idea that not all statements can be said to be true or false, but rather, some utterances exist to *do something*;[7] Searle pushed this a step further by arguing that, truly, *all* utterances do something.[8] Kiowa writer N. Scott Momaday argues that questions of truth and fiction are subsumed beneath the life and actions of the story itself:

> Stories are true to our common experience; they are statements which concern the human condition. To the extent that the human condition involves moral considerations, stories have moral implications. Beyond that, stories are true in that they are established squarely upon belief. In the oral tradition stories are told not merely to entertain or to instruct; they are told to be believed. Stories are not subject to the imposition of such questions as true or false, fact or fiction. Stories are realities lived and believed. They are true.[9]

The mainstream and Dene narratives about the discovery of the Port Radium mine reflect sharply contrasting attitudes about the relationship between people (both Dene and non-Dene) and the landscape; these are the attitudes that have shaped, and continue to shape, northern colonialism. At the same time, the metonymic relationship between the mine and the broader experience of colonialism imbues Port Radium with a homeopathic power: to heal Port Radium properly, with due attention given to the values and personhood of the Sahtúot'ı̨nę and the Sahtú landscape, is to take a great leap toward healing the damage of colonialism more broadly.

Intimate and multifaceted relationships between mining, colonialism, and indigenous cultures exist throughout the Americas. Délı̨nę's story finds its closest cognate in the story of the Navajo, whose ancestral land became home to a thousand uranium mines in the early twentieth century, and who lost countless elders to the effects of radiation exposure. Among the Navajo, uranium is a monster, *Leetso*, born from the ground and delivered by the Navajo miners.[10] Similarly, tin mines in Bolivia, which began to appear almost immediately following colonization of the region, are homes to a syncretic Devil whose growth in power corresponds to growth in labour alienation.[11] June Nash has discussed how Bolivian tin miners' insistence on foregrounding local cosmology as a framework within which to understand both the mines' internal functions and their ongoing social impacts has empowered miners to resist imposed models of modernization.[12] In Délı̨nę, neither the mine nor its ore has the agency attributed to monsters or devils, as they do in the Navajo homeland and in Bolivia. Rather, the land itself has a kind of personhood; the natural, social, and cultural worlds often disaggregated in urban societies and in analyses of indigenous societies[13] remain unified here, so the relationship between people and the land is governed by guidelines of interpersonal ethical conduct. Julie Cruikshank has discussed how Aboriginal and non-Aboriginal narratives about the Klondike gold rush construct the categories of "individual" and "society" in starkly contrasting ways that index radically different understandings of what qualifies as good, valued, or justified behaviour.[14] Like Cruikshank, I seek to describe how different narratives about Port Radium reflect values of the teller that relate to conceptions of personhood and the relationship between people and the environment. Like Nash and Taussig, I seek to uncover the ways in which these narrative differences create opportunities for local Aboriginal empowerment in the face of colonial pressure to continue to assimilate to mainstream Canadian norms.

I spent a total of thirteen months in Délı̨nę between June 2009 and August 2011, researching the way the community negotiates the pressure to assimilate to Euro-Canadian norms while retaining a sense of local Sahtúot'ı̨nę identity. To that end, I interviewed various community members of different ages; I travelled on the land around the lake on trips geared toward trapping, hunting, fishing, and community education; and I participated in diverse community gatherings and activities. When I discussed culture change and community adaptation with friends in Délı̨nę, Port Radium came up as a recurrent theme in conversation. The ongoing impact of the mine comes not only from its material footprint on the landscape of the Sahtú: it comes also from the knowledge spread about the mine and its history through the narratives that circulate about it in different spheres. The varied stories about the mine—with its deceptive beginning, controversial existence, and devastating outcome—stand as both icons of and cautionary tales about the broader experience of colonialism.

PORT RADIUM: A BRIEF HISTORY

The Port Radium mine was built to excavate a vein of pitchblende ore that was staked by Gilbert Labine in 1930. Various conflicting narratives account for how he came to discover the ore; those will be discussed below. Following the discovery, Labine's Eldorado Gold Mines Ltd. built a camp in a protected cove called Cameron Bay. Other companies set up small ventures nearby, at Contact Lake and on the Camsell River. The goal of most of these ventures was to unearth and sell silver, copper, iron, and especially radium, which was being used in everything from phosphorescent wristwatches to newly developed cancer treatments. Uranium, which was considered virtually worthless at the time, was dumped with other tailings into the lake water. Prior to Labine's discovery, Belgium had held a monopoly over the global radium supply, extracting the element—which was valued at $75,000 per gram in 1930—from its colony in the Belgian Congo.[15] In the mid-1930s, the Canadian government established a post office and RCMP outpost at the Cameron Bay site to serve all the workers in the area. By the late 1930s, the global radium market had become saturated, and Eldorado found itself in a pricing war with the Belgian Union Minière; facing expensive extraction processes, the northern Canadian companies found themselves priced out of the market. To avoid shutting down, Eldorado negotiated with the Union Minière to delineate the geographical boundaries of their respective markets. But when World War II broke out in 1939, all participating countries imposed powerful trade restrictions; when Canada joined the Allies in September of that year, Eldorado lost its market in Germany. In 1940, the mine closed temporarily.[16] The government closed its offices there, and all non-Aboriginal residents returned to their homes farther south. In 1941, the American atomic bomb project began seeking sources of uranium oxide, a mineral found in high concentrations in pitchblende. The Nazi naval military and its U-boats held the North Atlantic in a stranglehold. The transportation of uranium across the Atlantic from Belgian-controlled mines in Africa became prohibitively dangerous. In 1942, the American government asked to purchase sixty tons of Canadian uranium oxide for use in the Manhattan Project, its secret task force working to develop a new, extremely powerful bomb.[17] The order was enough to inspire the Canadian government to purchase a controlling share of Labine's Eldorado Gold Mines Ltd., rechristening it Eldorado Mining and Refining Ltd. and re-opening the Port Radium site as a Crown corporation.[18] Most of the uranium used in the Manhattan Project came from previously purchased stock imported from Belgian-controlled Congo and stored in Staten Island, New York; demands beyond that were met with uranium from Port Radium. The Eldorado Mine at Port Radium continued to produce uranium ore until 1960, when it closed down again. In 1964, it reopened as a silver mine. The mine shut down for the final time in 1982, and most of its associated buildings and structures were dismantled.

While Eldorado's archival records are largely closed to the public, all available evidence suggests that before 1977, the Aboriginal people were not hired to work underground in the mine.[19] Aboriginal people were hired informally to provide other services around the Port Radium site: many Délı̨nę residents describe gathering wood for use as fuel and building material. Aboriginal people were also directly employed by the Northern Transportation Company Ltd. (NTCL) to work as ore carriers and barge pilots along the uranium transport route that moved the ore from its extraction site to refineries farther south. But even those workers with jobs that did not involve handling ore report noticing the effects of the radioactive dust that coated everything, and the fuel and oil that would find its way into the water from the boats and machinery. To get water, people would have to break the oily sheen that clung to the lake's surface. Ducks and fish, cut open, smelled like fuel, and sometimes they grew tumors.[20]

The rise in cancer cases among Sahtúot'ı̨nę became noticeable after the mine closed in the 1980s. The community pressured the Canadian government to undertake studies of the history, epidemiology, and environmental impact of Port Radium. Délı̨nę secured an agreement after initial resistance to conduct joint research with the federal government in these areas. In 2000, the two governments formed the Canada-Délı̨nę Uranium Table (CDUT), a committee that aimed "to address concerns about the human health and environmental impacts of Port Radium."[21] The CDUT's final report, issued in 2005, stated that "according to risk modelling based on the [radiation] dose reconstruction, 1.6 excess cancers (more than baseline) are theoretically predicted in a group of 35 individuals with ages and radioactive doses the same as the ore transport workers,"[22] but they found no conclusive evidence that uranium transport workers experienced any direct health impacts from radiation exposure. Many community members and some independent researchers dispute this claim.[23] The CDUT's research did, however, identify long-lasting tears in the community's political, cultural, and psychological fabric that are directly traceable to the community's connection with Port Radium. Their final report outlined twenty-six recommendations for action toward remediating the mine's lingering impacts. The recommendations included mandates for traditional knowledge research and the establishment of a traditional knowledge centre in the community, local job training and capacity building, the protection of Sahtúot'ı̨nę interests in all future research in the community, staffing the community health centre with health-care workers sensitive to Sahtúot'ı̨nę culture and to the mine's health impacts, and remediating the environment and landscape as quickly as possible.[24]

The CDUT process was controversial throughout its execution. Most people, in my experience, were pleased with the list of twenty-six recommendations that emerged from the process, but have been displeased with what they perceive to be insufficient follow-through on the part of the federal government. The remediation of the landscape has been underway for several years, but the components of cultural and psychological healing have received comparatively little attention. Local efforts to do this work have proven difficult to fund; traditional knowledge research has largely come about on the impetus of independent researchers visiting from universities and non-governmental organizations who can draw on scholarly and arts-based funding. Furthermore, the environmental cleanup, upon which great progress has been made in recent years, is fraught with tension: many community elders dislike that barrels of waste are being buried on-site, rather than transported elsewhere for disposal. Many people still desire compensation for the cancer deaths that they attribute to radiation exposure on the land. The mine may have been closed for more than three decades, its openings sealed over and most of its buildings dismantled, but its story, in Délı̨nę, is far from over.

CONFLICTING DISCOVERY NARRATIVES: GILBERT LABINE

The Port Radium mine has two distinct and radically different origin stories: one told in Délı̨nę, and another told in publications and formal documents throughout the rest of Canada. These narratives are mutually exclusive: if one is wholly true, the other cannot be. That said, even if one narrative is true and the other is false, the secret of the discovery of the pitchblende vein at Port Radium remains with the people who were present at the time, none of whom are alive anymore. For people today, these discovery narratives reflect strongly contrasting perspectives on Aboriginal disenfranchisement and the fair use of northern land.

The commonly known narrative about the origin of the Port Radium mine originates with Gilbert Labine, the prospector who identified the site and whose company, Eldorado Gold Mines Ltd., established the mine there. This "mainstream" story has been reproduced in several major Canadian publications.[25] According to that story, Gilbert Labine took a prospecting trip to the Northwest Territories, possibly inspired by a 1900 report by J. Macintosh Bell, a geologist for the Geological Survey of Canada, which referred to a sighting of cobalt bloom at the site where Port Radium was later established.[26] In a 1934 speech on the history of the subject, Labine describes having used maps of the Northwest Territories secured from the Department of the Interior and having received guidance from the 1900 Bell report.[27] In 1960, however, Labine published a contradictory history of his expedition. In an early publication about the history of the mine, Labine denied having seen the Bell report:

> There is one fact I would like to point out: I had no geological maps and any of my early mapping of the structures was made by my own reconnaissance and not by the Geological Survey. It has been stated that the Bell Report of 1900 [. . .] was responsible for my going to Great Bear Lake. I would like to state here that this is absolutely false, as I did not read the Bell Report until the following year when I obtained a copy from the archives in Ottawa, after I had already been in that country. Further, I had travelled with Bell. As a matter-of-fact I was his first assistant in Canada

when he returned from New Zealand and in all of our discussions never once did he mention to me that he had seen anything of interest in the Great Bear Lake field.[28]

At least one rendition of the narrative indicates that Labine made a first prospecting trip to the Northwest Territories in 1929 without having read the Bell report and spotted the cobalt bloom during his return flight south. This inspired him to seek out the Bell report on his return home and then to head back to the region the following year, for further prospecting.[29]

Labine travelled with a partner, Charles St. Paul, and for a long time found nothing. One morning, shortly before they were about to give up, St. Paul was struck with snow blindness—a temporary, painful affliction of the eyes caused by overexposure to ultraviolet light without the use of eye protection, commonly triggered by sunlight reflecting off vast expanses of snow—and had to spend a few days recovering in the dark. Labine went for a walk on his own, where he stumbled across an ore vein that may have been pitchblende or may have been silver showing indications of pitchblende, depending on the version of the narrative. Labine found a plum-sized sample of the black rock, which he brought back to St. Paul in their tent. He nursed St. Paul's eyes back to health until St. Paul could confirm Labine's suspicions: the surface of the rock—which appeared black, shiny, and bubbly, like it had solidified while boiling—looked like pitchblende.

The two men returned to the south, where their hunch was validated at a lab. Armed with this new knowledge, Labine staked his claim at the site of the vein, and then set about the arduous task of securing venture capital to fund his goal of building a mine a thousand miles from the nearest railroad, which was in Edmonton. Through perseverance, backed by the extremely high price of radium at the time, he succeeded, and the mine was built on the site of that vein of pitchblende.

While my purpose in this analysis is not to evaluate the validity of this narrative, it contains inconsistencies that bear mentioning. As others have pointed out,[30] it is extremely unlikely that a prospector as experienced as St. Paul would have made the kinds of basic errors that would have led to snow blindness. Labine and St. Paul spent an entire season prospecting together, and yet Labine's discovery happened almost immediately following the onset of St. Paul's affliction (snow blindness typically heals in less than seventy-two hours). As well, pitchblende is not a common ore in Canada, making it questionable that the two men would have been able to identify the rock so readily. Different sources have told different stories about how Labine had previously been able to see a sample of the ore.[31]

Perhaps Labine intentionally lied in one version of this narrative. More likely, the narrative evolved over the twenty-five years between these tellings, curated according to the demands of different audiences and how Labine wished to understand his own story. A reader can hardly help but notice how Labine's personal narrative has Horatio Alger qualities, embracing every trope of the American myth of the self-made man, who typically comes from poverty and, through hard work, perseverance, and strategic risk taking, achieves great financial success.[32] Lest this individualistic formula seem too American to have been the unconscious result of a narrator's interaction with more collectively minded Canadian audiences, it bears noting that shortly after the pitchblende strike, the Eldorado mining company enlisted the support of a New York City–based PR firm in shaping its corporate image,[33] and that the trope of the self-made man has also been pervasive in Canada, particularly in the late nineteenth and early twentieth centuries,[34] and especially with reference to prospectors. Julie Cruikshank has referred to Horatio Alger as the Yukon prospector's "prototype."[35] By actively excluding the Bell report from later versions of the narrative, Labine frees himself of any obligation to share credit for the mine's discovery. Even the assistant he paid to accompany him plays only a supporting role in the narrative: St. Paul's snow-blindness not only sets up the conditions for Labine's discovery, but also illustrates his weakness and lesser competence as contrasted with Labine's robust success (after all, the two men were travelling together in the same conditions, but only one went snow-blind). Key tropes of Canadian literature find their way into the story, as well. Sherrill Grace describes a "northern narrative," in which a white, male hero is thrown into a northern landscape and must struggle to survive until he finds salvation "through endurance in this harsh yet potentially transforming landscape."[36] In lieu of the wolves and bears that most frequently symbolize northern danger, Labine's narrative has snow blindness; in lieu of explicit spiritual salvation, Labine finds financial salvation. Labine has ventured into that wilderness and not only survived but vanquished it (which, in turn, calls to mind Margaret Atwood's rhetoric of "survival" as connected to Canada's fascination with, and terror of, the north).[37]

At the same time that Labine's narrative positions him as a survivor of the North's many challenges, it also feeds into the rhetoric of opportunism surrounding northern expansion that was popular in Canada in the first half of the twentieth century.[38] The turn of the century saw expanding foreign and domestic markets for Canadian natural resources and a federal government, under Sir Wilfrid Laurier, that was keen to enable national industry to capitalize on these opportunities.[39] This attitude grew even more powerful in the 1940s, when Lester B. Pearson—then ambassador to the United States—published an article in *Foreign Affairs* celebrating the opportunity offered by the "great wealth in the Land of the Midnight Sun," saying that "a whole new region has been brought out of the blurred and shadowy realm of northern folklore and shown to be an important and accessible part of our modern world."[40] Labine's story combines the thrill of risk with the promise of opportunity, elevating his own stature while promoting broader Canadian motives to further colonize the North.

DISCOVERY NARRATIVES: ʔƎHTSÉO BEYONNIE

Unsurprisingly, the origin of the Port Radium mine is understood very differently in Délı̨nę than in the rest of Canada, and the Délı̨nę version includes Aboriginal people as prominent players in the mine's discovery and establishment. Like Labine's version, the Sahtúot'ı̨nę narrative comes in several versions that waver, slightly, around a common narrative centre. I quote it here from the version published in *"If Only We Had Known": The History of Port Radium as Told by The Sahtúot'ı̨nę*:

> Prior to 1930, a Dene man, Victor Beyonnie's dad was travelling to Caribou Point and camped at Port Radium. He noticed an unusual looking rock, which he showed a non-Dene prospector.
>
> The prospector took the rock to Edmonton and showed it to a prospector named Gilbert Labine. Being a geologist, Mr. Labine noticed that the rock possibly contained pitchblende. In 1930, Gilbert Labine began staking claims at Echo Bay on the eastern shore of Great Bear Lake. During 1931 he and his crew shipped ten tons of handpicked ore to Ottawa for further analysis.[41]

When the story is told in Délı̨nę, sometimes people say that Old Beyonnie met a prospector who gave the rock to Gilbert Labine, and sometimes people say that Old Beyonnie met Labine himself during his expedition to the Arctic. Everyone agrees that the prospector gave Old Beyonnie something tokenistic in exchange for the rock—coffee, rifle shells, a bag of flour—saying that the rock probably had no value. The ore proved to be pitchblende, of course, and its discovery not only saved Labine's company from the brink of bankruptcy, but made Labine himself very wealthy. As in stories of ox-hide purchases common in colonial North America, an agreement is made to exchange small items, but the colonizing party, with duplicitous intent, claimed far more than the spirit of the agreement allowed.[42]

The Délı̨nę narrative is widely known and commonly told in the community. It comes up as a metaphor or analogy in the context of other political conversations: an implicit cautionary tale. On the occasion that this narrative came up in one of my interviews, its structure and content were fragmented across three speakers (elder Andrew John "AJ" Kenny, his son Dennis who worked with me as an interpreter, and myself) and two languages (the Sahtúot'ı̨nę language and English):

> Sarah Gordon (SG): What do you think is the most important lesson for young people and future generations to learn from the Port Radium story?
>
> Andrew John "AJ" Kenny (AJK): [responds in Sahtúot'ı̨nę language].

Dennis Kenny (DK), interpreting: He said the government treated our elders really badly. Really badly. The way they treated them, there was no compensation, nothing. He said, look at my mom's grandfather. He's the one who found the . . . that stuff at Port Radium.

SG: The pitchblende?

DK: Yeah. He was the one who found it. And he never, her dad never even got compensated. He got a 25 pound flour. And they're talking about this guy who discovered it. Now he's rich, he's a millionaire, and there's a book about him. And no nothing about my mom's dad. My grandfather. He's the one who discovered. . .

AJK (in English): Victor. Victor Beyonnie's dad. Victor's brother is my wife's father.

DK: Yeah. Their dad. He's the one who found that.

SG: What was his name? It was Beyonnie, but. . .

DK: ʔəhtséo Beyonnie.[43]

AJK: Beyonnie.

DK: They just call him Beyonnie. He *[AJ]* said that story is, you know, it's important for people to learn it, how the government treated us. So badly. Not just that, but the explorers too. Like that guy who discovered that stuff.

SG: Labine?

DK: Labine, yeah. Some people said all the older people . . . should, you know, gather young people together and tell them exactly what happened and what they did for them and what happened to them. Now my Dad's talking about all this, how his brother's gone by cancer, my grandpa died by cancer, my grandpa was a really hard-working man . . . he was good hard-working man and he helped a lot of people just with, you know, giving them advice and stuff like that.⁴⁴

Virtually any time Port Radium is discussed in Délı̨nę, the narrative is tied to some concept of death. The Kennys' version of the origin narrative closes with a discussion of the cancer: the mine began with the unjust treatment of a Dene man, and it continues, now, with the unjust suffering of Dene people who lived or worked at or near the mine or the uranium transport route. But another narrative—less commonly told, but just as widely known in Délı̨nę—speaks to prior Dene knowledge of the mine site and of the lives it would take. Its focus, however, is not on the Dene lives lost, but the Japanese. The narrative, paraphrased based on my field notes, goes like this:

> The place where Port Radium is now, long ago, people knew it was a bad place. They said they should never sleep there or go near it. But long ago, some hunters were travelling and they stayed there one night. One of them was a prophet, and that night, he had a dream. He saw a large hole in the ground with white men walking into it. Then, he saw a large flying bird carrying a black stick. The bird carried the stick to a land far away and then dropped it; it made a giant, burning hole in the ground. Out of the hole, the spirits of thousands of people escaped, rising to the sky. The spirits looked like Dene people, but they were not Dene.

The versions of this narrative that I heard attributed the vision to an unnamed prophet who lived long, long ago, though they often cite Prophet Ayah (or Ayha)⁴⁵—an extremely important Dene prophet who passed away in Délı̨nę in 1920—as a teller of the tale. Most notable of these is the version recorded by Sahtúot'ı̨nę elder George Blondin in his book *When the World Was New*.⁴⁶ Other versions of the story, recorded by journalists and previous ethnographers, attribute the vision itself to Ayha, when he was a young man, before he had been instructed to share his visions with the world. One version, published in *News/North*—a regional newspaper that circulates throughout the Canadian territories—cites Leroy Andre and Joe Blondin, both Sahtúot'ı̨nę, as its sources:

Long before the Europeans came or any mines opened on the shores of Great Bear Lake, the Dene people learned Port Radium was deadly and many of them stayed away.

Délı̨nę didn't exist yet and the people still lived on the land, often travelling in groups of families to hunt and fish.

One day a group of Dene was passing through the area and they decided to camp near what would eventually become known as Port Radium. Among them was a powerful medicine man, the Prophet Ayha.

During the night, the others awoke to the prophet singing. He did not wake himself, but sang for most of the night in his sleep.

In the morning the people asked him why he was singing in his sleep, so he told them of his vision. He said he saw boats and many houses with smoke coming out of them. There were people with white skin going into a great hole in the ground and coming back out with rocks.

These people were carrying the rocks away and he decided to see where they were going. So in his dream state he followed them across Great Bear Lake and down along the river network to Fort McMurray and beyond there into the U.S. There the people made a long stick and put the rocks in it. They then loaded the big stick into a giant bird, which then took flight so he followed it as it flew over wide-open water.

When it came back over the land, the bird dropped the stick and it burst into a giant ball of fire and many people who lived there were burnt.

"Those people looked a lot like us," said Prophet Ayha. "I was singing for them."

He then told the people that all of this would happen after they died.

Many years later, in September of 1940, the Prophet Ayha passed away.

On Aug. 6, 1945, the U.S. dropped the first atomic bomb on Hiroshima. Three days later another fell on Nagasaki.[47]

Another print version of this story adds that long before colonization, the Dene knew that this place was dangerous; loud noises came from the rocks there, and it was "bad medicine" to pass nearby.[48]

CONTRASTING NARRATIVES

The mainstream Canadian narrative of the origin of Port Radium contrasts sharply with the Délı̨nę narrative with respect to where they situate power and agency. The mainstream story situates power with Labine: plucky, resourceful, down-on-his-luck prospector who challenges the frigid Canadian Arctic and survives, rewarded for his rigour and persistence by the discovery of an ore containing a wildly valuable element, radium. The Aboriginal residents of the area are, at best, background characters, part of the landscape; more often they are invisible. The Dene narrative, however, complicates these power dynamics. Because the rock is discovered by Ɂǝhtséo Beyonnie, a Dene man, the glory of its discovery (and associated power) should, according to the Sahtúot'ı̨nę, reside with him, but the prospector purchases the rock for a token sum so that he may assume its power for himself, accruing great wealth and prestige in the process. The Aboriginal version of the story focuses on the good faith with which the First Nations people generally greeted and supported the arriving Europeans, and on the unjust and inhumane treatment they received in response.

The differences in the roles played by Aboriginal people in the mine's discovery mirror the roles they play in narratives of northern colonialism more broadly: the mainstream narratives of Arctic colonialism erase Aboriginal sovereignty and personhood, especially in the face of industrial and governmental desires for land and resources on Aboriginal land.[49] The Crown recognized Aboriginal title to any land that had not been ceded through treaty, but as the land held value to the government and the Aboriginal people did not, the latter were conceptualized mostly as obstacles to overcome in order to secure the former.[50]

One constant remains through the different versions of Labine's story, in keeping with the colonial mentality: Aboriginal people do not appear. This noticeable absence is in keeping with subsequent descriptions of Port Radium. At Library and Archives Canada, I searched dozens of boxes of material related to the Eldorado, scanning for references to Aboriginal people at the mine sites. Boxes contained correspondence between the company's management staff, employment records, and file after file of media clippings—but little if any mention of Dene, as employees, traders, or even merely local residents. Julie Cruikshank has argued that "facts get established by enacting silences . . . there are things to be said and ways of saying them."[51] Whenever a narrative is shared, be it orally or in print, decisions of inclusion and exclusion are made based on culturally ingrained assumptions of what is or is not considered relevant. When southern, urban Canada hears stories of the founding of a new and successful mine, it expects Horatio Alger: independence and plucky self-reliance. When it hears of the country's movement to further colonize the North, it expects stories of challenge meeting opportunity. It does not expect stories about the local people of the area whose lives are integral to and affected (sometimes negatively) by the success of the story's lead character.

The picture of Aboriginal life at Port Radium emerged in the negative spaces of my archival scavenging. The archival material about Port Radium and the Eldorado corporation made no reference to the Aboriginal employees who lived at the mine and worked there casually, or who worked along the uranium transport route.[52] An unpublished essay called "Radium in Canada," by radiation expert Marcel Pochon, who was a high-level director at Eldorado for many years, says that "Great Bear Lake had had, in the past, very few visitors," going on to enumerate prospectors and missionaries who had travelled in the region, but giving no mention to the Dene and Tłı̨chǫ people who had travelled there for centuries; it goes on to describe the land as "uninhabited."[53]

A 1937 article in *Collier's Weekly* includes limited references to Aboriginal residents of the Port Radium area, but its inclusions and exclusions are telling. "Radium City," the topic of the article, has, in the author's description, only a single female resident—the wife of the mine manager—but nonetheless manages to have a "little gang of half-breed kids."[54] The author goes on to describe the warming effect of the smiles of young women he sees from afar, who, implicitly, are not the lone woman resident of Radium City.[55] Aboriginal people are present in the gaps of the story, in the spaces between and around its actual characters, who are, of course, white.

The background appearances of Aboriginal people in so many Port Radium stories raise the question that they may have appeared in the background of Labine's mine discovery narrative, as well, even if he never mentioned them or implied that they were there. Fred J. "Tiny" Peet, an electrician and miner who had worked at Port Radium, stated in an oral history interview that Labine mentioned having hired an "Indian" guide in his early expeditions to the area.[56] "Punch" Dickins, renowned bush pilot who flew in and out of Port Radium, said that Peet's story certainly would have made sense, because many Aboriginal people hunted in the area and would have been of great help with equipment and dog teams.[57] He also offered that, according to his understanding, prospectors' interest in the region had originally stemmed from the fact that the Dene and Dogrib people who lived there had copper arrowheads—a perspective that contrasts sharply with Pochon's assertion that the landscape of the Sahtú was devoid of human life. Just as northern colonization cannot escape the presence of Aboriginal people, no matter its attempts to work around them, neither can any discovery narrative about Port Radium completely and convincingly erase the presence and influence of Aboriginal people.

Differing perspectives on northern colonization, as illustrated by these contrasting narratives about the origin of the Port Radium mine, also reflect different attitudes about the nature of fair exchange and reciprocity. In both narratives, a piece of ore found its way from the earth on the shore of Great Bear Lake into the hands of a prospector. In the Sahtúot'ı̨nę narrative, that exchange was mediated by a Dene man; in Labine's narrative, he found the rock himself. In both cases, however, Labine profited enormously from his discovery, in terms of both finance and fame. In an urban Canadian narrative context, there is no problem with this arrangement. Labine gambled carefully and invested well in his trip, he made a discovery, and succeeded—any successful business, after all, is, at its core, about selling something for more than it cost to acquire or produce. Even if ʔəhtséo Beyonnie did find the rock and sell it to Labine for a token price, this may not inherently be unfair: when he bought it, he presumably was unaware of its value. But the nature of this exchange, and the resulting process of profiting from the gifts given freely by the land, violates fundamental moral codes in Délı̨nę.

During my time in Délı̨nę, I had two separate conversations, one in an interview and one informally, with individuals who drew parallels between the cultural prohibition against selling wild meat for money and the ethical problem with the mine. When a hunter or fisherman in Délı̨nę brings home meat, it is expected that he will share that meat with friends and relatives who ask, and that he will not request money in exchange. Morris Neyelle, a respected community leader, outlined in an interview the ethic of exchange that he inherited from his elders:

> You know, in my culture, like hunting, fishing, all those kind of things that my elders, even my parents always said, what's given to you free should be given back free. Don't take anything for it, especially money. So to this day, if somebody asks for meat—caribou meat, or fish—first thing they do is, especially outsiders, they go, "Well, I'll give you money." I say, "No!" I say, "It's given to me free, why should I take money for it? I didn't make it. It's given to me free, I should give it back free to whoever asks for it."
>
> But if you misuse it, the elders always said, if you misuse it by gaining from it, by accepting money for it, that way, everything will go, they always said. And I notice a lot of that happening in the other regions. I know there are a lot of stores where they were selling meat. I've seen that too. And it's not right. If they made it, sure. But they didn't make it, so that's why they're losing all the caribou.[58]

Neyelle and I often discussed the politics of selling meat, and how no person should ask for money for something like meat, which was given to them for free. In similar conversation, another elder told me, laughing, that if people were keen to give him something in exchange for meat, they could bring him five gallons of gas for his snowmobile when they knew he was heading out on a trip—not that anybody ever did!

On one occasion, while we were discussing these politics of exchange, Neyelle drew a poignant comparison between this ethic and the ethics of mining: the problem with mining, he said, was that humans extracted from the ground a rock that they did not make, and then sold it for profit. The land did not ask payment for the ore that they extracted. And thus, just as the caribou would retreat from the hunting grounds of people who disrespected them by selling their meat, so too was the land pulling away from the people. This idea of the land withdrawing could be interpreted in many ways: the growing sense, among elders, that younger generations are disconnected from the local landscape is one possibility; the idea that the resources of the land are becoming slimmer and harder to use, and its environment changing and becoming less hospitable, is another.

Implicit in Neyelle's story is an analysis of the nature of reciprocity and exchange in Délı̨nę, and the importance of not taking personal gain at the expense of another human or non-human being. The land, a living thing, creates all of its parts: the caribou, the fish, the minerals in the earth. When we, as humans, take any of those, we are receiving the gifts given by the land. Any travelling that a person may do to find the caribou herd, or any labouring the person may do to cut a hole in the lake ice, is simply the work that must be done in order to receive those gifts: it doesn't *make* anything. And if the land did not want people to receive those gifts, it would keep those gifts away from the people who seek them. So, to charge money for something given by the land is to take something for nothing, to profit without having done anything to deserve it.[59] That kind of empty profit is exemplified by money. Money is an abstraction that circulates between people without any obvious empirical roots in the world. Neyelle often told me that the elders said not to worry too much about money—it was, after all, only a thing of this world, of no real value.

When an elder says that he will accept five gallons of gasoline from anyone grateful to receive his meat, he is recognizing the difference between receiving money as interest and receiving gas as both a gift and an investment. Gas is a practical thing: without it, the skidoos, trucks, and boat motors will not run, and without those tools, the elder can neither travel to get more meat nor drive around town to deliver it to those who need it. He does not profit, per se, from a gift of gasoline: he cannot pocket five gallons of gas and use it to buy frivolous things for himself. Instead, he will use the gasoline—either that five-gallon can, or an equivalent five-gallon can that he buys for himself before his next trip—to do the work he must do in order to continue to provide meat to the community. And a person who gives him the gift of five gallons of gas also gives him the gift of time, saving him from having to drive to the gas station to buy those five gallons himself. All of this enables this elder to treat the land with the respect it deserves: by harvesting only the meat that the community needs at any given time, he does not have to overharvest on any given trip and risk harming the caribou herds or wasting meat.

REMEMBERING AN ABANDONED MINE

The buildings of the original Port Radium community have long since been dismantled, but the remnants of the mine itself still remain. As a part of the CDUT agreement, the land must be remediated as much as possible.[60] What constitutes a thoroughly remediated site remains a topic of contention. According to the federal government, and to the local officials in charge of liaising between local and federal authorities on the topic, the remediation work on the area of the Port Radium mine itself is complete; at the time of my visit to Délı̨nę in the summers of 2009 and 2010, remediation work had moved on to Sawmill Bay, with temporary workers being hired on three-week rotations to clean up that contaminated site. But the elders, in particular, continue to insist that the remediation work at Port Radium is incomplete. Morris Neyelle, for example, is dissatisfied with any remediation plan that involves disposing of any waste on-site—especially any toxic waste. That waste was made by people from the South, he says, so they should take it back south with them. The final agreement of the Canada-Délı̨nę Uranium Table stipulates that while tailings, abandoned machinery, and scrap may be disposed of on-site, any hazardous material should be removed for safe disposal elsewhere.[61] Sahtúot'ı̨nę employees of the clean-up process told me that they were responsible for burying material on-site. Nobody was able to say with any confidence whether that material was hazardous or not. Happiness that the remediation plans are moving forward, and that Sahtúot'ı̨nę workers are being employed as part of the process, is tempered by distrust in the overall process. Port Radium was built on deception and misinformation: why should any Dene believe that its deconstruction should be any more honest?

The parallel of the colonial experience and the Port Radium experience extends into the future. While the Sahtúot'ı̨nę work on remediating the mine site, they are also working on remediating themselves, not only from the damaging impacts of their experiences living and working on the uranium transport route, but also from the impacts of abusive residential schools[62] and tuberculosis hospitals,[63] of Indian agents, of living as wards of the state, of manipulative treaty processes.[64] The mine is both a part of this process and powerful metonymy for it: like many of those other colonial events, its development brought with it the promise of some good things (jobs for Dene people and new opportunities for trade, for example), but those good things have been outweighed by their negative consequences, which never could have been predicted by the Sahtúot'ı̨nę people. This metonymy certainly emphasizes the mine's cognitively destructive power, but it also imbues the mine, and its associated narratives and effects, with a kind of homeopathic power: to address, and heal from, the impacts of the mine is to address and heal from the impacts of colonialism more broadly. And planning to keep the problems of the mine from repeating themselves means also planning to assert control over the ongoing colonial relationship between Délı̨nę and Canada.

This is not to say that the presence of Port Radium within living memory of many Sahtúot'ı̨nę somehow simplifies the decolonization process. If anything, it highlights the tensions that exist between different generations and personalities within the community regarding how the community should assert its independence and negotiate its relationship between the First Nation government and economy, and the federal government and economy. For example, many community leaders, especially of older generations, express significant reservations about the prospect of any future natural resource development on their land, citing Port Radium as an example of how projects that may seem beneficial at first can have unanticipated long-term consequences, particularly when outsiders are the primary stakeholders. But other community leaders, particularly younger ones, recognize Port Radium as a cautionary tale for what can happen if the community is denied a seat at the decision-making table, but also argue that opening parts of the Sahtú to natural resource exploration, and potential extraction, can provide the community with much-needed cash inflow and employment opportunities. The community is concurrently working to implement self-government, geared toward asserting Sahtúot'įnę sovereignty over local governance, land management, and education. Just as resource development must proceed cautiously, taking into account the perspectives and needs of various community members, so too must the self-government process.

In her work on narratives of colonial encounter in Alaska, Cruikshank reminds us that "ideas have material consequences."[65] The narratives surrounding the origin, life, and afterlife of Port Radium *do things*: they situate power; they illustrate Dene and non-Dene understandings of personhood, agency, responsibility, and modernity; they assign meaning to Port Radium relative to a larger colonial context and use Port Radium as an icon for the deception inherent in colonial processes more broadly. The mainstream Canadian narrative of the mine's origin, with its erasure of any Aboriginal presence and its foregrounding of the mine's discovery as the achievement of a lone ambitious and resourceful person, reflects the values and interests of its largely white, urban audiences and tellers. The Dene narrative indexes concerns regarding the community's broader colonial context and concurrently shapes the community's recovery from the mine's impacts and the broader impacts of colonialism.

Cruikshank has argued that "viewing encounters of ideas historically shows how indigenous peoples continue to face a double exclusion, initially by colonial processes that displace them from land and ultimately by a neocolonial discourse that hastens the transformation of sentient and social spaces to measurable commodities called 'lands and resources.'"[66] In the words of Dennis Kenny, speaking on behalf of his father Andrew John Kenny: "He [Andrew John] said that [the Port Radium] story is . . . important for people to learn . . . How the government treated us. Not just [the government], but the explorers, too."[67] "The explorers" are the prospectors and any other adventurers who travelled through the North desiring and seeking out means for personal gain. The story of Port

Radium is, for the people of Délı̨nę, a cautionary tale that contains moral and political implications that influence the community's strategies for self-governance and cultural preservation. Beyond its impact on the environment, Port Radium has given an accessible face to the overarching ethos of colonial mistreatment, and in that respect, it gives Délı̨nę the strength of a visible adversary against which it may chart its course, demarcated by local traditions and values, into a self-determined future.

NOTES

1. Aboriginal Affairs and Northern Development Canada, "History of Giant Mine," accessed January 31, 2014, http://www.aadnc-aandc.gc.ca/eng/1100100027388/1100100027390; Natural Resources Canada, "Canada: A Diamond-Producing Nation," http://www.nrcan.gc.ca/minerals-metals/business-market/3630.

2. Délı̨nę Uranium Team, *If Only We Had Known: The History of Port Radium as Told by the Sahtúot'ı̨nę* (Délı̨nę: Délı̨nę Uranium Team, 2005).

3. Robert Bothwell, *Eldorado: Canada's National Uranium Company* (Toronto: University of Toronto Press, 1984); Liza Piper, "Subterranean Bodies: Mining the Large Lakes of North-West Canada, 1921–1960," *Environment and History* 13 (2007): 155–86, Peter C. van Wyck, "The Highway of the Atom: Recollections along a Route," *Topia* 7 (2002): 99–115. Peter C. van Wyck, *The Highway of the Atom* (Montreal: McGill-Queen's University Press, 2010).

4. Peter Blow, *Village of Widows* (Toronto: Lindum Films Inc., 1999); David Henningson, *Somba Ke: The Money Place* (2006).

5. See, for example, Julie Salverson, "They Never Told Us These Things," *Maisonneuve*, August 12, 2011; Andrew Nikiforuk, "Echoes of the Atomic Age: Cancer Kills Fourteen Aboriginal Uranium Workers," *Calgary Herald*, March 14, 1998.

6. Julie Cruikshank, *The Social Life of Stories: Narrative and Knowledge in the Yukon Territory* (Lincoln: University of Nebraska Press, 1998), 92.

7. John L. Austin, *How To Do Things with Words* (Cambridge, MA: Harvard University Press, 1975).

8. John R. Searle, *Expression and Meaning* (Cambridge, UK: Cambridge University Press, 1979).

9. N. Scott Momaday, *The Man Made of Words* (New York: St. Martin's Press, 1997), 3. See also Thomas King, *The Truth about Stories: A Native Narrative* (Minneapolis: University of Minnesota Press, 2008).

10 Esther Yazzie-Lewis and Jim Zion, "*Leetso*, the Powerful Yellow Monster," in *The Navajo People and Uranium Mining*, eds. Doug Brugge, Timothy Benally, and Esther Yazzie-Lewis (Albuquerque: University of New Mexico Press, 2006), 2.

11 Michael T. Taussig, *The Devil and Commodity Fetishism in South America* (Chapel Hill: University of North Carolina Press, 1980).

12 June Nash, *We Eat the Mines and the Mines Eat Us* (New York: Columbia University Press, 1979).

13 Julie Cruikshank, *Do Glaciers Listen?* (Vancouver: UBC Press, 2005), 4.

14 Cruikshank, *Social Life of Stories*, 72–97.

15 Bothwell, *Eldorado*, 8.

16 Ibid., 72–77.

17 Ibid., 97–98; van Wyck, *Highway of the Atom*, 117.

18 Bothwell, *Eldorado*, 117–54.

19 Canada-Délįnę Uranium Table, "Canada Délįnę Uranium Table Final Report" (Délįnę: Délįnę First Nation, 2005).

20 Délįnę Uranium Team, *If Only We Had Known*.

21 Canada-Délįnę Uranium Table, "Final Report," ii.

22 Ibid., 40–41.

23 Blow, *Village of Widows*.

24 Canada-Délįnę Uranium Table, "Final Report," ii–x.

25 Bothwell, *Eldorado*; Gordon C. Garbutt, *Uranium in Canada* (Ottawa: Eldorado Mining and Refining Ltd., 1964); W. O. Kupsch, "From Erzgebirge to Cluff Lake – A Scientific Journey through Time," *The Musk-Ox* 23 (1978): 7–87; Gilbert Labine, "Great Bear Lake" (Library and Archives Canada [hereafter LAC], 1934); Gilbert Labine, "Port Radium: The Story of Its Beginnings," *The Refiner* (1960); Peter C. Newman, "Gilbert Labine: Adventurous Bushwhacker," *The Beaver* (1959); Leslie Roberts, "Living on Radium," *Collier's Weekly*, June 5, 1937.

26 J. Mackintosh Bell, "Report on the Topography and Geology of Great Bear Lake and of a Chain of Lakes and Streams thence to Great Slave Lake," in *Annual Report* (Ottawa: Geological Survey of Canada, 1901).

27 Labine, "Great Bear Lake."

28 Labine, "Port Radium: The Story of Its Beginnings."

29 Kupsch, "From Erzgebirge to Cluff Lake," 53.

30 See, for example, van Wyck, *Highway of the Atom*, 106.

31 Ibid., 108.

32   See Irvin G. Wyllie, *The Self-Made Man in America: The Myth of Rags to Riches* (New York: The Free Press, 1966 [1954]).

33   van Wyck, *Highway of the Atom*, 109.

34   Allan Smith, "The Myth of the Self-Made Man in English Canada, 1850–1914," *Canadian Historical Review* 59, no. 2 (1978): 189–219.

35   Cruikshank, *Social Life of Stories*, 92.

36   Sherrill Grace, *Canada and the Idea of North* (Montreal and Kingston: McGill-Queen's University Press, 2001), 169.

37   Margaret Atwood, *Survival* (Toronto: House of Anansi Press Ltd., 1972); Margaret Atwood, *Strange Things: The Malevolent North in Canadian Literature* (London: Virago Press, 2004 [1995]).

38   See Piper, "Subterranean Bodies"; Morris Zaslow, *The Opening of the Canadian North, 1870–1914* (Toronto: McClelland and Stewart, 1971).

39   Zaslow, *Opening of the Canadian North*, 131–37, 279.

40   Quoted in Grace, *Canada and the Idea of North*, 9.

41   Délı̨nę Uranium Team, *If Only We Had Known*, 9.

42   Jason Baird Jackson, "The Story Of Colonialism, or Rethinking the Ox-Hide Purchase in Native North America and Beyond," *Journal of American Folklore* 126, no. 499 (2013): 31–54.

43   Literally, "Grandfather Beyonnie" or "Elder Beyonnie." "ʔəhtséo" is an honorific used to refer to elders and revered ancestors.

44   Andrew John Kenny and Dennis Kenny, "Personal Interview – 12 January" (2011).

45   Both spellings are generally acceptable in Délı̨nę.

46   George Blondin, *When the World Was New: Stories of the Sahtú Dene* (Yellowknife, NWT: Outcrop, 1990), 78.

47   John Curran, "Dene Medicine Man Foretold Horrific Future" (Yellowknife, NWT: Northern News Services, 2007).

48   Salverson, "They Never Told Us These Things."

49   See, for example, Rebecca Hall, "Diamond Mining in Canada's Northwest Territories: A Colonial Continuity," *Antipode* 45, no. 2 (2013): 376–93.

50   See René Fumoleau, *As Long as This Land Shall Last* (Toronto: McClelland and Stewart Ltd., 1975); Zaslow, *Opening of the Canadian North*, 7–9.

51   Cruikshank, *Social Life of Stories*, 95.

52   The final report of the Canada-Délı̨nę Uranium Table confirms that no employment records exist for Délı̨nę Dene people at Port Radium. It also asserts that "no Déline [sic] Dene were ever directly employed by Eldorado at the Port Radium mine or mill," though they did trade wood and meat at the village and

worked along the uranium transport route as ore carriers, deckhands, and pilots. Canada-Délı̨nę Uranium Table, "Final Report," 5, 27.

53 Marcel Pochon, "Radium in Canada," n.d., Eldorado Nuclear Ltd. fonds, LAC.
54 Roberts, "Living on Radium," 17.
55 Ibid., 28.
56 Fred J. (Tiny) Peet, "Interview by Jane Mingay," 1978, LAC.
57 "Punch" Dickins, "Interview by Jane Mingay," 1978, LAC.
58 Morris Neyelle, "Unpublished personal interview, 12 October" (2010).
59 Hall describes this as a kind of "accumulation by dispossession," a term developed by Harvey that builds on Marx's concept of primitive accumulation. Hall, "Diamond Mining in Canada's Northwest Territories."
60 See Canada-Délı̨nę Uranium Table, "Final Report," 83.
61 Ibid., x–xi.
62 See, for example, Alice Blondin-Perrin, *My Heart Shook Like a Drum: What I Learned at the Indian Mission Schools, Northwest Territories* (Ottawa: Borealis Press Ltd., 2009).
63 Little has been published on the trauma suffered by Inuit and Dene people who were forcibly relocated to tuberculosis hospitals during the early and middle parts of the twentieth century. See Joe Betsidea, "Unpublished interview, 14 April" (2011); Pat Sandiford Grygier, *A Long Way from Home: The Tuberculosis Epidemic among the Inuit* (Montreal: McGill Queen's University Press, 1994); Neyelle, "Unpublished personal interview, 12 October."
64 See Fumoleau, *As Long as This Land Shall Last*.
65 Julie Cruikshank, "Nature and Culture in the Field: Two Centuries of Stories from Lituya Bay, Alaska," in *Research in Science and Technology Studies: Knowledge and Technology Transfer*, ed. Marianne de Laet, Knowledge and Society 13 (Oxford: Elsevier Science Ltd., 2002), 37.
66 Ibid., 38.
67 Kenny and Kenny, "Personal Interview – 12 January."

| CHAPTER 3

"It's Just Natural":
First Nation Family History
and the Keno Hill Silver Mine

Alexandra Winton and Joella Hogan

For Yukon First Nation people, family history is intertwined with the history of the land. As the late Yukon First Nation elder Kitty Smith once told anthropologist Julie Cruikshank, "I belong to Yukon. I'm born here. I branch here. My grandpa's country, here. My grandma's. That's why I stay here . . . My roots grow in jackpine roots."[1] This sentiment holds true for the Na-Cho Nyäk Dun, or Big River People,[2] of the central Yukon, even though their relationships to the land and to each other have been transformed by over one hundred years of silver mining within their traditional territory.

In this chapter, we share the story of Herman Melancon, a member of the First Nation of Na-Cho Nyäk Dun and an underground miner. Using indigenous methodologies, historical research, and oral history generously shared by Herman, we parallel Herman's personal and family history with that of the Keno Hill mine, offering an intimate view into the complexity of relationships between northern First Nation peoples and industrial development.

Herman's family connection with the mine dates back to before he was born, to the mine's origins in the early twentieth century. At the heart of the Na-Cho Nyäk Dun traditional territory is the McQuesten River watershed, a region rich in fish and wildlife that sustained the Northern Tutchone-speaking people on their seasonal round throughout this rugged landscape.[3] It was near the headwaters of the McQuesten that high-grade galena, or silver ore, was discovered by Jacob A. Davidson in 1903. Hoping for gold, Davidson abandoned the find, unaware that he had stumbled upon one of the largest and richest silver deposits in the world, soon to be known as the Keno Hill mining district. Two years later and approximately one hundred kilometres downstream from Davidson's discovery, Dave Hager was born in McQuesten Village, a seasonal Na-Cho Nyäk Dun settlement at the mouth of the McQuesten River. Although it was not immediately apparent, Davidson's discovery would bring great changes for Hager, his family, and for the Na-Cho Nyäk Dun.

By virtue of their location, the Na-Cho Nyäk Dun had managed to escape some of the most severe impacts of the Klondike gold rush of 1898—the onslaught of thousands of gold seekers, the devastation of local wildlife populations, and the destruction of traditional hunting and fishing lands (to name a few). As the Klondike gold rush began to wane, however, increasing numbers of prospectors flooded into Na-Cho Nyäk Dun territory in search of the next bonanza.[4] There was gold in the area, but as more silver discoveries were made, many with ore averaging 300 ounces of silver to the ton, most prospectors abandoned their placer gold claims in search of silver. The first major staking rush did not occur until the 1920s, but when it did, development came quickly to the area. Mining techniques rapidly progressed from individual miners collecting float, or surface chunks of galena, to mechanized underground mining with corporate financing from the likes of the Guggenheim brothers. As the mines proliferated below ground, so too did the new communities above: the wild boom town of Keno City sprang up at the foot of silver-rich Sourdough, Galena, and Keno hills; while mining camps like Calumet, Wernecke, Elsa, and Bellekeno clung to the tops of these windswept peaks. Down in the valley, the more sedate community of Mayo developed into a regional centre on the banks of the Stewart River, where the silver-lead-zinc ore could be shipped out on sternwheelers (Fig. 1).[5]

Traditionally, Na-Cho Nyäk Dun people had travelled throughout this area in order to take advantage of seasonal harvests and lessen their impact on the land. In the summer months, Na-Cho Nyäk Dun families gathered on the banks of the Stewart and McQuesten Rivers to fish for chinook and chum salmon; in the fall, they moved into the mountains to hunt gophers and caribou, pick berries, and gather medicinal plants; and in winter, they returned to the lowlands to hunt for moose and ice-fish in the region's many lakes.[6] As the Keno Hill mining district developed, the expanding network of mines, roads, and communities bisected these traditional hunting and gathering grounds, contaminating water sources and ruining delicate plant and animal habitats. The Na-Cho Nyäk Dun were exposed to new diseases, religions, and societal pressures transported by newcomers. In keeping with the colonial policies of the day, the Na-Cho Nyäk Dun were forcibly settled into a sedentary community two miles downriver from Mayo, now known as the Old Village, where, despite being subject to curfews and segregation, they managed to live a semi-traditional lifestyle for many years.[7] As Na-Cho Nyäk Dun elder Dave Moses described it, his people adapted their seasonal round and subsistence patterns in order to accommodate the newcomers and their needs:

> After the boom, lots of people comin' to this place . . . Indian go get rifle to shoot moose and sell meat to Whiteman and make his living that way. When they find some rock in Keno, [First Nation] people move to Mayo and work on steamboat, cut wood, pile wood and sold wood on the barge . . . Pretty soon they use machine to grind the ore so it comes to a flour and put it in a sack . . . I been around Keno Hill when they first start. When I was a little boy people worked around Keno, Elsa, and hauled ore with horse team and caterpillar in the wintertime . . . Later they put in the highway and we haul ore back from Keno, back and forth with truck.[8]

As Moses explained, many First Nation people joined in the new economy, cutting wood for the steamships, selling meat, fish, and berries to the miners, and eventually, working in the mines themselves. For the Na-Cho Nyäk Dun, the way that they related to their environment began to change, as they adapted their hunting and harvesting patterns to help feed and shelter these newcomers and as they became involved in the extraction industry that quickly transformed their land.

By the late 1940s, most of the silver claims in the area had been consolidated by United Keno Hill Mines Ltd. (UKHM) and large-scale industrial mining had taken hold in the Keno region. UKHM operated up to nine different mines in the Keno Hill district, most of which consisted of a series of deep underground shafts, where miners worked in tandem, blasting out veins of ore to be hauled away by narrow-gauge railways. All ore was then transported by truck or aerial tramway to the company town of Elsa, the hub of the district, where a 250-ton capacity mill would separate out silver, lead, and zinc concentrates to be shipped to a smelter in Trail, British Columbia.

The expanding UKHM operations required massive amounts of electricity, which was provided by the Mayo hydro dam, built by the Yukon government in 1952 at the urging of UKHM. This dam drastically changed the hydrology of the Mayo River and raised the level of Mayo Lake by six metres. Transmission lines and roads were completed between Keno, Mayo, and Whitehorse, shifting the mode of transportation away from the rivers and damaging moose and caribou habitats.[9] Also in the 1950s, the Na-Cho Nyäk Dun were asked to relocate once again, this time back into the town of Mayo, where they were subject to discrimination, and many of their children were taken away to residential schools.

After the boom years of the 1950s and 1960s, the Keno Hill district went into a slow decline. Aging technology and low silver prices prompted cutbacks, strikes, and eventually, the indefinite closure of the mines in 1989. Most UKHM employees left the Yukon, Elsa was abandoned, and there was virtually no reclamation of the mine sites. As life in the region slowed, many Na-Cho Nyäk Dun people returned to traditional pursuits, such as hunting and trapping, and the environmental destruction left by the mining industry became more apparent.

Throughout this cycle of boom and bust, Na-Cho Nyäk Dun citizens became adept at operating in two new worlds: both the dark, cavernous world of the underground mines and the capital-based world of

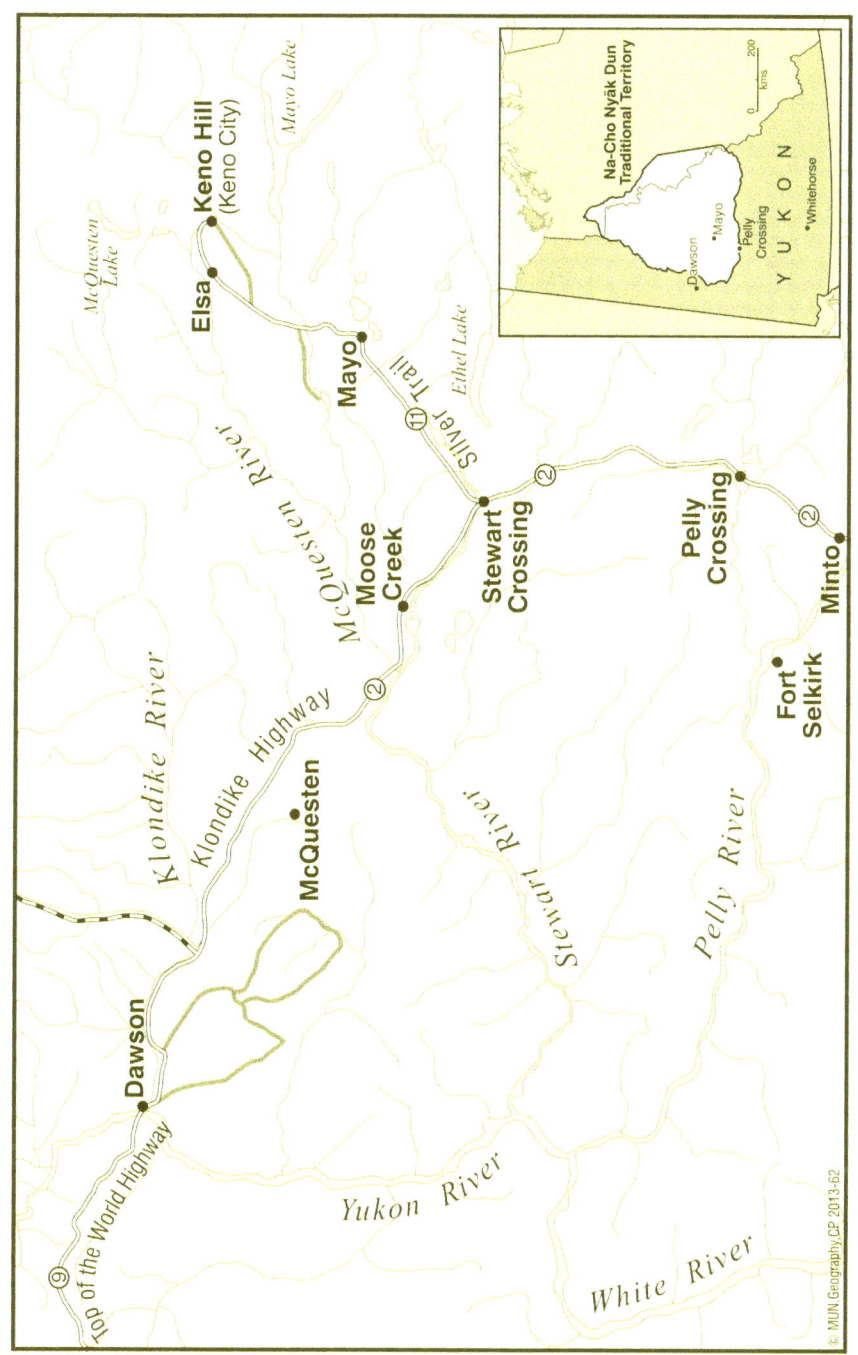

FIGURE 1: The Keno Hill, Yukon, mining district. Map by Charlie Conway.

FIGURE 2: Herman Melancon. Photo by Evan Rensch.

the newcomers who built them. The Na-Cho Nyäk Dun weathered environmental, economic, and social changes, while struggling to maintain their traditions, language, and culture. Perhaps the most severe impact was the newcomers themselves—an influx of hundreds of single Euro-Canadian men, who streamed into the region to work at the mines. Many of the newcomers entered into relationships with local Na-Cho Nyäk Dun women, and, whether they were considered legitimate or not, these relationships forever altered the cultural makeup of the region.

This complex reality is embodied by Dave Hager's grandson, Herman Melancon—the son of Dave's daughter Irene and Maurice Melancon, a non–First Nation miner from Quebec. At age fifty, Herman Melancon has spent more than half his life underground working as a miner. When asked to introduce himself, Herman quickly acknowledges his mining background:

> I'm Herman Melancon, from Mayo. I lived here most of my life, but I also lived in Tungsten, Northwest Territories for a couple

years, it's another mining town. It's running right now, that mine, and I been mining for 26 years. I first started in the Keno Hill Mine . . . I learned just from, probably from, coming from my Dad being a miner, see? It's just natural. I grew up around it, eh?

Just as Kitty Smith likened herself to the Jack pine—"I branch here," she said, her family and her history growing and diverging like the southern Yukon tree[10]—Herman Melancon's family grew and expanded with the development of the Keno Hill silver mine. Like the roots of a tree seeking water, the shafts, caverns, and tunnels of the mines have spread underground, following the silver-rich mineral veins. So too have Herman's family and the Na-Cho Nyäk Dun culture grown, so that both are now irrevocably intertwined with the history of the mine. Knit together in a complex pattern of mutual involvement and unequal impacts, the story of the Na-Cho Nyäk Dun and the Keno Hill mine is illustrative of how Aboriginal people across Northern Canada have been both affected by and involved in one of the most destructive forms of industrial development in their traditional lands. Our goal as researchers, authors, and community members is to share Herman's version of this story, which may help us to understand this complex relationship.

In order to help tell this story, we employ anthropological, oral history, and indigenous methodologies. The work of anthropologist Julie Cruikshank has influenced our approach to oral history and ethnography. Cruikshank's early work with Yukon First Nation women helped to set a new standard for oral history. By using long, uninterrupted interview excerpts, Cruikshank allowed, as much as possible, for stories to be told by the tellers themselves.[11] We make use of this style of storytelling here.

Indigenous academics such as Shawn Wilson, Lianne Leddy, and Margaret Kovach have also informed our approach to research.[12] As Wilson suggests, "We cannot remove ourselves from the world in order to examine it."[13] Indeed, this would be difficult for us, as we both live in or very near to the communities in which we have conducted research. Indigenous methodology also stresses the need for researchers to locate themselves during and within their research;[14] therefore, we will take a moment to do so now. Alexandra met Herman while conducting research for her master's degree. In the summer of 2011, Alexandra and a young Na-Cho Nyäk Dun citizen, Kaylie-Ann Hummel, interviewed Herman about his experiences with the Keno Hill mine. Joella (herself a Na-Cho Nyäk Dun citizen) met Herman through his wife Bobbie-Lee. Joella cannot remember exactly when she met him, as Herman has always been there, in the background, working in his garage or as the main character in Bobbie-Lee's stories. In the winter of 2013, Joella and Bobbie-Lee conducted another interview with Herman for this book chapter.

We have also been influenced by our own life experiences, storytelling cultures, and family histories, which counsel us to listen well, give credit to storytellers, and ask for permission before sharing their stories. Therefore, we have attempted to work as collaboratively as possible with Herman as we conducted research for and wrote this contribution. We are very much aware that our personalities and backgrounds have had an impact on our relationship with Herman and the way in which we portray his story; as such, we have included our own questions and responses in Herman's interview excerpts. We have used initials (HM, AW, JH, KH) to indicate who is speaking in these excerpts. As much as possible, we hope to allow Herman to tell his own story, fostering a relationship with the reader, who may then see the connections between Herman's community and his or her own.

According to Cruikshank, such narrative connections are vital to understanding change in modern First Nation communities:

> Yukon storytellers of First Nation ancestry frequently demonstrate ability to build connections where rifts might otherwise appear. They use narratives to dismantle boundaries rather than erect them . . . narrative storytelling can construct meaningful bridges in disruptive situations.[15]

In a complex era of mineral redevelopment, coupled with new Aboriginal self-government, we believe that Herman's story can serve as one such bridge. By sharing his narrative, we hope to create a broader understanding of the modern relationship between the Na-Cho Nyäk Dun, their land, and the processes that have affected it. Na-Cho Nyäk Dun are not alone in this situation; indeed many northern Aboriginal people are now faced with the difficulties of engaging in a modern economy,

FIGURE 3: Dave Hager standing on the sternwheeler "Keno" ca. 1930s. Yukon Archives/Dave Hager fonds/#8875.

while balancing new self-government responsibilities and struggling to preserve traditional ways of being. We hope that by sharing this story, we will shed light on the complexities of this modern era for northern Aboriginal people.

DOCUMENTING CHANGE

On September 27, 1905, Dave Hager was born to Jenny Jimmy at McQuesten Village. Known his entire life as "Big Dave," Hager was an amateur photographer, who both took part in and recorded a time of great change for the Na-Cho Nyäk Dun. His photographs can be found in the homes of his children as well as in the Yukon Archives. Dave documented his life and those around him through the lens of his camera; photographs of him show a young man in the early days of mining, when he was working on the steamboats (Fig. 6). These wooden paddlewheelers pushed barges loaded with ore from the Keno mines down the Stewart River toward Whitehorse. Ore would then be shipped on the White Pass Railway to the port of Skagway, Alaska, and then out to a smelter in Southern Canada. The Stewart and Yukon Rivers were lined with wood camps, where many First Nation men worked, providing the thousands of cords of wood needed to fuel the boats. Big Dave worked as a deckhand, loading the wood from shore onto the boats. Later, when the highways replaced the steamships, Hager worked for the Yukon territorial government, conducting highway maintenance between Mayo and Whitehorse.[16]

Dave, like many Na-Cho Nyäk Dun men, engaged in the new economy but did not catch gold—or in this case, silver—fever. He managed to keep a foot in both worlds by maintaining traditional pursuits, such as hunting, trapping, and raising his family on the land as much as possible. An anecdotal account, which appears in *Hills of Silver*, the only book dedicated to the Keno Hill silver mine, reveals Dave's attitude toward mining. After watching a crew of miners struggle for three days to move a large piece of mining equipment just one thousand feet, Dave, who was repairing a nearby bridge, is purported to have remarked, "those white men will do anything for money."[17]

In 1929, Big Dave married Alice Louise, a First Nation woman from Fort Good Hope, Northwest Territories. Together, Dave and Alice had six daughters: Martha, Rosie, Jenny, Irene, Laura, and Mary, all of whom were raised in the Mayo area. The Hager girls would all marry non–First Nation miners, or men involved with Keno Hill in some way. As Herman's wife Bobbie-Lee said, "they were all attracted to miners." One of these daughters, Irene, met and married Maurice Melancon, originally from Quebec. From a mining family himself, Maurice came north to work as an underground miner for UKHM. While Maurice was transplanted to the Mayo area, like many others, he gained acceptance in the community and knowledge of the land from his First Nation wife and her family. The skills to operate in both worlds were then passed on to their son Herman, who learned to hunt and trap from his grandfather and learned about the mining industry from his father.[18]

For a brief time, Maurice worked at another mine, called Tungsten, located in the southern Yukon. Tragically, Maurice was killed in a car accident while driving to work on the Tungsten road. Herman recalled the accident:

> He was driving to work in Tungsten, after he got called back there to go to work. And he was driving back there and there was only about, I don't know, bad pass there and he went over the bank. He drove right from Mayo too, that day, it was too far . . . I was nineteen. I went mining after he died, right after.

Born to a miner, Herman became a miner himself at the age of twenty. While Herman is definitely the miner in the family, his brothers have all been involved with mining either directly or in the support industries. When asked why he continues to mine, despite the dangers of underground work, Herman replied: "Well, it's part of me, it's in my blood. My Dad, probably to, in some way, make him proud a' me. And then, plus for the money, I like the money."

Herman is just one of the many Na-Cho Nyäk Dun citizens in the modern community of Mayo who are of mixed ancestry, due to the influx of young, Euro-Canadian men who came to work in the mines. These men came alone, and many of them married local Na-Cho Nyäk Dun women. According to the enfranchisement policies of the 1876 Indian Act, these women and their children then lost their status as Indian people and the few privileges it provided. Years later, they fought to regain their status and take part in the land claim agreements that were occurring across the territory.[19] The women were also isolated from their families, as many of them went to live in Elsa with their new husbands. Of course, these struggles did not change who they were, or their ability to connect with their culture or the land. Debbie Buyck, a Na-Cho Nyäk Dun woman whose husband worked in Elsa, has said that even while living in a predominantly non–First Nation company town, built on the standards of southern Canadians, Na-Cho Nyäk Dun women managed to practise many of their traditional activities:

> We did continue in our traditional ways outside of work, we went berry picking, hunting and fishing. We would go out on the weekend for drives or day hikes and would scout out old places our parents had taken us when we were kids . . . the mothers

and kids would go berry picking and fishing while our husbands worked in the mines, we took lots of picnics and some of the old First Nation ladies in camp taught us our traditional ways.[20]

Many of the Na-Cho Nyäk Dun women who lived up at Elsa talk of picking berries, hunting, and snaring small animals in the region, all attempts to maintain aspects of the seasonal harvest. The increased mining activity in the area and a more sedentary lifestyle affected the Na-Cho Nyäk Dun's ability to hunt, fish, and pursue traditional activities; however, these pursuits did not disappear entirely. Men and women both found ways to balance this new life with traditional ways of living, which were passed down to the generation currently in Mayo.

"IT'S HARD TO IMAGINE THESE HILLS WITHOUT MINING"

For Herman's grandparents and parents, the mine was a new development, an oddity, which altered their environment, brought newcomers into the area, and drastically reshaped their relationship to the land and to others. For Herman, however, the mine has always been there, and his involvement with it began early. As Herman's father was a non–First Nation man who worked for UKHM, his family was able to live in the company town of Elsa, where UKHM provided housing, recreation facilities, free steam heat, and discounted groceries. While Herman has fond memories of a busy youth in the community, not everyone was able to live there. Most First Nation men who worked for the mine lived a few kilometres down the road, in a small cluster of houses called Millerville. It is unclear whether they were forced to live there rather than in Elsa, or whether they received the same benefits as residents of Elsa.[21] For a young man of mixed ancestry, like Herman, life in such a racially divided mining camp must have been somewhat difficult, but, like many people now living in Mayo, he is reluctant to speak about the segregated past.

When Herman's father got better-paying work at the Tungsten mine, the Melancon family moved back to Mayo. After his father's sudden death, Herman began working for UKHM himself. His first job was digging trenches, but, being a tall, strong man, Herman quickly moved up to the position of trammer. As he explained, there was a natural progression in the training of an underground miner and a gradual increase in pay:

> HM: Trammer . . . that's where you drive around little trains for the miners, go outside and dump it [the ore], then you come back and help them, have to help them . . . And then you go to miner helper, later, from that. And then eventually you can do your own minin'. But now, it's all trained differently, they're all trained on heavy equipments, eh? Right off the bat, usually.
> AW: So they didn't have those sorts of training programs when you were young?
>
> HM: No, you learn different . . . I went underground, they get you diggin' ditch, eh? And they start you off from the bottom.

Herman has worked as a miner ever since, and now, as he said, "I run everything underground. Yeah, anything that moves, I run it, eh? I have to." While he is known throughout the community as a good miner, his career has not been without its dangers. When Herman was twenty-seven, he was in a horrific accident that occurred when a co-worker, drilling into the rock face beside him, accidentally hit a hole pre-loaded with dynamite. Joella asked Herman about the incident:

> JH: Do you want to tell me the story about how your arm got blew up?
>
> HM: . . . Well, we were drillin', I was in Bellekeno . . . I was workin' three weeks by myself . . . And then they hired young guy, [name omitted], used to live in Elsa. And the night shift crew, they bootlegged the face, so bootleg is like a socket of holes that they built, you know, couple feet deep, or a foot deep, whatever and then the shift boss spray paint them all blue, fluorescent blue, so you could see them, eh? . . . it's the law to drill away from them, six inches minimum, away from all those holes. Because if there's power still to light it, you can't see inside, it's frozen eh, the holes, all crushed. . . . So I was just pulling that drill down off

the backhoe on the left side, I only had three holes left to go and then it went off . . . just when my arm was in the air, I guess. And it blew me back or wherever, by the time I woke up I was probably about thirty feet from the face, landed about thirty feet away. Banged my arm. I could see something happened, 'cause I was laying on the ground and it was all quiet, you couldn't hear nothing. And then I could hear [Herman's co-worker] moaning away, way up at the face there. But I was lookin' around for my arm, 'cause I couldn't see it on my side. And something was hurting on my right side. And I was looking around, bending over, look around on the ground for my arm, 'cause I thought it was blown off, like that bending over and then all of a sudden, my arm . . . come flying, fell off the back of my neck and come, was hanging there. Then I was happy, 'cause it was still there. But I couldn't do nothing, I had to hold it up, it was broken, broken right off, eh? The bone was busted right off.

Fortunately, both men survived the explosion, and, after just two months off work, receiving workers compensation, Herman went back underground. "I had to," he said, "'cause workers comp bother me. . . . I still got pins in here, pins there and screws. . . I was scared for a while, to work underground, but it actually was a good therapy. . . . It helped me become stronger, after a year or two it didn't bother me no more."

This accident and others have left physical imprints on Herman's body. After so many years of working with loud, heavy machinery in a poorly ventilated, underground environment, his hearing is diminished, his lungs are damaged, and he bears the scars from potentially deadly accidents. These corporeal impacts are easy for Herman to discuss. Intangible impacts on his community and his culture are more difficult to express. While Herman is quick to acknowledge the history of the mine, he struggles to articulate how it may have affected his ancestors:

KH: How do you feel about the mining history in the area?

HM: I uh, there's a lot of history there. I think it goes back right to 1900's, so there's still silver up there. There's still silver there for a while. And there is a lot of history there, you can just tell by that old museum in Keno, eh? Just by going through there . . .

AW: And how do you think that mining history has affected the Na-Cho Nyäk Dun First Nation?

HM: I'm not sure how it affected. . . I know they're getting more um, native people involved in mining, with their training, eh?

Instead of discussing the past impacts of the mine, Herman speaks about the future, pointing out that there are more First Nation people working within the mining industry. Perhaps this is because Herman has never known the area without mining—as Joella says, "for this current generation [of which she and Herman are a part] . . . it is hard to imagine these hills without mining." The history of the mine and Herman's family story are so intertwined, it is impossible to separate one from the other in order to examine it.

Although Herman's family history has revolved around the mine, mining is just his work, and there is much more to his life. Herman also learned how to hunt and trap from his grandfather, Dave Hager, and he still employs these skills, as he hunts for moose each fall and maintains a trapline near Mayo. Herman was happy to speak with us about this aspect of his life:

AW: . . . And what about trapping, what sort of animals do you trap?

HM: Usually marten and wolverine, link [lynx] usually. Usually the one that pay, eh? 'Cause now it's pretty well . . . pretty low right now.

AW: And where do you trap?

HM: I got a little line ten miles out of Mayo, it starts on the, it's called the Kurtz River and the Ridge Trail. I have a little, I got a little trapping shack on it too.

Hunting and trapping are important activities that, along with mining, tie Herman to his traditional territory and keep him connected to the land. Hunting and trapping are also important skills that Herman was able to rely on to keep him afloat when the mining industry went into an inevitable decline.

MINE CLOSURE AND SELF-GOVERNMENT

In January 1989, as many employees were returning from Christmas holidays, UKHM suddenly announced that, due to low mineral prices, the company was closing the mine indefinitely. All residents of Elsa were given two weeks to vacate the town, which was quickly abandoned. Most UKHM workers left Elsa immediately, finding work at other mines in Southern Canada, but about half of them remained in the Yukon, some moving to Mayo and a few others settling in Keno or the surrounding rural areas.[22] The mines were shuttered, with just a small crew of loyal employees left to treat run-off water from the mines and guard the buildings from vandals. Herman was not immediately affected by the mine closure, as he had already moved on to contract mining work and was able to support his young family, while supplementing his wages with trapping. Kaylie-Ann asked him about this time:

> HM: 1989? When it shut down? I quit before it shut down. I worked there for six years and then I went to work for contractors, so I was already gone outta there. Like I say, I guess it shut down, everybody moved out, all to B.C., all the families moved right outta there. And they just had a skeleton crew there, eh? That's what I remember from the closure, how it shut down there.

Eventually, Herman also had to leave his home community to seek work in mines in the southern Yukon and British Columbia. Herman hinted at the difficulties of this lifestyle: ". . . you're always gone, eh?" When the mine shut, most Na-Cho Nyäk Dun citizens moved into Mayo to be with their extended families, but this was a quiet time for the Mayo region. While there was some employment to be found with the territorial government and the Local Improvement District,[23] there was little other work to sustain the population. However, long-awaited land claims and self-government processes were finally beginning to take shape.

When the mine was in its final years, Na-Cho Nyäk Dun elders and a team of land use technicians and negotiators were working on a monumental self-government and land claims agreement. Historically there were no treaties negotiated with Yukon First Nations. In order to address this, in 1973 the Yukon Native Brotherhood delivered a landmark document to Prime Minister Trudeau, *Together Today for Our Children Tomorrow*, which was Canada's first comprehensive land claim.[24] This claim eventually developed into the Yukon Umbrella Final Agreement, a modern-day treaty under which the majority of Yukon First Nations negotiated individual land-claim and self-government agreements.[25] In 1993, after decades of negotiations, the Na-Cho Nyäk Dun became the first Yukon First Nation to sign its land-claim and self-government agreements with the federal and territorial governments.[26]

During the land claims process, blocks of land were selected in a large range of sizes throughout the Na-Cho Nyäk Dun traditional territory. These lands were selected for traditional pursuits, harvesting, protecting heritage sites, and future housing and development, as well as for economic development. Much of this land was selected with future generations in mind. Many young Na-Cho Nyäk Dun men were charged with surveying the lands, interviewing elders, and documenting these sites for land claims. Those who were involved with self-government had hope for a brighter future, while others were able to maintain a subsistence lifestyle by cutting wood, trapping, or working as hunting guides.

As Na-Cho Nyäk Dun people began the long process of reclaiming the land, both on paper and in person, they became more aware of the environmental degradation caused by years of mining. There had been very little reclamation or remediation work done at the mine sites; as a result the relics left from mining activity were a danger to wildlife, such as moose, which were found tangled in wire left over from the mine. People became cautious about hunting and fishing near the shuttered mines, and concerns were raised about the quality of water and the impacts on fish and wildlife.

Na-Cho Nyäk Dun began the slow process of self-government, taking over services formerly operated by the federal government, such as administration of housing and health and social benefits. Self-government ushered in a new era for the Na-Cho Nyäk Dun, who again have a measure of control over their traditional lands. The First Nation now owns approximately 4,700 square kilometres of settlement land, within a traditional territory spanning 162,465 square kilometres.[27] The First Nation of Na-Cho Nyäk Dun now must be consulted regarding development within that region. In order to ensure that their citizens benefit from future development, the First Nation created the Na-Cho Nyäk Dun Development Corporation, a business arm of the government, which can enter into agreements with mining and exploration companies.[28] Now, the First Nation of Na-Cho Nyäk Dun must act as stewards of the environment and the economy, keeping an eye on the activities of new mining companies, while at the same time supporting work and training opportunities for its citizens.

A NEW ERA OF MINING

Throughout these years, Elsa remained abandoned, Keno City shrank to approximately twenty residents, and Mayo experienced an economic downturn: with the need for services greatly diminished, the town fell to 800 residents, or half of its former population.[29] Tourism was regarded as an important economic alternative to mining, so the Yukon government and local tourism organizations marketed the Mayo area with a mining theme. The name "Silver Trail" was given to the Yukon Highway 11, which links Mayo, Elsa, and Keno with the Klondike Highway, and wilderness tourism was suggested as a complementary attraction to the region's mining history.[30]

But in 2006, after sitting abandoned for nearly two decades, the Keno Hill mine site was purchased by Alexco Resource Corporation, a junior mining and reclamation company based in Vancouver. The Yukon government awarded Alexco a contract to remediate the mines and the Elsa townsite, while simultaneously allowing the company to conduct its own exploration work and assess the possibility of redeveloping the mine. In 2010, the Bellekeno Mine, one of United Keno Hill's top producers, was reopened by Alexco, and the first trucks of silver, lead, and zinc ore in over twenty years began to run through the Keno Hill region.

With a settled land claim and the implementation of self-government, the Na-Cho Nyäk Dun Development Corporation was eager to create economic opportunities for its citizens and the region of Mayo and has done so by supporting the redevelopment of the Keno Hill mine. For Herman, this new relationship was inevitable for his First Nation, but it also creates a new form of responsibility:

>AW: What do you think about Na-Cho Nyäk Dun being involved in the mine up there?

>HM: Well, I guess they probably have to keep an eye on them, what they do up there. I dunno, make sure they keep the area clean and everything, and treat the water. I guess they gotta keep treating the water steady there, eh?

>AW: So how do you feel about the First Nation being involved in the mining industry as a whole?

>HM: Well, if they're, eventually they will have to, eh?

>AW: Yeah?

>HM: They'll have to uh, know about mining and have more people trained.

The mine redevelopment was not of immediate concern for Herman; however, he soon returned to work at Keno Hill:

>KH: How did you feel when you heard that the mine was reopening?

>HM: At the time I didn't mind, 'cause it was close to home. But I wasn't worried about it, 'cause I had a job somewhere else anyway, but then I got on with . . . I went to work for Alexco for a while . . . then I got on with the contractor, soon as they started mining, eh? Right away.

Even though twenty years had gone by, Herman found little difference between his work for UKHM and his work for Alexco; however, he easily draws contrasts between the communities, the old Elsa and that of the new Alexco operation. UKHM Elsa was a real town, with families and an air of permanency, whereas the new incarnation of Elsa is simply a work camp:

> HM: Elsa, when it was a town site, when people, families lived there, it was more busy. Now it's more like a work-camp, eh, basically. It's pretty quiet up there now. Like, this is just a working camp, eh?
>
> AW: And what's it like for you to be working around Elsa, now, a town that you sorta grew up in, that you spent some time in? When you see the old school and buildings like that, how does it feel?
>
> HM: Well there's probably lots 'a memories in there, yeah, when you're a kid. I remember playing hockey there all the time and that. And it was a pretty busy little town, eh?
>
> AW: Yeah. Do you miss it? Like, do you miss living there, or having that town around?
>
> HM: I dunno, yeah, a little, in a way, but I just like quiet places, eh? Quiet towns, I don't like very big places. I could stay in a big place, but not very long. 'Cause see here, some days you still, you don't even have to lock your door.

For Herman, there was an emotional connection to the town of Elsa, but with eight young daughters to care for, Herman is less concerned with the changes in Elsa, an abandoned mining town, than with the possible changes in Mayo, where his family now lives:

> AW: So how do you think the mine reopening there, how do you think it's affected the community of Mayo?

> HM: It probably bring a little more money into the community, more than anything, and some jobs. Yeah. But the majority jobs always go to outside, eh?
>
> AW: Why do you think that is?
>
> HM: Um, they're more qualified.

Herman is one of the few experienced Na-Cho Nyäk Dun miners working at Keno Hill. As in the previous era of mining, most of the well-paid management and technical trades positions are awarded to people from outside the Yukon, who work on a fly-in, fly-out basis. Much like the UKHM era, the influx of young, predominantly male workers at the mine has created concerns about alcohol, drugs, and social changes in the community. However, it was just such an influx of young men that brought Herman's father to the Yukon and created a comparable situation for Herman's grandfather, Dave Hager. Both Herman and his grandfather Dave managed to balance traditional subsistence work while engaging in a modern economy, and both were living in a time of cultural and social change—for Dave it was an influx of newcomers and industrial development, while Herman has witnessed the revival of such development in a new era of Aboriginal self-government. Like his grandfather, Herman also has a large, primarily female family to look after—Herman has eight daughters, and Dave had six. Herman continues to drill away at the same mineral veins his father worked on, expanding the underground maze of tunnels and shafts, some of which are supported by timbers cut by his grandfather. But while his family history may be intrinsically linked to that of the mines, Herman appreciates the world above ground much more than the dark, damp mines below:

> AW: . . . So what's it like working underground, like what are your days like there?
>
> HM: If you keep busy, it go by fast. Yeah, I like that. Sometimes you get tired of the dark, I like to come out in the daytime when it's nice outside, I'll come all the way outside to eat lunch, yeah, like if it's real nice.

AW: Just to get some sunlight?

HM: Yeah, to come see some daylight, eh?

[Laughter]

HM: 'Cause you know, I don't wanna spend all day down there, like to come out at least once, eh? . . . So, you see a lot 'a dark, eh? Feel like a mushroom.

Herman also sees an important distinction between his work life and his home life. Mining may pay the bills, but most important for Herman is the rural, northern lifestyle, which he is able to maintain despite physically demanding shift work:

AW: . . . Well, when you're not working up there, what other sorts of activities do you do out on the land and in the area around here?

HM: Well, I like hunting. I used to trap and all that before, but prices are way too low, I don't bother anymore, in the winter.

AW: Where do you go hunting?

HM: Usually up towards Elsa, up that way. Up in that area, or up river.

AW: Is it for moose, mostly?

HM: Yeah. Sometime I go hunting for sheep too.

AW: Have you noticed any changes with the moose habitat or behavior, or anything around the mine sites?

HM: Yeah, there's less moose around. There's more activity, eh, with this mining going on.

AW: Right. So how does that affect your hunting?

HM: Well, usually people have to go further out, eh, for their moose.

Herman, a man who has spent more than half his life mining—who sustains his family and his traditional pursuits through mining—is still critical of such development. While he is not given to public speaking, his wife, Bobbie-Lee, has represented him at community meetings, speaking out about certain mineral development projects to which Herman is opposed, such as a proposed hardrock mine near Mayo:

> HM: . . . But see they're gonna open that Victoria Gold too, and that's in huntin' country there, boy. Yeah, that's out McQuesten Flats there, and all that, that area.
>
> AW: How do people feel about that one?
>
> HM: I'm not sure. Me, myself, I don't care much for open pit mining, 'cause it makes a big hole, eh? And too, when they start using heap leach, eh?
>
> AW: Right. So do you think that kind of mining, up at Keno Hill is better for the environment?
>
> HM: In a way, eh? Yeah, because it's one hole in the side of the mountain, not compared to a open pit. Open pits are huge. When they, they can affect the area way more, yeah, so but they still, it still goes on all over, eh?
>
> AW: Do you feel like you have a say? Like as a First Nation person in the area, like your First Nation has sort of a say in that kind of mining?
>
> HM: Uh, yeah, yeah, they must have some kinda say, 'cause that's all Band land [First Nation settlement land] over there, on this side of it. But they usually go ahead anyway, eh?

Both dependent on the mining industry and critical of it, Herman personifies the complexity of the mining and development debate in the Yukon. In spite of the sometimes fierce rhetoric surrounding contemporary mineral development in the territory, few Yukoners are completely against or completely in favour of mining. Instead, there is a spectrum of what people view as acceptable. Herman, like all of us, is mired in the modern, industrial world, which is still dependent on non-renewable resources, such as metals and fossil fuels. While Herman toils underground to uncover these metals—silver, lead, and zinc—many of us benefit from his work and in turn, partake in the industry.

"LONG AS IT DOESN'T CHANGE THE YUKON TOO MUCH"

The United Keno Hill Mine has left its mark on the Na-Cho Nyäk Dun traditional territory. The largest environmental scars are hidden deep within the Wernecke Mountains, where there are hundreds of kilometres of hollowed-out mining shafts and, as people say, more timber below ground than above. The social and cultural effects of the mine, however, can still be seen, heard, and felt in the community of Mayo and among the Na-Cho Nyäk Dun. For Herman Melancon, who has worked at the mine nearly his entire adult life, this work has not only shaped his family history, it will also leave a permanent, physical impact on his life. Indeed, with each scar and injury, the story of decades of underground mining is slowly being inscribed on Herman's body. Near the end of our interview, Herman did confess that he was concerned about his health and would eventually like to stop working underground:

> AW: Well, I think that's pretty much all of our questions, do you have anything else you wanna add, or . . .?
>
> HM: I'd probably like to quit, uh, stop from it, you know, stop mining eventually, try something else, maybe placer, maybe placer mining. 'Cause sometimes all the diesel smoke underground, I get tired of it, coughing that black stuff up.
>
> AW: Right. You're worried about your health?
>
> HM: Yeah, I'm starting to get worried about my lungs, actually. I start trying to wear mask more, you know? I wear mask more now, 'cause some equipment, you know, you got lots a' equipment moving underground and the ventilation is not good

enough then you, you get too smoky and breathe too much of that diesel fumes in, eh?

    AW: Right, right. Do you notice any effects, any like physical effects yet, from working down there, or . . .?

    HM: . . . my hearing is less, eh? Percentage. It's still pretty good, but it's, as years go by it's gonna get less, eh?

Herman is critical of the environmental degradation caused by mining and of the physical impacts on his own body, but for him, mining is a part of his life. A career change would mean shifting from one form of mining (underground) to another (placer). Herman has been a miner since he was twenty years old, and while he is realistic about the boom and bust life cycle of the industry, he is confident there will always be another mine where he can find work:

    AW: . . . So who do you think should regulate that [the mining industry]?

    HM: Probably government, eh? Don't open it up too much. There's a lot a' nice country, eh, up in the Yukon. But I dunno how long the mining boom will last, eh? Maybe it'll last ten years.

    AW: Do you think about that, like, what you'll do if the mine shuts down again?

    HM: Keno Hill? I'd probably just go out, more out in B.C. again, eh? But there's another mine open, there's another one at Minto anyway, Minto too. There's how many mines going in the Yukon—one, two, three, eh? I think.

There may always be another mine at which Herman can work, but with each new mine, or redevelopment, there are social and environmental consequences to bear. One of the most significant social impacts of the Keno Hill redevelopment has been a polarization of opinions between people in the surrounding communities. For the Na-Cho Nyäk Dun, differences in opinion about the mine have served to widen the gap between elders and the younger generation who, like Herman, grew up with mining. For many elders, protection of the land is of the utmost importance, while for a younger generation, who are now running the First Nation government, economic development is also a high priority. Herman spoke about this gap:

> AW: . . . Do you think there's a difference between the way elders in Mayo feel about the mining and the way younger people feel about mining?
>
> HM: Yeah, there's probably a difference. Elders never like it very much, eh? . . . in the old days, elders didn't bother with mining very much. That's 'cause long time ago, Indian people, they used to find gold in the river and they just threw it back in there, didn't they?
>
> AW: Yeah.
>
> [Laughter]
>
> HM: Didn't bother, but now, I'm not sure how elders feel about it now. But you see more younger people working at mines, now, eh? . . . Long as it doesn't change the Yukon too much, eh? The Yukon should be left the way it is, see. You uh, shouldn't overpopulate too much, here, shouldn't change very much, 'cause they'll just ruin it.

Herman demonstrates an important similarity between the opinions of elders and those of younger Na-Cho Nyäk Dun citizens. While they may strive for economic development and recognize their historical connection and economic dependence on the Keno Hill mine, Na-Cho Nyäk Dun citizens, old and young, value the land and feel that they belong to it. As Na-Cho Nyäk Dun elder Helen Buyck has asserted, "The land was their teacher, and the knowledge they have of it is far greater than most people can appreciate."[31] This land and the Na-Cho Nyäk Dun culture have been shaped by over one hundred years of underground mining, but those hundred years represent just a fraction of their story. While some of that history will remain hidden deep within the silver-laden mountains, much of it will be told both through the stories and the bodies of the Na-Cho Nyäk Dun themselves.

We have shared elements of Herman's story, in an attempt to bridge the burgeoning gaps between generations of Na-Cho Nyäk Dun citizens and between First Nation and non–First Nation Yukoners. However, as Julie Cruikshank points out, "that a culture is shared does not mean that all individual interpretations will be the same."[32] Indeed, this is not a definitive representation of contemporary Na-Cho Nyäk Dun opinions or culture, it is simply our interpretation of Herman's story, which is one we find particularly poignant and illuminating. The very act of sharing stories such as Herman's may serve to create understanding between generations and, we hope, demonstrate the ability of narrative to unravel the complexity of modern relationships between northern indigenous peoples, industrial development, and the land that they share.

AUTHORS' NOTE: At the time of editing in 2013, Alexco Resource Corporation announced that, due to decreasing silver prices, the company would be laying off up to 25 per cent of its employees at the Keno Hill silver mine, with tentative plans to reopen in the early 2015.[33] Herman was laid off and has been exploring other work opportunities at mines down south. Unfortunately, fur prices have been too low for Herman to profitably work his traplines.

ACKNOWLEDGMENTS

Joella and Alexandra wish to express their gratitude to Herman Melancon for his generosity and bravery in sharing his story. Mussi Cho Herman. Mussi to Bobbie-Lee Melancon and Kaylie-Ann Hummel for their assistance, and to the First Nation of Na-Cho Nyäk Dun for its support of this project.

NOTES

1 Julie Cruikshank, *The Social Life of Stories: Narrative and Knowledge in the Yukon Territory* (Lincoln: University of Nebraska Press, 1998), 1.

2 This literal translation omits much of the meaning held within the name. Northern Tutchone elders speak of a much more nuanced definition, in which "Na-cho" translates to "our elders," thus Na-Cho Nyäk Dun means "flowing from our elders," demonstrating the direct connection between the words of the elders and the Stewart River, which has sustained the Northern Tutchone people.

3 Northern Tutchone is an Athabascan language, traditionally spoken by First Nation people of the central Yukon, whose descendants still identify as Northern Tutchone people. Although they are now divided into three separate communities and First Nations—Mayo (First Nation of Na-Cho Nyäk Dun), Pelly (Selkirk First Nation), and Carmacks (Little Salmon Carmacks First Nation), there are still many historical, cultural, and familial ties between the Northern Tutchone. Na-Cho Nyäk Dun was the name chosen by the Northern Tutchone people of the Mayo area for their government during the land claim and self-government process. Here, we use the term Na-Cho Nyäk Dun to represent the Northern Tutchone people living in the Mayo area, both historically and today.

4 Ken Coates, *Best Left as Indians: Native-White Relations in the Yukon Territory, 1840–1973* (Montreal and Kingston: McGill-Queens University Press, 1991), 86–95; Lynette R. Bleiler, Christopher Robert Burn, and Mark O'Donoghue, *Heart of the Yukon: A Natural and Cultural History of the Mayo Area* (Mayo, YT: Village of Mayo, 2006), 99; Dominique Legros, "Oral History as History: Tutchone Athapaskan in the Period 1840–1920," in *Occasional Papers in Yukon History* No. 3 (2) (Whitehorse, YT: Yukon Cultural Services Branch, 2007).

5 Lynette R. Bleiler and Linda E. T. MacDonald, *Gold and Galena: A History of the Mayo District, with Addendum* (Mayo, YT: Mayo Historical Society, 1999), 60–70.

6 Bleiler, Burn, and O'Donoghue, *Heart of the Yukon*, 86–89; Legros, "Oral History as History," 244–308.

7 Yukon Archives, Mayo and Area 1/2, John Hawksley, *Report on Mayo Band of Indians* (Dawson City, YT: Department of Indian Affairs, 1916), 1–3; Bleiler, Burn, and O'Donoghue, *Heart of the Yukon*, 90–91.

8 Dave Moses, in *Heart of the Yukon*, 91.

9 Bleiler, Burn, and O'Donoghue, *Heart of the Yukon*, 119.

10 Quoted in Cruikshank, *Social Life of Stories*, 1.

11 See Julie Cruikshank, Angela Sidney, Kitty Smith, and Annie Ned, *Life Lived Like a Story: Life Stories of Three Yukon Native Elders* (Lincoln: University of Nebraska Press, 1991).

12 Shawn Wilson, *Research Is Ceremony: Indigenous Research Methods* (Winnipeg: Fernwood Press, 2008); Lianne Leddy, "Interviewing Nookomis and Other Reflections: The Promise of Community Collaboration," *Oral History Forum d'histoire orale* 30 (2010): 1–18; and Margaret Kovach, *Indigenous Methodologies – Characteristics, Conversations, and Context (Toronto*: University of Toronto Press, *2009)*.

13 Wilson, *Research Is Ceremony*, 14.

14 Kathy Absolon and Cam Willet, "Putting Ourselves Forward: Location in Aboriginal Research," in *Research as Resistance: Critical, Indigenous, and Anti-oppressive Approaches*, eds. Leslie Allison Brown and Susan Strega (Toronto: Canadian Scholars' Press/Women's Press, 2005), 97.

15 Cruikshank, *Social Life of Stories*, 3–4, 24.

16 Bleiler and MacDonald, *Gold and Galena*, 275.

17 Aaro E. Aho, *Hills of Silver: The Yukon's Mighty Keno Hill Mine* (Madeira Park, BC: Harbour Publishing, 2006), 58.

18 Much of this information came from personal conversations between Joella and local elders throughout 2012 and 2013.

19 Coates, *Best Left as Indians*, 83, 89, 239.

20 Bleiler, Burn, and O'Donoghue, *Heart of the Yukon*, 92.

21 We only learned about Millerville through oral history, as it is not mentioned in any of the written documents about the Keno Hill area.

22 Yukon Territory, *Silver Trail Tourism Development Plan* (Whitehorse, YT: Yukon Department of Tourism, 1989), 8.

23 This became the municipality of Mayo.

24 Coates, *Best Left as Indians,* 162, 231.

25 These agreements are collected on the Aboriginal Affairs and Northern Development Canada website, accessed January 5, 2015: https://www.aadnc-aandc.gc.ca/eng/1100100030607/1100100030608.

26 Bleiler, Burn, and O'Donoghue, *Heart of the Yukon*, 129.

27 First Nation of Na-Cho Nyäk Dun, "History," Retrieved from http://nndfn.com/history/.

28 Bleiler, Burn, and O'Donoghue, *Heart of the Yukon*, 91.

29 Ibid.,122.

30 Yukon Territory, *Silver Trail Tourism*, 1.

31 Bleiler, Burn, and O'Donoghue, *Heart of the Yukon*, 87.

32 Cruikshank, *Social Life of Stories*, 43.

33 Alexco Resource Corporation, News Release, *Alexco Implements Cost Savings Measures* (Vancouver: Alexco, May 31, 2013), 1–2; "Alexco Wants to Head Back to Work at Keno Silver Mine," CBC News, December 6, 2013, http://www.cbc.ca/news/canada/north/alexco-wants-to-head-back-to-work-at-keno-silver-mine-1.2453849.

| CHAPTER 4

Gender, Labour, and Community in a Remote Mining Town

Jane Hammond

The quickest way to travel from the island of Newfoundland to Labrador City is by airplane. A small Provincial Airline plane does a twice-daily milk run from St. John's to Labrador City, stopping in Deer Lake and Goose Bay before reaching its final destination. In early May 2011, as the plane approached western Labrador, I was struck by the scale and extent of over five decades of open-pit iron mining on the landscape. Mountains of red earth sat next to large craters sculpted by years of mass-mining for iron ore. Labrador City, originally built by the Iron Ore Company of Canada (IOC), made up the majority of the western Labrador's population and continually underwent construction to keep up with expanding iron production, as contractors cut large sections of the surrounding wooded area to make room for more temporary housing. Less visible than these environmental changes, however, is the history of unequal gender relations that have accompanied the extraction of iron ore in western Labrador, as women struggled to enter the mining workforce and gain independence and opportunity in this male-dominated company town.

The study of women's place in industrial towns, and more specifically mining communities, has blossomed since the 1980s as scholars moved from focused studies of gender in industry to a more all-encompassing approach to place, community, and industry. This shift in focus reflects changing historiographical understandings of gender and labour history. By the 1980s, scholars such as Angela John recognized that past historical accounts reflected inaccurate, stereotyped gender roles and ideologies.[1] Several studies examined traditional gender divisions and explored the significance of women's positions as housewives to the success of the town and industry. For example, Luxton and Fox used a Marxist approach to reinterpret the work of housewives, arguing that "women's work in the home is one of the most important and necessary labour processes of industrial capitalist society."[2] Similarly, Vicky Seddon's public history of the British coal miners' strikes of the 1980s determined that most women became involved in the strike because they viewed it as an extension of their duties as housewives.[3] By the late 1990s to 2000s, scholars transitioned in their approach from women's studies to gender studies, and the focus moved to understanding relationships between men and women. Recognizing the complexity of gender and labour history, these studies used interdisciplinary, mixed-methods approaches.[4]

Drawing from these models, this study of Labrador City women's history also used an interdisciplinary framework and multi-method techniques. The primary research for this chapter consists of twenty-five oral history interviews and group workshops conducted in May 2011.[5] Oral history is used to reveal the hidden voices and complex divisions among working-class residents in the town, since these might not be readily apparent using other research methods. To obtain multiple points of view, interviewees included current or retired employees of the mine and residents of the community ranging in age from twenty-four to eighty-two. Following the interview process, interviews were carefully analyzed since "only slowly do underlying strands of a community's culture reveal themselves, as interview after interview sounds the same themes."[6] From oral histories, historians can learn not just the facts, but also the experiences, feelings, and insights of the people being interviewed. The study was further supported with research into archival sources and published documents by IOC and the provincial government.

This research reveals that women's entry into the mining workforce was halting in Labrador City. From the early 1970s to the 2000s, top-down forces, such as the influence of the company, along with bottom-up social pressure from other working-class residents made it difficult for women to enter wage labour comfortably or in large numbers. According to Karen Beckwith, change in traditional gender roles in an industrial community requires three major preconditions: women must have the desire to act, the opportunity to change their position, and the strength of group support.[7] Certainly, in broad terms the advent of the 1970s provided a potential opportunity for women in Labrador City to make employment gains, as the second wave of feminism took hold, and women throughout North America began to enter wage employment. This was no less true in Newfoundland, where feminist groups throughout the province encouraged women to take a more active role in society, forming women's unions with a common goal of "lobbying the governments to influence policies that regulate women's lives."[8] One result was employment gains: the number of women working in the province's paid labour force increased by 56.9 per cent between 1970 and 1980.[9] In fact, according to interviewees in Rick Rennie's studies, women of other mining towns in the province, notably Buchans and St. Lawrence, were active participants in the workforce.[10] Despite these gains in the workplace and toward gender equality in the province and in North America at large, at Labrador City, IOC policies continued to structure and maintain an unequal division of labour at the Carol Lake Mine. Women struggled to obtain positions at the mine and, once they entered the mine workforce, faced social and occupational discrimination. Through a study of Labrador City from the 1970s to 2000s, this discussion will reveal that despite women's ability to obtain positions in the mine, the social life and work culture of the company and town maintained gendered inequality until the 2000s. Labrador City's gender history can be divided into three periods. After an initial period between the town's founding in 1958 and 1975, when the company worked to build not only a stable town but also to create an ideal image of the community by encouraging traditional gender roles, women began to resist actively the imposition of unequal gender relations. From 1975 to 1989, women grew increasingly discontented, even as they tentatively entered the mine workforce, mainly in supporting positions. Throughout the 1990s and 2000s, women's activism, changing attitudes, and strong demands for labour led to greater gender equality. While this chapter focuses on the latter two stages, it is important to first give a brief history of the mine and town's conception.

After the early success of the Iron Ore Company of Canada's (IOC) mining project in Schefferville, Quebec, in 1958 the company began developing the Carol Lake Mine and accompanying mining town, Labrador City. As in many mining towns developed in the 1960s, the founding company invested a great deal of time and money into community development and transformed the region from a land shared with moose, bears, and other wild animals into a modern, suburban-style Canadian town.[11] To maintain this atmosphere, the company controlled the type of people who could live in the community. By hiring only male employees and hand-picking "satisfactory" residents, IOC wielded power not only over the economy of the town, but also influenced the people's way of life and their connection to place.[12] As a result, the town was an environment where women often felt trapped and forced to depend upon their husbands for survival. It was this feeling of inequality that shaped women's level of engagement in community life and their labour opportunities.[13]

In its hiring policies, the company sought to emulate the idealized social structure of the 1950s. Through the 1960s and even as late as the 1970s, IOC's unofficial policy was to hire women only for office work because the mining industry was seen as a man's world. As Nichole Churchill noted, "For a very long period of time, Labrador West was a male-dominated community . . . and you know breaking into the male workforce itself was a problem."[14] In giving priority to married men, it was clear that IOC was still enforcing the nuclear family lifestyle.[15]

By the 1970s, however, some changes made it possible for women to enter non-traditional occupations. The federal Royal Commission on the Status of Women in 1969 recommended that more women be employed in Canadian companies. In response to this report and its corresponding need for employees, IOC began hiring a limited number of women in low-paying, untrained positions in the company. These positions were previously seen as masculine, and included labourers, janitors, office workers, and heavy equipment operators.[16] Despite this initial action, it was not easy for women to enter these jobs. Elizabeth Andrews was one of those women who struggled to get an entry-level position: "I wrote all the tests, passed them and just waited to be hired. And you're hearing all these people getting hired and you're still. . . . Well what it was is you had to keep going back all the time." Andrews finally broke into the mining industry workforce in August 1974 and was among the first group of women to get hired.

While the industry's high salary was important, women did not endure the hardships associated with entering the mining industry simply to get extra money. There was more behind this action. A woman's entrance into the mining industry in the 1970s was a political statement, whether or not this was her primary intention.[17] The desire for equal rights and women's independence was, quite possibly, stronger in the mining districts throughout Newfoundland and Labrador since, as expressed in histories of other towns, the women of the province were used to greater opportunities:[18]

> I just didn't get what the problem was. Back home [Stephenville], if there was fifteen women in the field, there was probably fifteen men. Nobody thought nothing less of it. It was just the normal thing to do.[19]

After many years of being trapped in their homes with nothing to do, or stuck in abusive relationships, some women in Labrador City saw work at the mine as an opportunity to achieve financial independence. However, they still lacked strength in numbers and the support of the town and company.

The economic downturn in the early 1980s highlighted women's tenuous place in the mine's workforce. Between 1978 and 1982, the mine went through a bust cycle, resulting in mass layoffs. Since the majority of IOC women workers held entry-level jobs, they were the first to go. The very few women who successfully avoided the layoffs were harshly criticized by some IOC men and their wives. Many in the community felt that it was unfair for women to take the few remaining positions when men needed the money to support their families.[20] Clearly, some people in the town did not believe women were or should become breadwinners. As well, many still believed—as they had when women had first entered the mine—that this was not a serious long-term venture but simply an attempt by women to prove their ability to make extra money. At the end of the recession, IOC was quick to hire back many of the male workers who had been laid off during the bust cycle. However, women were less likely to be rehired by the company. Suddenly women found themselves in the same battles they had struggled through in the 1970s.[21]

WORKING WOMEN AND GENDER ROLES IN THE COMMUNITY

It wasn't long before the mine and town again saw prosperity. With this boom cycle came mass hiring, and despite IOC's primary loyalty to men, the company once again needed women. Thus, by the mid-1980s a larger number of women began working in the mining industry. Women's changing place in the workforce had implications for gender relations and social life in the town. Social divisions emerged between IOC-employed and non–IOC-employed women. Even when IOC women were accepted into housewives' social groups or clubs in town, shift work made it practically impossible to become a member as they missed too many meetings. People developed very strong opinions of wives and mothers working in the mine. One IOC woman remarked:

> I didn't go to the hair dressers. I went a couple of times but you don't go back. Oh, no. You get in there and you're listening to all these women talking about the women working in IOC: "Nothing but a bunch of whores in there screwing everything that's in there." One day I turned over and I said: "Listen, you know who we're screwing? Your husbands, cause it's all your husbands we're working with."[22]

Unable to socialize with the women of the town, some IOC women tried socializing with other IOC employees, but this came with its own problems. Just as IOC men had been accustomed to their male-only work environment, they also enjoyed a male-only social life. When IOC women entered the bars where IOC employees commonly gathered after a shift, the men ignored them:

> They didn't want you in there because they figured you didn't belong. You go in here at happy hour at the clubs in the evening after work, even though it was a union centre and the women should have been allowed just as much as the men, but if I walked in with my husband everything just went dead . . . it was obscene for you to walk into the club because that was where the men went after work, not the women.[23]

Women were not given the cold shoulder just at the clubs. While the men would talk to women at the plant, once in town they refused to even acknowledge women's presence. The small group of IOC women suddenly felt like outcasts in their own community. As some interviewees noted, it took a "different breed of women" to last in the mining workforce because they had to endure hardships both on and off the project.[24] The few IOC women who were willing to endure the work and town harassment built strong and lasting friendships with each other. As Elizabeth said, "I knows some of them more than I knows my family, right because I was nineteen when I started. I mean, I knew them for thirty years."[25] These friendships were essential in preventing depression and social isolation among the women.

In spite of the gains made by women in the 1980s, in many ways Labrador City retained many of the gendered social roles of the previous generation. At this time, 73 per cent of the adult residents were married. Of these, 87 per cent lived in a nuclear family setting where women remained responsible for domestic labour. Most often, there were four to five members per household.[26] This lifestyle was strongly encouraged by company housing policies. Mining companies in the region urged residents to buy housing since company officials believed that home ownership encouraged long-term residency, a stable workforce, and lower costs for the company.[27] IOC provided subsidies for its employees, but these subsidies came with a number of unwritten rules. Priority went to married men, preferably those with families, followed by single men. IOC women, married or single, could not obtain any housing subsidies.[28] While technically a home could be owned by anyone who could afford it, the majority of women did not qualify for a mortgage. This forced some women to remain in abusive relationships, and it caused others to seek marriage when they had no interest in it.²⁹ Ultimately, IOC women were deprived of the independence and freedom they were fighting for when originally entering the industry.

Even as late as the 1980s, there was a clear socially acceptable path for Labrador City women to follow. As single, young adults, women worked until they found husbands. They could take jobs as babysitters, nurses, schoolteachers, office workers, or service industry workers.³⁰ They were then expected to take the position of housewife until their children reached maturity. Social pressure and particularly a lack of child-care facilities encouraged women to remain home with younger children. The town, which prided itself on all its modern amenities, did not have a day-care centre until the 1990s.³¹ As one resident recalled:

> Mom didn't work for most of while we were growing up. I can remember her coming to us and saying she was getting a job and we were going to have to take care of ourselves at lunch. So we'd have to get our own lunch. I was probably about thirteen or fourteen.³²

Once the children reached their teen years, women could choose to go back to work part-time or even full-time in positions categorized as women's work. It appeared that the community embraced and in many cases encouraged women to work in the town's service industry. In fact, some women even gained town support as they became entrepreneurs. When Alison Wiseman became a widow in 1987, she decided to open the Dollar Plus Souvenirs and Convenience store in the town mall. When Wiseman faced financial hardships, the community gave her monetary and emotional support.³³ Women were also supported when opening home-based businesses such as Linda Cassell's sewing venture.³⁴ Much like their entrance into the mining industry, some women saw this as an opportunity to gain independence.

While working was acceptable for those women without young children, some interviewees reserved harsh judgments about the effect that working mothers' absences had on their children. One informant believed that staying home when her children were young had made them better behaved and educated. She stated:

> I stayed home ... when she got home from school, to give her a ride, to do different things with her ... Some kids were left from the time they got out of school in the evening to the time their parents got home in the evening. God only knows what they were at ... Mine were supervised and I found a difference, not only in their behaviour but even in their attitude and their language ... I carted around a lot of kids and I picked them up and dropped them off and I could see a difference. Every day as time went by some of them got worse.[35]

She further argued that there was a marked difference in school grades. Since working mothers did not have as much time to devote to their children and could not always find a babysitter, children were left to do their homework as they pleased. The parents would find out how their children were doing through the report cards, and by then it was too late to fix the problem. The housewives, on the contrary, "kept an even keel on it," which yielded better grades.[36]

On the opposing side, women who did work often believed that less doting on the child created a more independent, responsible, and well-rounded young adult who was ready to move out and experience life. Elizabeth Andrews was one working woman who had strong opinions on the matter:

> When the mother stays at home, [the children] had a harder job leaving the home, they had a harder job going to university; a lot of them don't know how to take care of themselves ... I must say mine are three quite independent go-getters. Determined they were ... and I say: "Parents, do they think they're doing a wonderful job if their kids never had to cook, never had to clean, never had to do for themselves?"[37]

The working mother believed that by taking on the added pressure of working while raising a family they were also able to provide more for their children and meet the demands of the consumer world. Keeping them busy in extracurricular activities and giving them things such as skidoos prevented boredom and created a well-rounded child.[38]

Interviewees suggested that working women were sometimes blamed for changing marital relations and domestic breakdowns.[39] Working alongside men at the mine led to a perceived increase in extramarital affairs. As men and women spent sixteen-hour shifts together, they were often seeing more of their work partners than they were of their spouses, ultimately leading to family breakups.[40] From a different perspective, women also used mine employment as a means to secure financial independence and escape unsatisfactory or abusive relationships.[41] In households where both spouses worked in wage labour, families needed to adopt a new division of domestic labour, sharing the household responsibilities as a result of opposing shift work and limited access to babysitters and cleaning services.[42] But changing one's domestic practices was not easy for some families, and in Labrador City it triggered resentment from some of the men who were used to being "tended on" by their wives.[43]

In the late 1980s and 1990s, both IOC and the town's women organized to push for gender equality in Labrador City. These efforts led to the establishment of the local Women's Centre run by Marion Atkinson and Barbara Doran. This was designed to provide a safe place for women in the area and to promote equal rights.[44] In response to growing rates of sexual and marital problems (such as abuse, infidelity, and divorce) in the town, the Labrador West Status of Women Council and the newly formed Women's Centre created the Labrador West Status of Women Committee to study gender relations in the town and in the workforce.[45] Finding that working women felt that they were not supported, the Women's Centre established coffee groups, children's play time, skills-training programs, and emotional support groups. The centre was also a safe house for victims of domestic abuse.[46] The IOC publicly supported the Women's Centre and the Labrador West study. The company also joined the federal government's voluntary affirmative action program, which was launched in 1978 as a means of establishing gender equality in the workplace.[47] IOC's actions helped regain the image they once had of being supportive and involved in the success of the community as well as the industry.[48]

By the 1990s, despite IOC's outward enthusiasm for equality in the workplace and town, the company and its male employees still drew clear lines separating feminine and masculine work at the mine. A study on women in the workforce of Newfoundland and Labrador revealed that while 46.8 per cent of women worked in paid labour, women represented less than 10 per cent of the workforce in trades, technology, and operations fields.[49] Those women who worked with IOC still most often occupied roles consistent with stereotypes about women's abilities: they worked as cleaners, or were funnelled into other jobs that were not as physically demanding. Their positions usually did not pay enough relative to the community's cost of living. Therefore, the majority of women still depended on marriage for financial security.[50]

Nevertheless, in the 1990s, there was a sudden increase in female mine workers at IOC. At this time, the town and company noticed a large number of second- and third-generation children leaving the small town to seek education or jobs, which meant a loss of potential employees. Thus, IOC joined forces with College of the North Atlantic's local campus and the local steelworkers' union to develop the two-year mining technician diploma as part of the "Employee of the Future" program.[51] Students were almost guaranteed a job with the company upon successful completion of the program.[52] As anticipated, the program effectively encouraged a number of young men to stay in the community. It also unexpectedly raised the percentage of female employees. Since both IOC and the college followed the affirmative action program, their education program needed to be accessible to everyone, and women of all ages took advantage.[53] Noreen Careen, the director of the Labrador West Women's Centre, noticed the program resulted in a significant change in the lives of working women in the community. As she argued, "it's the greatest move that was ever made because it gave women an opportunity for their independence that otherwise they would never have had in this community." Just as IOC strongly influenced women of the 1960s and 1970s to embrace the role of housewife, it was now one of the main forces encouraging women's paid employment. Now the majority of Labrador City women were working, if not at IOC then in the supporting services in the town.[54]

The growing participation of women in the Labrador City workforce did not make up for the labour shortfall in the booming community. In fact, some secondary industries and volunteer organizations began to suffer from a lack of available employees. IOC had an unwritten rule that it would not steal employees from other businesses; however, if an employee quit a position before applying to IOC, that person was considered fair game. With limited choices, stores began hiring people at younger ages, and soon they were hiring children as young as thirteen and fourteen. Other stores and restaurants were forced to close earlier in the evening, as early as six o'clock. Many businesses that should have succeeded, given the average family income in the town, were forced to close permanently.[55] By the early 2000s, businesses began hiring temporary workers from developing countries.[56] Tim Hortons was the first company to do so, and soon Walmart, McDonald's, and other local businesses followed. These companies hired groups of foreign workers, most often from the Philippines, for two-year contracts that included one shared house for approximately ten to twenty employees.[57]

Like the secondary industries, most of the volunteer roles in the town were originally filled by women. As noted earlier, women used these roles to give them something to do throughout the day. Yet with full-time paid positions, women no longer needed or had time to participate in volunteer work. While children took on some responsibilities in the town, few were willing to become involved in volunteer work when they could easily find paid employment. This resulted in fewer social events in the community, and some residents felt the town spirit was fading.[58]

As a result of IOC's new policy toward hiring women, the company achieved parity in the number of male and female employees by the early 2000s.[59] The company added women's locker rooms, washrooms, and lunchrooms, and removed derogatory pictures from the walls. IOC and the union developed an anti-harassment policy with zero tolerance for bullying.[60] As IOC increasingly sought female employees, men encouraged their wives to enter the workforce, and mothers encouraged their daughters to avoid depending on male breadwinners. As one mother recalled:

> I tell my girls, you get your education, get a job, be independent, look out for yourself, then find someone that you can stay with and be with. I've seen so many people I know my mother's age, like I have a friend whose parents were in a bad relation but

the woman couldn't leave because she has nothing to go to: no education, no job, no skills, but now it's so different.[61]

Increasingly, women in Labrador City regarded paid employment as a means of achieving financial independence and personal empowerment.

Still, the achievement of gender parity in the town's main industry did not eliminate conflict over gender roles in the workplace. Many mining-industry employees noted that the positions in the workforce have simply been divided along gender lines. While a few women filled trade positions such as mechanics and electricians, they were a very small minority. Most women remained in less physically demanding positions such as truck driving. Some women did feel that they fit into this work environment well. One remembered:

> I worked with the guys and the guys treated me like I was one of the boys. And one of the guys even said that to me, "Oh you're just one of the boys," and I think that was one of the best compliments I ever had . . . I thought that was perfect. I got along great with the guys. There was none of this sexual suggestiveness or anything like that. Every now and then they'd tell an off-colour joke, but I mean you just laugh at it. As long as it's not geared towards me personally, I don't mind an off-colour joke or dirty joke or whatever. No big deal to me.[62]

It is clear, however, that in order for women to fit in, they needed to present themselves as "one of the boys." If women appeared weak, male employees did not appreciate working with them. In fact, some men believed that women were given "easy jobs" because they were often unable to do jobs requiring strength. This left all the physically demanding jobs for the men. If a man wanted to rotate to an easier job for a few weeks, there were often none available.[63]

❊❊❊❊

As the town of Labrador City celebrated its fiftieth year in 2008, the residents took pride in their generations of hard work in the mines and marvelled at the vast changes to the region's landscape. Yet, the same mine that evokes such strong pride was also at the centre of the cracks that emerged in the town's gender relations, a fraught history revealed through oral histories of this fifty-year period. Despite women making inroads into paid labour in the 1970s, company policy and a masculine workplace culture constrained the number and quality of positions that were available to women. Oral interviews also suggested that working women experienced social pressure and at times outright condemnation for entering the workforce. Moreover, through the period of low commodity prices in the 1980s, it was female workers who were most vulnerable to layoffs, showing that their foothold in a male-dominated workforce remained tenuous.

If the call for gender equality in the Carol Lake Mine began with the broader feminist call for greater access to wage employment in the 1970s, progress on the issue was slow. Indeed, gendered inequalities maintained a strong hold over Labrador City until the 2000s, when a significant increase in women's employment and more tangible social and workplace equity policies from the town and company (child care, anti-harassment, etc.) led to significant gains for working women. Other studies have suggested that, as in Labrador City, a combination of policy changes in the workplace and fundamental cultural and social change within the local community is a prerequisite for the advancement of gender equity in single-industry towns.[64] In Labrador City, policy innovation and cultural change proceeded more slowly than was typical elsewhere in Canada. The memories of women who lived and worked in the town suggest, moreover, that the intransigence of management and cultural resistance among other working-class residents worked in concert to reinforce male dominance and gender inequity at the mine.

NOTES

1. Angela John, *By the Sweat of Their Brow: Women Workers at Victorian Coal Mines* (London: Croom Helm, 1980).
2. Meg Luxton, *More Than a Labour of Love: Three Generations of Women's Work in the Home* (Toronto: Women's Educational Press, 1980), 13; and Bonnie Fox, *Hidden in the Household: Women's Domestic Labour under Capitalism* (Toronto: Women's Educational Press, 1980).
3. Vicky Seddon, *The Cutting Edge: Women and the Pit Strike* (London: Lawrence and Wishart, 1986). See also Sheila Rowbotham, "More Than Just a Memory: Some Political Implications of Women's Involvement in the Miners' Strike, 1984–85," *Feminist Review* 23 (Summer 1986): 407–21; and Jill Liddington, "Gender Authority and Mining in an Industrial Landscape: Anne Lister 1791–1840," *History Workshop Journal* 42 (October 1996): 58–86.
4. See Mary Murphy, *Mining Cultures: Men, Women, and Leisure in Butte, 1914–41*, Women in American History (Urbana: University of Illinois Press, 1997); Janet Finn, *Tracing the Veins: Of Copper, Culture, and Community from Butte to Chuquicamata* (Berkeley: University of California Press, 1998); Elizabeth Jameson, *All That Glitters: Class, Conflict, and Community in Cripple Creek* (Urbana: University of Illinois Press, 1998); Thomas Miller Klubock, *Contested Communities: Class, Gender, and Politics in Chile's El Teniente Copper Mine, 1904–1951* (Durham, NC: Duke University Press, 1998); and Kathryn McPherson, Cecilia Louise Morgan, and Nancy Forestell, *Gendered Pasts: Historical Essays in Femininity and Masculinity in Canada*, Canadian Social History Series (Toronto: University of Toronto Press, 2003).
5. "All interviewees are given pseudonyms, except those representing specific groups or organizations."
6. Linda Shopes, "Oral History and the Study of Communities: Problems, Paradoxes, and Possibilities," in *The Oral History Reader*, ed. Robert Preks, 2nd ed. (New York: Routledge Taylor and Francis Group, 2006), 267; and Donald A. Ritchie, *Doing Oral History: A Practical Guide*, 2nd ed. (Oxford: Oxford University Press, 2003).
7. Karen Beckwith, "Lancashire Women against Pit Closures: Women's Standing in a Men's Movement," *Signs: Journal of Women in Culture and Society* 21, no. 4 (Summer 1996): 1042.
8. Sharon Gray Pope and Jane Burnham, "Change Within and Without: The Modern Women's Movement in Newfoundland," in *Pursuing Equality: Historical Perspectives on Women in Newfoundland and Labrador*, ed. Linda Kealey (St. John's, NL: Institute of Social and Economic Research, Memorial University of Newfoundland, 1993), 207.
9. Labrador West Status of Women Council, "Submission to the Employment Practices Commission," 1989.

10   Rick Rennie, *The Dirt: Industrial Disease and Conflict at St. Lawrence, Newfoundland* (Halifax: Fernwood Publishing, 2008), 108–12; Ed Hamilton, Buchans Historic Research Group, and Red Indian Lake Development Association, *Khaki Dodgers: The History of Mining and the People of the Buchans Area* (Grand Falls-Windsor, NL: Red Indian Lake Development Association, 1992), 57–59; and Labrador West Status of Women Council, "Submission to the Employment Practices Commission."

11   Arn Keeling, "'Born in an Atomic Test Tube': Landscapes of Cyclonic Development at Uranium City, Saskatchewan," *Canadian Geographer* 54, no. 2 (2010): 228–52.

12   Richard Geren, Blake McCullogh, and Iron Ore Company of Canada, *Cain's Legacy: The Building of Iron Ore Company of Canada* (Sept-Îles, QC: Iron Ore Company of Canada, 1990); Jacqueline Jacques Driscoll, "Development of a Labrador Mining Community: Industry in the Bush" (PhD diss., University of Connecticut, 1984); and Iron Ore Company of Canada, "Historical Data," St. John's, 1967.

13   Elizabeth Andrews, Personal Interview, May 15, 2011; and Nichole Churchill, Personal Interview, May 9, 2011.

14   Nichole Churchill, Personal Interview, May 9, 2011.

15   Linda Ann Parsons, "Labrador City by Design: Corporate Visions of Women," in *Their Lives and Times: Women of Newfoundland and Labrador: A Collage*, eds. Carmelita McGrath, Barbara Neis, and Marilyn Porter (St. John's, NL: Killick Press, 1995), 214–17.

16   Elizabeth Andrews, Personal Interview, May 15, 2011; and Nichole Churchill, Personal Interview, May 9, 2011.

17   Beckwith, "Lancashire Women against Pit Closures," 1996.

18   Valerie Hall, "Differing Gender Roles: Women in Mining and Fishing Communities in Northumberland, England, 1880–1914," *Women's Studies International Forum* 27, no. 5/6 (November 2004): 521–30; Linda Kealey, ed., *Pursuing Equality: Historical Perspectives on Women in Newfoundland and Labrador* (St. John's, NL: Institute of Social and Economic Research, Memorial University of Newfoundland, 1993); Richard L. Loder, "Changing Familial Patterns in a Newfoundland Mining Town: A Town of Widows and Orphans," paper presented to Professor Louise Charimonte (Centre for Newfoundland Studies, Memorial University of Newfoundland, 1972); and Rennie Slaney, *More Incredible Than Fiction: The True Story of the Indomitable Men and Women of St. Lawrence, Newfoundland from the Time of Settlement to 1965: History of Fluorspar Mining at St. Lawrence, Newfoundland* (St. John's, NL: Confederation of National Trade Unions, 1975).

19   Elizabeth Andrews, Personal Interview, May 15, 2011.

20 Parsons, "Labrador City by Design," 215.
21 Geren, McCullogh, and Iron Ore Company of Canada, *Cain's Legacy*; Driscoll, "Development of a Labrador Mining Community"; Parsons, "Labrador City by Design," 214–17.
22 Elizabeth Andrews, Personal Interview, May 15, 2011.
23 Ibid.
24 Margaret Carlton, Personal Interview, May 7, 2011.
25 Elizabeth Andrews, Personal Interview, May 15, 2011.
26 Linda Ann Parsons, "Passing the Time: The Lives of Women in a Northern Industrial Town" (MA thesis, Memorial University of Newfoundland, 1987), 36–47.
27 John H. Bradbury, "Declining Single-Industry Communities in Quebec-Labrador, 1979–1983," *Canadian Studies* 19, no. 3 (1984): 132–33.
28 Parsons, "Labrador City by Design," 216–20.
29 Ibid.
30 Labrador West Status of Women Council, "Submission to the Employment Practices Commission."
31 The daycare facility was provided by the Women's Centre; however, with limited volunteers, the program ended a few years later and there is no longer daycare available (Noreen Careen, Personal Interview, May 9, 2011).
32 Andrea Locke, Personal Interview, May 8, 2011.
33 Alison Wiseman, Personal Interview, May 15, 2011.
34 Andrea Spracklin, "No Business Like Sew Business (re: Cassell's Sewing Business in Labrador City)," *Downhomer* 14, no. 10 (March 2002): 35.
35 Laura Fewer, Personal Interview, May 13, 2011; individuals who agreed included: Lorraine Carolynn, Personal Interview, May 13, 2011; Ashley Fewer, Personal Interview, May 11, 2011; and Gerard Martin, Personal Interview, May 7, 2011.
36 Laura Fewer, Personal Interview, May 13, 2011.
37 Elizabeth Andrews, Personal Interview, May 15, 2011. Individuals who agreed included: Alison Wiseman, Personal Interview, May 15, 2011; Andrea Locke, Personal Interview, May 8, 2011; and Nichole Churchill, Personal Interview, May 9, 2011.
38 Debbora McDonald, Personal Interview, May 12, 2011.
39 Parsons, "Labrador City by Design," 222–24.
40 This was a common consensus among the interviewees. Some people who specifically focused on this subject included: Ashley Fewer, Personal Interview, May 11, 2011; Lorraine Carolynn, Personal Interview, May 13, 2011;

Gerard Martin, Personal Interview, May 7, 2011; Laura Fewer, Personal Interview, May 13, 2011; and Noreen Careen, Personal Interview, May 9, 2011.

41 Noreen Careen, Personal Interview, May 9, 2011; and Elizabeth Andrews, Personal Interview, May 15, 2011. Ashley Fewer, Personal Interview, May 11, 2011; Lorraine Carolynn, Personal Interview, May 13, 2011; Gerard Martin, Personal Interview, May 7, 2011; Laura Fewer, Personal Interview, May 13, 2011.

42 Paul Rickards, Personal Interview, May 9, 2011.

43 Labrador City Questionnaire, 25 Personal Interviews, May 6–16, 2011.

44 Pope and Burnham, "Change Within and Without," 180.

45 Labrador West Status of Women Council, 1–3; and Pope and Burnham, "Change Within and Without," 180.

46 Pope and Burnham, "Change Within and Without," 180, 185–86, and 198–89.

47 Bobbie Boland, "At a Snail's Pace: The Presence of Women in Trades, Technology, and Operations in Newfoundland and Labrador" (St. John's, NL: Women in Resource Development Committee, April 2005), 15.

48 "NFB of Canada: Too Dirty for a Woman," film (Labrador City, NL: IOC, 2011), and "Labrador City: From Tailings to Biodiversity," film (Labrador City, NL: IOC, 2010).

49 Boland, "At a Snail's Pace," 5.

50 Parsons, "Labrador City by Design," 215.

51 Boland, "At a Snail's Pace," 25.

52 Robert Tobin, Personal Interview, May 12, 2011.

53 Noreen Careen, Personal Interview, May 9, 2011

54 Nichole Churchill, Personal Interview, May 9, 2011.

55 John Shmuel, "Labrador City's Huge Worker Shortage Threatens Small Businesses," *National Post*, December 3, 2012.

56 Ibid.

57 Labrador City Questionnaire, 25 Personal Interviews, May 6–16, 2011; Gerard Martin, Personal Interview, May 7, 2011; Ashley Fewer, Personal Interview, May 11, 2011; and Lorraine Carolynn, Personal Interview, May 13, 2011.

58 Personal Interviews; comparison of Driscoll, "Development of a Labrador Mining Community," and the Labrador West website, http://www.labrador-west.com/default.php?ac=changeSite&sid=1.

59 Personal Interview with Marilyn Currie, the co-chair of J.O.S.H.E. and the United Steelworkers, May 2011.

60 Margaret Carlton, Personal Interview, May 7, 2011, and Personal Interview with Marilyn Currie, the co-chair of J.O.S.H.E. and the United Steelworkers, May 2011.

61 This was a common consensus among the interviewees. The quote was from Margaret Carlton, Personal Interview, May 7, 2011.

62 Andrea Locke, Personal Interview, May 8, 2011.

63 Lorraine Carolynn, Personal Interview, May 13, 2011.

64 See Griselda Carr, *Pit Women: Coal Communities in Northern England in the Early Twentieth Century* (London: Merlin Press, 2001), and Rowbotham, "More Than Just a Memory."

| CHAPTER 5

"A Mix of the Good and the Bad": Community Memory and the Pine Point Mine

John Sandlos

Mining brings massive transformation to lives and landscapes. Almost inevitably, the people who worked in and lived near historic mines are compelled to tell stories about these changes. Whether it is recollections of hardships and good times within the mining camps, memories of large-scale environmental change, reminiscences of social life within a mining town, or remembrances of work on or under the ground, the process of telling stories can generate multiple and sometimes contested interpretations of local mining heritage within a particular landscape. Was the mine a good place to work? Was the town a good place to raise a family? Did pollution arise from the mine? What kind of landscape changes (open pits, roads, tailings ponds, etc.) did the mine produce, and how well were these physical reminders cleaned up? How did mining change pre-existing forms of natural resources use? The answers to these questions provide rich source material that can help us to understand the complex social, cultural, and ecological memory and meanings associated with mining activity in small communities that are typically located in remote regions.

This has been particularly true for indigenous communities who must balance the inherently short-term benefits of mining development with long-term residency in particular localities and regions. And indeed, there is a large body of writing on the social, economic, and environmental inequities associated with large-scale mining development in subsistence-oriented indigenous communities throughout the globe.[1] Oral history research methods have sometimes been used as a means to capture these local voices and stories in places where mining has led to environmental injustices such as acute pollution and the exposure of indigenous communities to chemical and radiological toxins.[2] Such community-generated stories can provide a powerful corrective to boosterish histories (often commissioned by companies or published by mining heritage societies) that celebrate mining as a historical gateway for the extension of capital, settlement, and development in remote regions.[3]

Similar to the community-generated stories, the historical narrative of the Pine Point Mine, at first glance, seems to pit the environmental and economic impacts of development against the subsistence economies of First Nations communities. The Consolidated Mining and Smelting Company (later Cominco) operated the mine on the south shore of Great Slave Lake from 1964 to 1988 through a subsidiary company, Pine Point Mines, Ltd. As a massive lead-zinc mine, the Pine Point Mine was also central to the Canadian government's post–World War II colonial agenda in Northern Canada. Government records are replete with references to the mine as a gateway development that would stimulate additional mines throughout the North and quickly catapult northern Aboriginal people from the moribund fur trade economy to more modern forms of industrial wage labour.[4] So great was the political consensus in favour of northern development in the late 1950s and early 1960s, the federal government provided subsidies of nearly $100 million for a railroad, a highway extension, and hydroelectric development to support Pine Point, infrastructure that was meant to kick-start further mineral development and modernization throughout the Northwest Territories.[5]

Even at the early stages of the mine's operating life, however, several critics began to argue that the economic promise of Pine Point for adjacent First Nations communities, particularly for the Chipewyan and Métis community of Fort Resolution roughly sixty kilometres to the east, had gone largely unfulfilled. As early as 1968, the political economist Kenneth Rea invoked Harold Innis's staples theory to criticize Pine Point as another in a long line of export-oriented northern development projects that contributed little to local economic development.[6] Sociologist Paul Deprez's 1973 report for the Winnipeg-based Centre for Settlement Studies adopted much the same tone, highlighting how the federal government's failure to provide local skills training, a viable housing policy for northerners, and an extension of the Pine Point highway further east to Fort Resolution meant the mine workforce was mostly imported and the local economic benefits of the mine severely limited.[7] In 1977, Justice Thomas Berger's report on hearings into the proposed Mackenzie Valley gas pipeline quoted liberally from Fort Resolution testimony, suggesting that Pine Point demonstrated the negative impact of industrial mega-development on local hunting and trapping activities.[8] Almost simultaneously, the Canadian Arctic Resources Committee (CARC—an environmental NGO) released a report highlighting the negative social, economic, and environmental consequences of the mine, particularly for Aboriginal hunters and trappers in Fort Resolution.[9] The sudden closure of the mine in 1988, the subsequent collapse of the town of Pine Point, and the abandonment of the mine's forty-seven open pits in an unremediated state further cemented the idea that Fort Resolution and other nearby First Nations communities had derived little from the mine other than the mess that was left behind.

In recent years, a very different parallel story has begun to emerge about the legacy of Pine Point. In 2011 media artists Michael Simons and Paul Shoebridge released an interactive web documentary about the town of Pine Point through the National Film Board. Inspired by the efforts of former Pine Point resident Richard Cloutier to keep the abandoned and demolished ghost town alive through the web memorial titled "Pine Point Revisited," the remarkable web documentary traces the overwhelmingly positive memories of life in the town of Pine Point. In a broader sense, the Pine Point website and documentary provide a powerful example of what

Figure 1: Abandoned waste rock pile, Pine Point Mine, NWT, 2009. Photo by John Sandlos.

many historians and geographers have identified as a close identification with local place and landscape in remote mining towns, even after the cessation of mining activity.[10] The Simons and Shoebridge documentary declares (as looped video of a figure-skating performance of the Wizard of Oz at the Pine Point annual winter carnival plays in the background) that "most Pine Pointers think their home town was the best place on earth to have lived."[11] With a close-knit community and seemingly endless recreation activities, the Pine Point experience evokes for many waves of nostalgia for the demolished community.

How are these competing stories—the mine as colonial disruption versus the mine work and the mine town as a near-paradise—interpreted within the First Nations communities adjacent to the Pine Point Mine? A complicated answer can be found in the thirty-nine oral history interviews the Memorial University–based "Abandoned Mines in Northern Canada" project conducted mostly in Fort Resolution, but also in the largely Slavey K'atl'odeeche First Nation (Hay River Reserve) and among members of the Hay River–based North Slave Métis Alliance.[12] The most striking and surprising feature of these interviews is that they feature parallel stories about Pine Point as a source of economic and environmental disruption, but also memories of Pine Point as a great place to live and work. Negative memories of local racism or environmental change are often juxtaposed *within the same interview.* This mix of positive and negative memories is all the more surprising because other oral history projects on historical large-scale northern mines have suggested unambiguously negative consequences for adjacent First Nations.[13] The process of selecting interviewees—local research assistants largely chose individuals who had spent time living in the town or working at the mine, as opposed to those with no association with the town or mine, who might have been more critical—may have biased comments toward those nostalgic for a past life in Pine Point. Nonetheless, as Emilie Cameron has recently argued, stories about Northern Canada are not always subject to the binary categories—north versus south, mining versus communities, colonial versus indigenous, industrial activity versions traditional economies—that scholars have often chosen to highlight.[14] In the more than two decades since Pine Point has closed, people in adjacent First Nations communities have retained memories that that speak both critically and nostalgically about the mine and the town, a reflection of their ability to accommodate *and* resist that massive changes that that the federal government's development agenda and private capital brought to their region in the 1960s.

SPEAKING OF PINE POINT: THE IMPACTS

A long tradition of testimony and storytelling about the impact of Pine Point exists in nearby Aboriginal communities, and has circulated for decades in public hearings about the mine or other development projects. Beginning with Justice Berger's landmark Mackenzie Valley Pipeline Inquiry in the 1970s, residents of Fort Resolution in particular have told an overwhelmingly negative story about the Pine Point Mine. The basic storyline resonates with the descriptions of adverse effects in the previously mentioned academic literature: local First Nations were never consulted about the mine, they received few economic benefits (including jobs) from the mine, development at the mine severely impacted trapping activities, and the community was left with nothing but the huge mess of abandoned pits and the tailings pond.[15] Chief Robert Sayine, speaking through an interpreter at the Berger Inquiry in 1975, described Fort Resolution's experience with mining development:

> He [Sayine] says you should see our own Pine Point there, he says for about ten miles radius around Pine Point he says you'll never see no green trees around there for about ten miles radius around Pine Point. He says everything is just—all the dead trees, that's all you could see around there.
>
> He says look at that water around there because it never freezes during the winter, and you could smell it even when you are in a car passing through there, you can smell that water.
>
> Yes, he says right at the meeting wherever they're going to have a stockpile for these pipes for the pipeline, he says, you told us there was going to be about 400 people is going to be employed there, and he says that's the same kind of promise we got from Pine Point in 1960 when we sat in the meeting with them. There was going to be lots of jobs for natives there, but what we get today, he says there's nothing for natives over at Pine Point.[16]

Local First Nations' criticism only hardened after the mine closed in 1988, as people realized that Cominco's abandonment and restoration plan did not extend beyond covering the tailings pond with gravel, removing the houses and buildings at Pine Point townsite, and treating remaining water discharge to mitigate zinc discharge. The pits, the roads, and some infrastructure (the power facility) were left as long-term reminders of mining's impacts on the landscape. Speaking for the Fort Resolution Hunters and Trappers Association, Cecil Lafferty testified to local objections to the limited reclamation plan at a Northwest Territories Water Board hearing on renewal of the mine's water licence in 1990:

> Presently the operation may be terminated, except for the retreatment program and minor clean-up, however, we have lost that whole area for future utilization, economically. With the long term lease on the area it would be unworthy to even consider the area in our land selection. This was once a pristine traditional harvesting area for the people of Fort Resolution. The Pine Point Mines Limited have yet to implement an adequate restoration program that will be satisfactory to the people concerned. The operation over a period of twenty years, had numerous open pits that were left as is to fill up with groundwater. The waste dumps were not covered over with soil that could at least enhance growth, and what little land that was not dug up flooded over by the pit dewatering programs, which in turn killed all the trees and surrounding vegetation in the area.
>
> During its short lived operation, Pine Point Mines Limited had also put in hundreds of miles of haul roads that are all about four times as wide as our normal highways. Ironically, these roads impeded proper drainage of the land and therefore flooded . . . The tailings pond with its high level of metal, particularly zinc in a soluble form, is being flushed out annually into the environment . . . We strongly believe that our waters are polluted and that the fish are absorbing the metals.[17]

In 1993, Bernadette Unka, the chief of Fort Resolution's Chipewyan band government, the Deninu K'ue First Nation, summed up the environmental and economic dislocation associated with the mine for a national audience at the hearings of the Royal Commission on Aboriginal Peoples:

> Pine Points [sic] Mines nor Canada have never compensated the Dene people that used those areas in their hunting, fishing and trapping. They have never been compensated for their loss or for the land devastation. When I say land devastation, if you are to fly over Pine Point Mines you would look down and you would think you were flying over the moon with the craters and open pits that are left open. The people have never been compensated for the hardships and the heartaches induced by mineral

development. While the company creamed the crop at $53 million during their peak years, we got very little jobs and what we did get were very low-paying jobs.[18]

Although comments such as those above were produced for public consumption, the large volume of public testimony about the negative impacts of the mine on First Nations represents a significant source of oral history recollections about the mine. Taken together, they suggest that feelings about the Pine Point Mine ran raw during the operational phase and in the years after closure, with broken promises of jobs and economic development consistently juxtaposed with the negative impacts on Aboriginal harvesting and fishing, and on the land more generally.

In 2010, many people we interviewed recounted many similar stories describing the extensive impact of the mine.[19] Even before the mine opened, respondents noted, several traplines were destroyed in the area as line cutting and seismic exploration proceeded in the 1950s and early 1960s. As Angus Beaulieu described it, "People were trapping in that area, and they bulldozed people's traplines, and many people lost their traps and everything. Every 900 feet they had a bulldozer go in there in the winter . . . A lot of people lost their traps. People in those days, they didn't say too much."[20] Kevin Fabien remembered, "All the animals disappeared; they went farther away because of all the construction going on and all the equipment running." Other interviewees explained that pollution and habitat change also meant that game animals disappeared in the area. As George Balsillie put it, "The mine killed everything around there."[21]

As with the testimony from earlier sources, several interviewees suggested that these lost opportunities associated with trapping were not replaced with chances to earn a living at the mine. Angus Beaulieu claimed that only an average of eight people from Fort Resolution were hired at Pine Point at any given time in the 1960s and 1970s, a number he claims did not increase dramatically until thirty-eight additional Fort Resolution residents were hired on just prior to Justice Berger arriving to conduct hearings and then subsequently laid off shortly afterwards.[22] For those local Aboriginal people who did get jobs, several interviewees suggested they were often confined to menial low-skilled jobs, particularly cutting

Figure 2: Abandoned mine pit, Pine Point Mine, NWT, 2009. Photo by John Sandlos.

seismic exploration lines in the forest or the repetitive work of shovelling ore in the dusty environment of the loading shed. As Sam Bugghins from K'atl'odeeche First Nation stated in regard to Aboriginal people from Hay River, "mostly they cut lines" when they worked at the mine.[23]

In broad terms, some interviewees resented the lack of consultation and compensation in advance of the mine development, which is now standard practice and takes the form of impact and benefit agreements between mining companies and First Nations. Greg Villeneuve recalls that "they didn't even inform Res [Fort Resolution] about anything that time. Like now, other mines they pay out these, you know, like Res used to get IBA money for everybody, to give it out. And back then, I guess I'm not sure how it worked, but Res never did get a cent out of Cominco. And that's I think where, you know, we lost out on lots in Fort Res. We could'a had lots. It could'a been a rich little town."[24] Lloyd Cardinal claimed that the lack of consultation was possibly the most significant source of local resentment about the mine in Fort Resolution: "The biggest negative impact that Pine Point brought was that they didn't get our consent."[25] Whether it was a lack of employment and training opportunities or the absence of consultation and other financial benefits, the notion persists that in economic and political terms the mine bypassed nearby Native communities.

Many interviewees noted—in contrast to the ephemeral economic benefits—the more lasting environmental impacts of the mine. Indeed, the abandonment of the open pits and waste rock piles with little attempt at remediation is still a major source of discontent in adjacent Native communities. Leonard Beaulieu's concerns about the landscape changes and safety issues associated with the pits are representative of many interviewees: "Look at the way they left Pine Point. Goddamn place is full of holes, about forty big goddamn holes full of water . . . Yeah, you know, they left the mess like that."[26] In a similar comment, Henry McKay criticized Cominco directly for the lack of remediation at the mine site:

> The way I look at it, those people that make the mill and mine and make all that money and are gone, they don't care. 'Cuz they made their money, and they're gone, and they just leave everything to us. Big holes, you know? We can't look after that. They should make us a promise that they'll do something about it at the end, you know? Like fill out that hole they're makin'. And, you know, why do they leave that big stockpile there? Put it back! It'd be safer for animals. It's only right. They shouldn't be leaving it like that.[27]

In addition to the pits, people continue to have widespread concerns about the health and environmental impacts of water pollution. Roy Fabian, chief of K'atl'odeeche First Nation, claimed that when he was younger his father told him not to drink the water in the area of the mine.[28] K'atl'odeeche resident Harold Moore claimed that the government lied to miners about contaminants in the drinking water supply, and many died after moving to Pine Point.[29]

Leonard Beaulieu summed up local concerns linking the spread of water-borne contaminants from the mine to deformities in fish and the increase of cancer among the local Native population in the 1970s:

FIGURE 3: Water Treatment Pond, Pine Point Mine, NWT, 2009. Photo by John Sandlos.

Until 1973, all the waste was pumped down the hill into the lake. Poison. All that shit that's sitting in the pond right now. You know, holding back in that dyke? And then, they had two big pits; the X-15 pit right behind the highway, and the W-17 right next to . . . I know all those pits. I used to haul out of there. The waste from the water, from the thing you know, there's always water pumping and pumping. They'd pump that across the highway. Every time you drove by there. In them days you'd get sixty below, up until 1975 you'd get sixty below. And if it freezes . . . stink. Oh shit. It used to run way up into the bush and it'd run through to Paulette Creek, back into my lake. And then people are wondering why there was cancer. Now today they don't know who to blame. They're blaming it on the tar sands. It's not the tar sands . . . Every elder that gets sick: cancer. Cancer. It's the goddamn water! There's no water treatment plant that's going to stop that cancer from, you know? That's why us here we don't drink the darn water. I buy water from the store. That has really impacted the lives of the people in the North West Territories; their health. That Cominco Pine Point Mine.[30]

Scientific studies in the 1970s suggested that heavy metals, particularly copper, cadmium, lead, and zinc, were leaking from the tailings pond during spring runoff, but spikes above safe levels were localized in nature. An Environment Canada report from the mid-1990s questioned the methodology of these earlier studies, and claimed that water and fish in the area were generally safe for human consumption.[31] Nonetheless, in our interviews and informal conversations in Fort Resolution and K'atl'odeeche First Nation, people persistently identified water quality and human health issues as one of their biggest concerns associated with the mine and one of the most pressing research needs within the community.

People who spoke with us also highlighted some of the negative social consequences associated with the mine. Some who moved to Pine Point reported difficulty adjusting to the new town, particularly when they experienced incidents of racism. Although only seven years old when she moved to Pine Point, Priscilla Lafferty recalls being "scared" due to the large numbers of outsiders, and remembers that "just because we were Native we were called down and what not."[32] Denise McKay remembered similar fears grounded in the fact that she did not speak much English.[33] Other interviewees reported memories of racial violence and division. Melvin Mandeville recalled, "We used to . . . especially the Newfoundlanders, the Newfies, whatever, we'd fight against them lots, and they'd be callin' us wagon burners and we'd be callin' them Newfielanders or whatever. We fought like that, as kids. The older adults too, they'd be drinkin' and fighting." Mandeville also suggested there was racism in the schools, describing Native children being singled out to read in front of the class when the teacher knew they were not good readers, and being told he could not wear moccasins to school because the teacher did not like the smell.[34] Gord Beaulieu remembered that the police could be rough on Native people living in Pine Point. Obviously, the town of Pine Point was not always the "best place on earth" for some of its residents.[35]

Interviewees suggested that the other major negative social impact was increased exposure to alcohol after a highway extension connected Fort Resolution with Pine Point and the rest of the world in 1972. Ronald McKay told us that Fort Resolution became known as "Little Vietnam" after the highway was extended to the town: "There was shootings, and fighting. Actually, it turned the whole community upside down with the boozing. It was kind of like the end of the road development thing, where boom! Everything just boomed and no rules. People just partied. There was a lot of money, you know."[36] Although interviewees suggested that the situation improved over time, some indicated that the long-term impact of the road was to undermine the close-knit nature of their previously isolated community. Angus Beaulieu recalled, "And it seemed like people were much closer before than after . . . Before it was like one big family, people got along much better. It seemed like about from the time they got that road in, people kind of . . . you know, I don't know how to explain it, but it was not the way it was before Pine Point."[37] Ron McKay similarly claimed that Fort Resolution was "really, really strong" before the road came in, but increased mobility made people more individualistic and less willing to help neighbours.[38] Leander Beaulieu noted that "people used to be more together" before the road and associated changes such as the introduction of electricity and television, but "now they're more in their own little world . . . more distant."[39] In broad terms, Tommy Unka stated that the introduction of the mine, the road, and associated southern influences "kind of dragged me away from my traditional lifestyle."[40] For Fort Resolution residents, the mine was a watershed event in their history, a development project that ultimately fostered closer links with the outside world and a move away from the bush life.

SPEAKING OF PINE POINT: A GREAT PLACE TO LIVE

Stories about the past can unfold in ways that defy our attempts to uncover singular meanings about social and environmental change. Certainly in the case of our interviews about Pine Point, people told us about experiences that challenge previous assessments of the mine as a wholly negative experience for nearby Native communities. As with non–Native Pine Pointers, there was overwhelming consensus among interviewees that the town of Pine Point offered an exceptional quality of life, that work at the mine often offered tangible monetary reward, and that rather than wholly displacing trapping, work at the mine offered Native people income that supplemented wildlife harvesting practices when fur prices were low. Powerful stories of Pine Point's environmental, political, and social impacts were thus very often tempered with accounts of the positive aspects of the mine. This was particularly true among younger interviewees, people who may be remembering the halcyon days of childhood, but for whom life in the community of Pine Point remained a positive and momentous part of their life histories.

If there was any theme that came through loud and clear in the interviews, even among critics of the mine's environmental and economic legacies, it was the fond memories for the town of Pine Point. Lorraine Mckay, the first child ever to be born at Pine Point, claimed simply that she "loved it there, because I was raised there and knew everybody."[41] Linda McKay asserted that "if that place would have opened up I'd be the first one to move back there . . . Oh, do I ever miss that place man. Sometimes I'd sit there, my mom and I would just sit there and talk; talk about Pine Point."[42] Garvin Lizotte remembers Pine Point as "a picture perfect town," where "every yard had flowers and grass, picket fence. It was a beautiful town." As Lizotte's comments hint, part of the affinity people feel for Pine Point stems from the fact that it resembled a modern suburb, with all the facilities, amenities, and activities one would expect in a southern small town. Citing what he felt were excellent schools and many opportunities for sports and recreation, Lizotte remembered that "it was the best of the best of everything."[43] As Dene and Métis began to visit or move to Pine Point, the quality of life in the more traditional Fort Resolution seemed diminished compared to the ultra-new and modern mining town. Eddy McKay recalled his growing perception of a stark contrast between the two towns:

> I guess it was totally different from Fort Res, as you know. It had a lot of things, you know. Stores, all kinds of stores, and everything was paved, and running water. You know, everything

was right up to the times. You had the best of pretty much everything for a small community....Oh yeah, there was a lot of sports there. That is where this arena came from. They moved it over here after it closed. The ball field went to Hay River. They had a soccer field right next door to the school. And then the high school had another big field, so there was a lot of green space, I guess, recreational space.[44]

As did McKay, many interviewees cited sports and recreation as the focal point of the town's social life, with memories of baseball tournaments, the Arctic Winter Games, and the Pine Days festival flowing into many of our conversations. Larry Dragon, a Métis from Hay River, described the town as "close knit" because "if you wanted something, like to get into recreation, they had the arena, which Cominco build 99 percent of it . . . You had a curling rink . . . towards the end you had a swimming pool. They had everything there. They had a golf course; the best golf course in the Territories back then."[45] Such testimony suggests that the efforts of federal and territorial governments to create a model northern mining town—a family-oriented community with a high quality of life that contrasted with rough mining camps or divisive company towns—was at least partly successful.

Certainly many interviewees confirmed Dragon's comments about the close social cohesion within the town. Many stated that everybody got along at Pine Point, and even some of those who cited incidents of racism suggested that for the most part outsiders embraced local Native people as friends. Ron McKay, who described racial tension between Aboriginal people and outsiders, also described how "the non-Native people were actually really nice to—like, my dad had some of the greatest friends there that were non-Native. They took care of him and everything, so they were good people."[46] Lorraine Mckay suggested that "growing up they used to get along, everybody from Res or Hay River who'd go to Pine Point, they were always welcome."[47] Several people noted that they had a particular affinity for Newfoundlanders, due to shared interests in hunting and fishing. And for some local people, one of the exciting aspects of moving to Pine Point was the opportunity to meet people from all over the world. As Garvin Lizotte explained, "We had at that mine guys from

Iran, my dad had good friends from Portugal, Argentina, you know, all over. I could just keep on naming them. It was just awesome to live there. Good culture, eh. Just because of the mine, people come in for all different trades. I enjoyed it. Like I lived beside a real Italian family; the mom and dad were both from Italy."[48] Melvin Mandeville likewise recalled that he "found it interesting, because coming from a community where it's all Chipewyan and Native and not too many white people, or Hispanics and coloured and stuff. So it was good in that sense, to meet different people and knowing that the world isn't just here. There's a big world, eh."[49] If the comments of many interviewees suggest that racism was part of the social landscape at Pine Point, other testimony suggests that residents were often able to create a cultural middle ground within the community.

Many people also invoked the idea of successfully accommodating change more broadly when discussing their embrace of southern cultural and economic norms. Ronald Beaulieu described the expanded entertainment and shopping opportunities in Pine Point, suggesting that the culture brought up from the south, "it's different than us, so to us it's exciting. Maybe to them it's a regular thing, but we see it different." Although Beaulieu repeated a common sentiment when he suggested that the new money that accompanied the mining jobs "screwed a lot of people up" with increased alcohol consumption, he and many others also cited the introduction of good-paying jobs as one of the best aspects of the new mining economy.[50] Gord Beaulieu recalls that

> We were working six days on, two days off. And the money was good. It was probably better money at that time than anywhere I've worked since, with the value of money back then. In 1979, '80, you could go to the store, and if you buy a hundred dollars' worth of groceries . . . we couldn't carry it out of the store; it was too much. Nowadays if you buy a hundred dollars' worth of groceries, and you can just walk out carrying it in one hand. So, you know, for the value of the money, and even vehicles were cheap back then. So I made good money. I had fun in Pine Point.[51]

Tommy Beaulieu recalled that many more Fort Resolution people were able to buy vehicles because of mine wages, suggesting in turn that increased mobility opened up the opportunity to buy cheap groceries in

Pine Point.[52] Indeed, many interviewees felt that access to cheaper food was one of the most significant positive impacts of the mine and the eventual road extension to Fort Resolution.

It is tempting to conclude that the introduction of modern wage labour and outside sources of food undermined local patterns of subsistence hunting and commercial trapping in the South Slave region. Many comments from interviewees suggest, however, that Native people in the area took a flexible approach to various economic opportunities, often moving between trapping and mine labour to take advantage of shifting prices and market conditions. Darin Mckay remembered his father's movement between two types of labour:

> I think he did trapping on the side, yeah. He always trapped, all his life he's been a trapper. Like before he moved to the mine here, that's what he did ... I guess when trapping wasn't the greatest, that was when fur was cheap. And that maybe, five, ten years after we moved there, or maybe five years after we moved to Pine Point, the fur price went up. Just when the mines were shutting down too they were laying off people. And my dad was a trapper, so we moved out of Pine Point because they were shutting down, you know, there was not money I guess in lead and things, startin' to get old. Then we moved to Res and he started trapping again, hunting and old times. Right 'til today he still traps and hunts. And he does trap in that area, Pine Point, right now.[53]

Leonard Beaulieu, who worked off and on at the mine and on the road crew from 1965 to 1974, claimed that trapping "was not worth it" when the mine opened in 1964, but by the mid-1970s lots of people quit their jobs because fur prices were very high. He recalled, "At that time, lynx averaged $400 apiece. Damn right, sport. In two months I made $21,000. Never make that working on a CAT."[54] Some interviewees indicated that settling in Pine Point and taking advantage of the associated wage labour opportunities drew them out of trapping for good. For others, however, movement between trapping and mining labour provided a means to cushion the blow from the international price swings that could cripple local economies associated with both these forms of primary resource production.

For still other interviewees, the transition from life in a modern town back to the more subsistence-oriented Fort Resolution proved difficult after the closure of the mine. Denise McKay said that she did not want to move back because there were no jobs (though she did find a job in the community hall), and her kids were sad to lose their friends at Pine Point.[55] Eddy McKay remembered a period of adjustment to life at the older town:

> I didn't like it at first, cuz, you know, there was no running water, and we had honey buckets I guess, and cramped housing; having to go to the school and shower over there. So it was totally different. And then, I don't know, I got (pauses), what would you say? I guess I accepted it more. And opened myself to the life in Fort Res, and then it wasn't too bad after a little while.[56]

Catherine Boucher similarly recalled that "for them [returning Pine Pointers] it was a big change for the families I guess because, even for me, when I came back it was different. Oh there was no pool table. You know, the things I liked to do when I was in Pine Point." Aside from missing the good life at Pine Point, Boucher suggested that one major source of difficulty for people moving back to Fort Resolution was the fact that there was no housing.[57] As well, according to Melvin Mandeville, the difficulty of adjusting to his return to Fort Resolution was compounded by the fact that many people labelled the Pine Pointers as outsiders.[58] In any case, memories of adjusting to the comparatively poor facilities at Fort Resolution point to a mixed legacy. As many attest, quality of life in Pine Point was quite high for some Native workers lucky enough to find work and housing in the town. But the juxtaposition of a modern town with another that (at the time) lacked basic municipal services also suggests the lack of lasting economic benefits that flowed to Fort Resolution during the life of the Pine Point Mine.

Such a mixed record is reflected in the ambivalent attitude of many Native people to the mine. While we did encounter some unequivocal Pine Point boosters and some who directed only harsh criticism at the mine among the many people we spoke with, most suggested in some way that the mine represented a mixed legacy for Native people in the surrounding communities. As mentioned previously, a remarkable number

Figure 4: Abandoned street, Pine Point, NWT, 2009. Photo by John Sandlos.

of people mixed stories of racism with stories of how the people in Pine Point were friendly, and pointed out how well they got along with many outsiders. Darin Mckay, for example, juxtaposed difficult memories of racism with broader recollections of the very positive social life in the town:

> Yeah, [the town was] a little bit rough. They kinda didn't like Natives, some of them. But lots of them were nice to us, you know, white people. "Come in and have cookies," or something. We had neighbours—yeah there was a few, the ones that didn't like us, I guess, had some kind of beef. But I didn't know; I was a kid, eh? I remember that. It was a good town, to tell you the truth. It was a nice place there; they had lots of good stores, a ball park. They had everything—lots of stuff going on for kids once in a while. They had parades, you know, carnivals. It was pretty good.[59]

Others mixed harsh criticism of the social, environmental, and economic impact of the mine while acknowledging the positive side of life and work at Pine Point. From K'atl'odeeche First Nation, Daniel Sonnefrere (speaking through an interpreter), asserted:

> Some places look bad, some places look good, because it's helped some people to work and there are still people working today. They learned a lot from there. But for us it was bad, because too much drinking... It was a good job, good work, you get good pay, you get to keep it. It's too much drinking [and] we had a problem with that.[60]

Tommy Unka maintained a similar perspective on Fort Resolution's experience of the mine:

> Well, like I said, it kinda brought the south to us, you know. So there's that impact, you know. But it was also a lot of good stuff like at Christmas time we had a little more stuff because of Cominco mines, because the stores were there and shit like that, you know. So there's the goods, you know, my family, my Dad had a little more rum and stuff like that. There's a lot of parties, you know, and I was young, so you know, I enjoyed these little perks. But also of course there was always a down side to a good thing. And some of the down side was some of the social problems that happened as the highway came in.[61]

Gord Beaulieu summed up his perception of the mine by stating simply, "There's a lot of good that came with it, but there's a lot of bad too. A lot of negative. I was young back then, and I had a lot of fun. I had fun at Pine Point."[62] A mix of the good and the bad: this idea came through time and time again in the interviews and challenged our initial assumption that Aboriginal communities simply regarded Pine Point as a blot on their collective historical experience.

Such an ambivalent view of Pine Point's history has deeply influenced local opinions on the recent plans of a smaller resource company, Tamerlane Ventures, to reopen the mine (but not the town) and remove the remaining economical deposits of lead and zinc. At environmental assessment hearings in Fort Resolution in 2008 about the project, people raised concerns about the negative impacts of Pine Point, particularly ongoing environmental concerns and the fact that the community derived little economic benefit from the mine.[63] Our interviewees raised many of the same issues, with many opposed to or ambivalent about the idea of reopening the mine. Catherine Boucher proclaimed an oft-repeated concern for the environmental impacts of new mining activity: "For me, I don't think it'd be good for our land. They've been taking things off our lands for so long; we don't get nothing back."[64] Angus Beaulieu echoed the latter part of Boucher's comments when he interpreted the lack of consultation prior to exploration work as a sign that history was repeating itself: "We're hitting the table so it's never going to happen to us again, and it happened again. These people come in and start drilling without even coming to Fort Res here."[65] Lloyd Cardinal was similarly critical that Tamerlane's bulk sampling and test mining program had proceeded without an impact and benefit agreement, and many in Fort Resolution and at K'atl'odeeche were firm that development should only proceed if the communities received employment, assurances that the site would be remediated, and an IBA.[66] Others juxtaposed environmental concerns with the pressing need for more employment and economic activity in the South Slave region.

Some, however, wholeheartedly welcomed the return of mining at Pine Point. When asked what he thought about the mine opening up again, Garvin Lizotte replied, "Well I'm just waiting. I'm a truck driver, eh. So I'm ready to go to work day one."[67] Gord Beaulieu felt that the community was more prepared than in the 1960s for a second Pine Point project: "So it's not like, if they open up a mine and everybody has all this money, it's not like this whole town is going to go back like it did again, like it did back in the '60s and '70s. We're already used to it, so that part won't change that much. But it will help the economy."[68] Lorraine Mckay claimed that if they did reopen the mine, she would move back and put up a house in the place where she was born.[69] Whether one is a supporter or critic of Tamerlane's Pine Point project, memories of the profound impact of the original mine continue to shape local responses to industrial development in the South Slave region.

CONCLUSION

People's life stories almost always proceed as a series of ups and downs. So why, then, is it important to suggest that Native people's experience with industrial mining in the South Slave region was mixed? Part of the answer lies in the fact that the rich and complex oral history of Pine Point expands beyond the common (though understandable) emphasis in northern oral history projects on preserving traditional stories and accounts of the pre-industrial fur trade and trapping life. Indeed, the Pine Point oral histories offer a rare glimpse, not at long-ago stories passed from generation to generation or the hunting and trapping life that dominated in South Slave communities prior to World War II, but at the various ways that Aboriginal people in the region resisted, accommodated, and in some cases embraced post-war industrial development. Interviewees provided trenchant and perceptive critiques of historical approaches to northern development, but their comments also problematize previous studies suggesting that northern Aboriginal people received no benefit from the mine and associated developments. Personal histories of individuals suggest that the mine did not simply bypass Aboriginal people; nor were they purely the victims of an externally imposed development project. If the oral interviews confirm that, in general, Aboriginal communities in the surrounding area realized very few social and economic benefits while having to live with a lasting legacy of environmental damage in their proverbial backyard, they also suggest that many individuals responded creatively to the social and economic opportunities associated with the mine and the modern community that came with it. Some took advantage of available wage labour opportunities while never completely abandoning hunting and trapping as a potential economic safety net. Others translated their experience working in the mine into employment in other development projects, whether mines, mineral exploration, or work on the road crew. Still others embraced the social life and economic opportunities associated with a modern town.

Acknowledging these stories should not be misinterpreted as an apologia for a development project that largely failed to fulfill the promise of sustained economic development in the South Slave region. Nor is it an attempt to flatten the variety of human responses to Pine Point into abstract social science concepts such as community resilience and adaptation. Instead, these oral histories should remind us that simple dualistic stories of traditional communities versus modern mines do not necessarily accord with the complex individual experiences of people who shaped, and were shaped by, the massive social, environmental, and economic changes that came with developments like the Pine Point Mine.

ACKNOWLEDGMENTS

The author would like to sincerely thank all those in K'atl'odeeche First Nation and Fort Resolution who participated in oral history interviews. Many thanks also to Arn Keeling, Frances Mandeville, and Catherine Boucher, who conducted many of the interviews that formed the basis for this chapter. Finally, thank you to Rosy Bjornson, IMA Coordinator, Deninu Ku'e First Nation, for tremendous help with local logistics and communications. Marsi cho!

NOTES

1 H. Ali Saleem, *Mining, the Environment, and Indigenous Development Conflicts* (Tucson: University of Arizona Press, 2003); Subhabrata Bobby Banerjee, "Whose Land Is It Anyway? National Interest, Indigenous Stakeholders, and Colonial Discourses," *Organization & Environment* 13 (March 2000): 3–38; Al Gedicks, *Resource Rebels: Native Challenges to Mining and Oil Companies* (Cambridge, MA: South End Press, 2001); Stuart Kirsch, *Reverse Anthropology: Indigenous Analysis of Social and Environmental Relations in New Guinea* (Palo Alto, CA: Stanford University Press, 2006); Marcus B. Lane and E. Rickson Roy, "Resource Development and Resource Dependency of Indigenous Communities: Australia's Jawoyn Aborigines and Mining at Coronation Hill," *Society and Natural Resources* 10 (1997): 121–42; Nicholas Low and Brendan Gleeson, "Situating Justice in the Environment: The Case of BHP at the Ok Tedi Copper Mine," *Antipode* 30 (1998): 201–26; Joan Martinez-Alier, "Mining Conflicts, Environmental Justice, and Valuation," *Journal of Hazardous Materials*, 86 no. 1–3 (2001): 153–70.

2 Déłı̨nę Uranium Team, *If Only We Had Known: The History of Port Radium as Told by the Sahtúot'ine* (Déłı̨nę, NWT: Déłı̨nę Uranium Team, 2005); Lianne Leddy, "Cold War Colonialism: The Serpent River First Nation and Uranium

Mining, 1953–1988" (PhD diss., Wilfrid Laurier University, 2011); Lianne Leddy, "Interviewing *Nookomis* and Other Reflections: The Promise of Community Collaboration," *Oral History Forum d'histoire orale* 30 (2010): 1–18; Doug Brugge, Timothy Benally, and Esther Yazzie-Lewis, *The Navajo People and Uranium Mining* (Albuquerque: University of New Mexico Press, 2006).

3. Robert Bothwell, *Eldorado: Canada's National Uranium Company* (Toronto: University of Toronto Press, 1984); Richard Geren, Blake McCullogh, and Iron Ore Company of Canada, *Cain's Legacy: The Building of Iron Ore Company of Canada* (Sept-Îles, QC: Iron Ore Company of Canada, 1990); Ryan Silke, *High-Grade Tales: Stories from the Mining Camps of the Northwest Territories, Canada* (Yellowknife, NWT: Ryan Silke, 2012).

4. John Sandlos and Arn Keeling, "Claiming the New North: Development and Colonialism at the Pine Point Mine, Northwest Territories, Canada," *Environment and History* 18 (2012): 5–34.

5. Peter Clancy, "Working on the Railway: A Case Study in Capital-State Relations," *Canadian Public Administration* 30 (1987): 450–71.

6. Kenneth J. Rea, *The Political Economy of the Canadian North* (Toronto: University of Toronto Press, 1968).

7. Paul Deprez, *The Pine Point Mine and the Development of the Area South of Great Slave Lake* (Winnipeg: Center for Settlement Studies, 1973).

8. Thomas Berger, *Northern Frontier, Northern Homeland: The Report of the Mackenzie Valley Pipeline Inquiry, Vol. 1* (Ottawa: Supply and Services Canada, 1977), 123–24.

9. Janet E. Macpherson, "The Pine Point Mine," in *Northern Transitions, Volume I: Northern Resource Use and Land Use Policy Study*, eds. Everett B. Peterson and Janet B. Wright (Ottawa: Canadian Arctic Resources Committee, 1978), 65–110.

10. Richard V. Francaviglia, *Hard Places: Reading the Landscape of America's Historic Mining Districts* (Iowa City: University of Iowa Press, 1991); Peter Goin and C. Elizabeth Raymond, *Changing Mines in America* (Santa Fe: Center for American Places, 2004); Peter Goin and Elizabeth Raymond, "Living in Anthracite: Mining Landscape and Sense of Place in Wyoming Valley, Pennsylvania," *The Public Historian* 23, no. 2 (2001): 29–45; John Harner, "Place Identity and Copper Mining in Sonora, Mexico," *Annals of the Association of American Geographers* 91, no. 4 (2001): 660–80; Ben Marsh, "Continuity and Decline in the Anthracite Towns of Pennsylvania," *Annals of the Association of American Geographers* 77 (1987): 337–52; William Wyckoff, "Postindustrial Butte," *Geographical Review* 85 (1995): 478–96; David Robertson, *Hard as the Rock Itself: Place and Identity in the American Mining Town* (Boulder: University Press of Colorado, 2006).

11 Paul Shoebridge and Michael Simons, *Welcome to Pine Point* (Montreal: National Film Board, 2011). For the Pine Point website, see "Pine Point Revisited," accessed February 21, 2013, http://pinepointrevisited.homestead.com/.

12 Interviews were conducted in Fort Resolution, Hay River, and K'atl'odeeche First Nation in May 2010 by Catherine Boucher, Arn Keeling, Frances Mandeville, Rosalie Martel, and John Sandlos.

13 Yellowknives Dene First Nation, *Weledeh Yellowknives Dene: A History* (Dettah, NWT: Yellowknives Dene First Nation Council, 1997); Déline Uranium Team, *If Only We Had Known*. The negative memories associated with these two prominent abandoned mines in the Northwest Territories, the Port Radium radium/uranium/silver mining complex and Giant Mine at Yellowknife, can be explained in part by the simple fact that they produced much more acute pollution than Pine Point. Keeling and Boulter's chapter in this volume suggests that the combination of political critique with a strong mining identity is not unique to communities near Pine Point.

14 Emilie Cameron, "Copper Stories: Imaginative Geographies and Material Orderings of the Central Canadian Arctic," in *Rethinking the Great White North: Race, Nature, and Whiteness in Canada*, eds. Andrew Baldwin, Laura Cameron, and Audrey Kobayashi (Vancouver: UBC Press, 2011), 169–90.

15 Sandlos and Keeling, "Claiming the New North."

16 A transcription of Robert Sayine's comments to the Berger Inquiry in Fort Resolution, October 1975, was found in a file marked DKFN, Cominco File, in the archive of the Deninu K'ue First Nation band office.

17 Cecil Lafferty's testimony was found in the transcript titled Northwest Territories Water Board Public Hearing on Pine Point Mines Limited's Application for Renewal of License N1L3-0035, June 12, 1990. A copy was found in the Deninu K'ue First Nation band office.

18 Presentation by Chief Bernadette Unka of the Deninu K'ue First Nation to the Royal Commission on Aboriginal Peoples, 17 June 1993, Native Law Centre Fonds, RCAP Vol. 167, Box 26, University of Saskatchewan Archives.

19 Interviews were conducted in a semi-structured format, with interviewers suggesting questions or themes for discussion, but also allowing the interviewee to take the lead in the conversation as well. Because of this, the interviews are not really for quantitative analysis in the manner of standardized surveys or focus groups, though some obvious trends are noted in the paper.

20 Angus Beaulieu, interview by John Sandlos and Arn Keeling, digital recording (Fort Resolution, NWT, May 19, 2010).

21 George Balsillie, interview by John Sandlos, digital recording (Fort Resolution, NWT, May 20, 2010).

22   Angus Beaulieu, interview by John Sandlos and Arn Keeling, digital recording (Fort Resolution, NWT, May 19, 2010). These numbers are difficult to corroborate with existing employment data from the 1970s because the numbers do not distinguish between Aboriginal people hired at Fort Resolution and those hired from outside the region in any given year. Macpherson has noted in her report that the number of Aboriginal workers hired from northern Alberta may have been substantial, and the mixing of the two groups in employment data was a major bone of contention in Fort Resolution. See Macpherson, "The Pine Point Mine," 89.

23   Sam Bugghins, interview by Arn Keeling and Rosalie Martel, digital recording (K'atl'odeeche First Nation, NWT, May 20, 2010). Twelve other interviewees mentioned that Native workers were confined to lower-skilled work, or that they had worked as line cutters. Some testimony in the interviews suggests there were more employment opportunities at the mine for Fort Resolution residents in the 1980s, but Aboriginal employment records for the 1980s were not found in the archives.

24   Greg Villeneuve, interview by John Sandlos, digital recording (Fort Resolution, NWT, May 20, 2010).

25   Lloyd Cardinal, interview by Arn Keeling, digital recording (Fort Resolution, NWT, May 18, 2010).

26   Leonard Beaulieu, interview by John Sandlos and Arn Keeling, digital recording (Fort Resolution, NWT, May 19, 2010).

27   Henry McKay, interview by Arn Keeling, digital recording (Fort Resolution, NWT, May 19, 2010).

28   Roy Fabian, interview by Arn Keeling and Rosalie Martel, digital recording (K'atl'odeeche First Nation, NWT, May 20, 2010).

29   Harold Moore, interview by Arn Keeling and Rosalie Martel, digital recording (K'atl'odeeche First Nation, NWT, May 20, 2010).

30   Leonard Beaulieu, interview by John Sandlos and Arn Keeling, digital recording (Fort Resolution, NWT, May 19, 2010).

31   Yves Berube et al., "An Engineering Assessment of Waste Water Handling Procedures at the Cominco Pine Point Mine," unpublished report, Ottawa: Department of Indian Affairs and Northern Development, April 1972; J. N. Stein and M. R. Miller, "An Investigation into the Effects of a Lead-Zinc Mine on the Aquatic Environment of Great Slave Lake," unpublished report, Winnipeg: Resource Development Branch, Fisheries Service, Department of Environment, April 1972; M. S. Evans, L. Lockhart, and J. Klaverkamp, "Metal Studies of Water, Sediments and Fish from the Resolution Bay Area of Great Slave Lake: Studies Related to the Decommissioned Pine Point Mine,"

Environment Canada, National Water Research Institute, Burlington and Saskatoon, NWRI Contribution No. 98-87, July 1998.

32 Priscilla Lafferty, interview by John Sandlos, digital recording (Fort Resolution, NWT, May 19, 2010).

33 Denise McKay, interview by John Sandlos and Frances Mandeville, digital recording (Fort Resolution, NWT, May 19, 2010).

34 Melvin Mandeville, interview by John Sandlos, digital recording (Fort Resolution, NWT, May 20, 2010).

35 Gord Beaulieu, interview by John Sandlos, digital recording (Fort Resolution, NWT, May 19, 2010).

36 Ron McKay, interview by John Sandlos, digital recording (Fort Resolution, NWT, May 19, 2010).

37 Angus Beaulieu, interview by John Sandlos and Arn Keeling, digital recording (Fort Resolution, NWT, May 19, 2010).

38 Ron McKay, interview by John Sandlos, digital recording (Fort Resolution, NWT, May 19, 2010).

39 Leander Beaulieu, interview by John Sandlos, digital recording (Fort Resolution, NWT, May 20, 2010).

40 Tom Unka, interview by Arn Keeling, digital recording (Fort Resolution, NWT, May 20, 2010).

41 Lorraine Mckay, interview by John Sandlos and Frances Mandeville, digital recording (Fort Resolution, NWT, May 20, 2010).

42 Linda McKay, interview by Arn Keeling, digital recording (Fort Resolution, NWT, May 20, 2010).

43 Garvin Lizotte, interview by John Sandlos and Frances Mandeville, digital recording (Fort Resolution, NWT, May 19, 2010).

44 Eddy McKay, interview by John Sandlos and Frances Mandeville, digital recording (Fort Resolution, NWT, May 19, 2010).

45 Larry Dragon, interview by John Sandlos, digital recording (Hay River, NWT, May 21, 2010).

46 Ron McKay, interview by John Sandlos, digital recording (Fort Resolution, NWT, May 19, 2010).

47 Lorraine Mckay, interview by John Sandlos and Frances Mandeville, digital recording (Fort Resolution, NWT, May 20, 2010).

48 Garvin Lizotte, interview by John Sandlos and Frances Mandeville, digital recording (Fort Resolution, NWT, May 19, 2010).

49 Melvin Mandeville, interview by John Sandlos, digital recording (Fort Resolution, NWT, May 20, 2010).

50 Ron Beaulieu, interview by Arn Keeling and Catherine Boucher, digital recording (Fort Resolution, NWT, May 19, 2010).

51 Gord Beaulieu, interview by John Sandlos and Arn Keeling, digital recording (Fort Resolution, NWT, May 19, 2010).

52 Tommy Beaulieu, interview by John Sandlos, digital recording (Fort Resolution, NWT, May 20, 2010).

53 Darin Mckay, interview by John Sandlos and Frances Mandeville, digital recording (Fort Resolution, NWT, May 19, 2010).

54 Leonard Beaulieu, interview by John Sandlos and Arn Keeling, digital recording (Fort Resolution, NWT, May 19, 2010).

55 Denise McKay, interview by John Sandlos and Frances Mandeville, digital recording (Fort Resolution, NWT, May 19, 2010).

56 Eddy McKay, interview by John Sandlos and Frances Mandeville, digital recording (Fort Resolution, NWT, May 19, 2010).

57 Catherine Boucher, interview by Frances Mandeville, digital recording (Fort Resolution, NWT, May 20, 2010).

58 Melvin Mandeville, interview by John Sandlos, digital recording (Fort Resolution, NWT, May 20, 2010).

59 Darin Mckay, interview by John Sandlos and Frances Mandeville, digital recording (Fort Resolution, NWT, May 19, 2010).

60 Daniel Sonnefrere, interview by Arn Keeling and Rosalie Martel, digital recording (Katlodeeche First Nation, NWT, May 20, 2010).

61 Tom Unka, interview by Arn Keeling, digital recording (Fort Resolution, NWT, May 20, 2010).

62 Gord Beaulieu, interview by John Sandlos and Arn Keeling, digital recording (Fort Resolution, NWT, May 19, 2010).

63 Mackenzie Valley Environmental Impact Review Board, Report of Environmental Assessment and Reasons for Decision on Tamerlane Ventures Inc.'s Pine Point Pilot Project, EA-0607-002, February 22, 2008.

64 Catherine Boucher, interview by Frances Mandeville, digital recording (Fort Resolution, NWT, May 20, 2010).

65 Angus Beaulieu, interview by John Sandlos and Arn Keeling, digital recording (Fort Resolution, NWT, May 19, 2010).

66 Lloyd Cardinal, interview by Arn Keeling, digital recording (Fort Resolution, NWT, May 18, 2010).

67 Garvin Lizotte, interview by John Sandlos and Frances Mandeville, digital recording (Fort Resolution, NWT, May 19, 2010).

68   Gord Beaulieu, interview by John Sandlos and Arn Keeling, digital recording (Fort Resolution, NWT, May 19, 2010).

69   Lorraine Mckay, interview by John Sandlos and Frances Mandeville, digital recording (Fort Resolution, NWT, May 20, 2010).

Section 2
History, Politics, and Mining Policy

| CHAPTER 6

The Revival of Québec's Iron Ore Industry: Perspectives on Mining, Development, and History

Jean-Sébastien Boutet

> Tes ancêtres t'ont conduit à moi pour me raconter les images de tes rêves.
>
> —Joséphine Bacon[1]

These words, borrowed from a remarkable Innu poet, introduced the implementation strategy for Québec's northern development plan—initially labeled *Plan Nord*—for the first five-year period (2011–2016). One can emphatically question the pertinence of associating this expression with a technocratic agenda conceived to engineer social, economic, and ecological progress for Québec society. Yet it is possible to ascertain elements of continuity with previous generations of policy-makers who gazed northward and nurtured bold dreams of resource exploitation in the subarctic hinterland.² Through repeated invocations of the necessity for renewal in a changing and more competitive world, the provincial government hoped that "the scope of the Plan Nord will make it in the coming decades what the [hydroelectric] development of La Manicouagan and James Bay were to the 1960s and 1970s."³ In the realm of northern development, innovative visions for the future thus meet the aspirations of yesteryear; for proponents of the resource industry, past megaprojects are always lurking, never fully erased or forgotten yet often simplified and reformatted to accommodate contemporary priorities.

Subarctic Québec is once again at the heart of a treasure hunt for the control of its "blooming iron ore scene."⁴ Targets for the exploitation of major deposits in the Labrador Trough (Fig. 1),⁵ a geological region "set to transform into a major force in the iron ore sector" worldwide,⁶ point to the revival of a decisive episode in the history of large-scale resource development in the province, a period that in fact predated the harnessing of Québec's most powerful rivers. This new iron ore rush is stimulated, not unlike the first round of mineral activities of the postwar period, by the developing needs of emerging world powers. Known around the globe for their impressive size and good ore content, the vast iron formations of northern Québec on which China and India are hoping to "feast"⁷ could comprise the long-term strategic reserves that these countries need in order to meet increasing domestic demand for finished steel. As competition for the control of iron deposits heightens globally, Québec proposes, once again, to hand over its best mineral reserves in the North to interests located outside of the region. At the same time, government administrators aim to integrate—on paper, at least—the virtues of profitable subsurface exploitation with improved environmental protection and substantial economic benefits for local indigenous communities and the province more generally.

This chapter argues that a retrospective synopsis of mineral activities in the Labrador Trough, at Schefferville in this particular case, can and should inform Québec's present ambitions for industrial growth in the north. *Plan Nord*, the province's latest twenty-five-year northern development plan, was formulated without comprehensive citizen engagement (and while bypassing the process of seeking approval from many indigenous communities) in a region that the government continues to treat

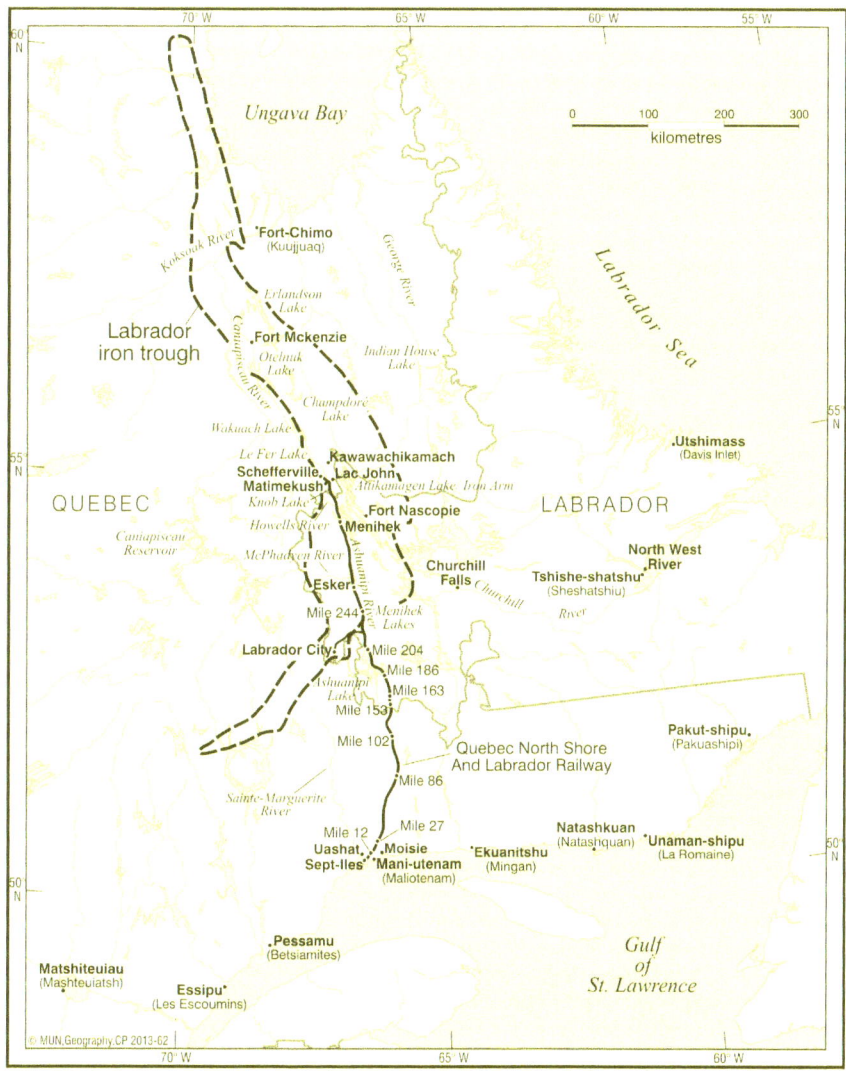

FIGURE 1: The Québec-Labrador peninsula, with the geological formation of the Labrador Iron Trough. Map by Charlie Conway.

primarily as a "bank" of natural resources where few benefits are derived for the rural peripheries and their inhabitants.[8] The first section adopts a state-corporate historical approach as it recounts the opening of the Ungava region to industrial interests through iron ore exploration (mid-1930s) and the beginning of operations at the Schefferville mine (1954). In the second section, the story of the mine closure shifts the perspective toward Innu and Naskapi individuals, who remember how indigenous and mining worlds ceased to interact after 1982 at Schefferville. Finally, the third section explores the contemporary context and provides an overview of the development projects that currently define the renewed quest for iron ore extraction in the Québec-Labrador borderlands.

The diverse, multi-layered interpretations of the mining past (section 2), understood in conjunction with the historical context that confirms the supremacy of state and corporate interests throughout the period (section 1), show the asymmetrical yet complex nature of the relationship of indigenous groups living in the area with industrial modernity. This relationship continues to be marked, informed, and shaped by the region's deeply contested mining history, a stark reminder for contemporary developers and policy-makers (section 3) that, despite mining's instability and apparent ephemerality, the mining past of the region lives on. Amidst these unresolved legacies, this analysis suggests that mining proponents' assertive guarantees of a durable, prosperous, and equitable industrial future for societies inhabiting the provincial north and the rest of Québec ought to be met with equal scrutiny and skepticism.

THE FIRST PHASE OF IRON MINING IN THE LABRADOR TROUGH

> Les Américains possèdent les capitaux qui nous manquent. Ils mettent nos ressources naturelles en valeur, réveillent une richesse endormie.
>
> —Robert Rumily[9]

During the late interwar years, Québec politicians sought to extract wealth from what they considered to be the subarctic hinterland's dormant mineral assets. In the Ungava region, the provincial administration subjected the geological formation known as the Labrador Trough to a certain preferential treatment, notably by granting exploration permits and large mining concessions to outside interests. However, this bold redefinition of geographical space in favour of foreign corporations collided heavily with the indigenous populations' prior occupation of their homelands.

In 1939, Québec conferred to McKay (Quebec) Explorers (MQEC) an impressive exploration concession of more than 10,000 square kilometres located in the Whale River and Swampy Bay River watersheds. A few years previously, explorers and geologists, typically accompanied by knowledgeable indigenous guides, had identified ore bodies with potential commercial value. At the time, the first government of Québec premier Maurice Duplessis (1936–39) imposed on this concession a modest annual rent of $1,000 and $2,000, for the first two years respectively, which could be renewed at a slightly augmented rate in subsequent years.[10] Following MQEC's failure to develop the concession, Jules Timmins, a prosperous mining entrepreneur heading the Toronto-based Hollinger Consolidated Gold Mines, entered a partnership in December 1943 with the influential American steel magnate George Humphrey, who was at the helm of the Cleveland mining firm M. A. Hanna.[11] Together these veritable "movers and shakers of mining empires"[12] controlled the Hollinger North Shore and Exploration Company (HNSE) and secured their access to the great property located at the heart of the Ungava Peninsula.

The recently formed HNSE venture received additional support from the Québec government via a special law passed in its favour three years later, in 1946.¹³ In the course of that year, the legislature introduced the *Loi pour faciliter le développement minier et industriel du Nouveau-Québec*, which defined a "mining exploitation lease valid for eighty years and covering an area of 300 square miles" on the shores of Knob Lake, the future site of the mining town of Schefferville.¹⁴ When shareholders incorporated Iron Ore Company of Canada (IOC) in Delaware in 1949, in order to assemble the capital essential for kick-starting the Schefferville operations, HNSE subleased these mining rights to the newly formed American-led iron and steel conglomerate.¹⁵

In the world of corporate executives, the region that was traversed by numerous Aboriginal trading and transportation routes remained at the time completely cut off from the large industrial markets and freight networks so crucial to moving the ore riches out of Nouveau-Québec. Knob Lake deposits somehow had to be linked to the St. Lawrence navigational waters (later on, the St. Lawrence Seaway) and thereby to the great manufacturing towns of Baltimore, Detroit, Cleveland, or Philadelphia. The railway engineers' answer to this technical challenge, 575 kilometres of steel rails and wooden ties to be laid over marshland, through mountains, and across forest, presented an enormous logistical and financial puzzle for IOC. Indeed, throughout the construction phase, the economic feasibility of this project often hung in the balance, as the anticipated profits from the mine (projected from identified ore resources) were constantly re-evaluated against the growing upfront capital expended on the infrastructure needed to transport the bulky, heavy, and relatively low-valued iron commodity to manufacturing markets.¹⁶

In order to support this major transportation program, in March 1951, the provincial government proceeded not through leasing but by means of direct sale of a thin strip of land stretching from Sept-Îles (on the Gulf of St. Lawrence) to Knob Lake. This band of territory, extending hundreds of kilometres deep into the boreal forest and bisecting the Innu homeland, would host the railway to Schefferville, to be built and operated by Quebec North Shore and Labrador Railway Company (QNS&L), one of the several subsidiaries formed by IOC. In exchange for $4,000, the railway company led by Jules Timmins became the "absolute owner" of this piece of land, free of any additional charges.[17]

In addition to guaranteeing the company woodcutting rights along the railroad right-of-way, Québec also conceded to QNS&L a shore lot near the coastal settlement of Sept-Îles for the establishment of a train terminal, a classification yard, and a modern seaport. Through one of its energy subsidiaries, Gulf Power Company, IOC gained control over sections of the Ste Marguerite River and the Menihek Lakes where it installed hydroelectric power plants destined to supply electricity to the terminal in Sept-Îles and the mine and townsite at Schefferville.

Several Innu labourers obtained casual work on these various construction sites, and played an especially crucial role in the resupplying of railway work camps.[18] This labour, if it remained very temporary and seasonal, nonetheless allowed a handful of families to confront the difficult conditions afflicting the trapping economy of the immediate postwar period. In general, however, mining officials engaged in the radical transformation of human and natural landscapes without the explicit consent of local populations. The newly damned Ste Marguerite River, for example, had constituted for decades a vital transportation axis for hunting groups based on the coast of the St. Lawrence who wished to reach the interior of the territory during their long winter hunts. Yet for federal bureaucrats employed by the Indian Affairs agency in Sept-Îles, this land takeover was clearly a matter for optimism. According to the prevalent paternalistic dogma, the employment opportunities that were generated through the development and operation of the Schefferville mine would consecrate the local communities' gradual retreat from the life on the land, thus encouraging a more sedentary lifestyle and the introduction of wage dependency. Going as far as threatening, in a few extreme cases, to suspend aid for families who refused to seek paid employment, the Canadian government urged the Innu residing near Sept-Îles—but in an even more decisive manner, the Naskapi of the Ungava region—to relocate to the town of Schefferville, a booming region that was traditionally part of a number of families' comprehensive harvesting activities and networks. There, federal bureaucrats promised, people would find "better housing, education, and so forth,"[19] with job opportunities "that will last one hundred years."[20]

In Québec City, the provincial government of Maurice Duplessis argued that the new rail link from Schefferville to the coast "opens the Ungava region to commerce and civilization, as well as to the industry."[21] For the premier, Québec's natural resources had for many years remained "underutilized," and this unfortunate state of affairs had engendered substantial losses for the province. By promoting portions of the hinterland to mining interests and welcoming the implementation of modern infrastructure, the government hoped to profit in return from this "marvelous industrial development that will contribute powerfully, not only to the complete valuation of the Nouveau-Québec region, but also to the progress and prosperity of the province in general."[22]

To what extent did the creation of public wealth from iron mines imagined by the Québec state in fact materialize? In general terms, the level of mining revenues flowing to the province—expressed as the percentage of government incomes drawn from the mining sector through taxes, royalties, mining rights and permits over the total value of mineral production—fell by roughly one-fifth during the second Duplessis mandate (1944–59).[23] Notwithstanding improvements of the taxation regime undertaken by the Liberal administration of Jean Lesage in 1965, which successfully increased returns on production to the public, revenues generated through mining rights, particularly in the iron ore sector, remained weak.[24] At the end of the 1960s and the beginning of the 1970s, $2.4 million generated through mining rights in the iron ore sector accrued to the Québec treasury every year; by comparison, the total value of iron production reached nearly $185 million, on average, during the same period, 1967 to 1972.[25] This relatively marginal amount obtained from mining rights was proportionally lower than that registered in the Québec mining industry as a whole, with consequences for government finances given that iron ore was one of the most important mining sectors in the province.[26] Yet at the same time, Hollinger and Hanna Mining were cashing in on substantial revenues received through the transfer of their mining rights (originally granted by the Québec state) to IOC. These monies, according to Paquette, were "worth more than five times that which was drawn [via mining rights] by the owner of the mining domain"[27]—the provincial Crown.

For Duplessis and his administration, engaged in the rhetoric of economic autonomy for Québec, liberalism in mining policy represented a somewhat paradoxical situation closely tied to the evolution of international markets. During the Second World War, the United States became aware of the impending exhaustion of its iron ore supplies located in the Mesabi Range (Minnesota). North American appetite for iron grew steadily as a consequence of the war—consumption increased two and a half times between 1935 and 1945[28]—and its aftermath. But by 1953, as domestic demand grew due to Cold War (and Korean War) militarism and the rise of postwar consumer society, US production was on the decline.[29] Starting in the 1950s, American (but also European and Japanese) steelmakers restructured their commercial operations and sought control over strategic ore deposits worldwide.[30] The American steelmaking industry gained access to some of the most attractive reserves by formalizing financial links with mining companies, thus ensuring a reliable access to iron ore inputs at a cost that it could better regulate. By contrast to its European counterparts, who were "relying mainly on European iron ore producers," American steelmakers were "procuring iron ore from their own captive mines."[31]

In this context, producers and investors saw in the Labrador Trough a good business opportunity, by virtue of the region's advantageous geography—a relative proximity to US steelmaking centres—which complemented the presence of stable political institutions and a favourable fiscal and regulatory environment. Bethlehem Steel and National Steel, in conjunction with four other major American steelmakers, eventually took majority ownership of IOC on February 1, 1962. The beginning of the company's operations, in 1954, had strongly helped to "reshape the Canadian iron ore industry," since merely half of a decade later, Schefferville ore accounted for more than 60 per cent of iron production in the country.[32] Notwithstanding the modest impact of such growth on direct and indirect labour opportunities—in 1966, the effect of economic expansion on employment in the iron ore sector was lower than for each of wood felling, the manufacture of women's garments, and the fabrication of newspaper in the province as a whole[33]—the iron ore region, in Schefferville and elsewhere, never really experienced the development of a diversified, mature industrial economy. The narrow, external linkages between the mining and steelmaking sectors (exemplified by the IOC venture) instead confined the region to an export-driven enclave of minimally processed iron ore.³⁴ Located downstream of extractive activities, yet institutionally dissociated from the provincial mines, Québec manufactures could not capture the fiscal benefits typically associated with corporate integration, nor were they in a position to secure primary inputs at a better cost.³⁵ Ultimately, the exploitation of iron ore did not generate significant multiplier effects for the regional and provincial economies, despite the fact that mineral production underwent a veritable explosion in the immediate postwar period, and even though the added value produced by this sector and the fabrication of primary steel in Québec jumped from $30 million to $188 million during the quarter century 1944 to 1975 dominated by both Unionist and Liberal administrations.³⁶

The birth of national steelmaker Sidbec-Normines was in fact engineered to remedy some of the structural deficiencies that plagued the provincial iron ore industry, notably by favouring domestic processing, but the launch of this partially state-owned entity in 1976 turned out to be poorly timed.³⁷ A crisis was about to hit global steel markets: in the period 1974 to 1977, worldwide production of steel fell by 5 per cent, and the stocks of iron ore began to accumulate (including in the Labrador Trough) as metal prices were depressed.³⁸ Québec iron operations, incapable of competing with Venezuelan and Brazilian counterparts who exploited higher-grade iron deposits and incurred cheaper labour costs, inevitably felt the effects. In 1985, the subarctic municipality of Gagnon, *"là où la chaleur humaine remplaçait l'astre du jour,"*³⁹ was bulldozed and effectively erased from the territory. The Schefferville mine, keenly recognized for the quality of its iron ore at the turn of the midcentury, also succumbed to the crisis. On November 2, 1982, IOC president Brian Mulroney announced to employees and authorities that the company would wind down operations,⁴⁰ an outcome that clearly showcased the precariousness of the community's reliance on a single export production. Despite their initial desire to shut down the municipality, provincial and federal authorities decided not to condemn Schefferville to the same fate that was about to descend upon Gagnon three years later.⁴¹ Between 1985 and 1998, town lots were in fact amalgamated to the Innu reserve of Matimekush through official Government of Canada purchases of Québec land register.⁴²

MEMORIES AND MINING LANDSCAPE AT SCHEFFERVILLE

> The mine closed in 1982. The company packed its bags and took off. Open pits, destruction everywhere; our animals driven away. We made the company rich. Where's my share? Where's my thank you? There was none of that.
> —A Naskapi individual[43]

Throughout the operations of the Schefferville mine, between 1954 and 1982, indigenous miners remained confined to very marginal positions within the company hierarchy. Relying on a variety of strategies, these labourers worked to adjust and maintain their own practices in order to combine the labour at the mine with their life on the land. Yet overall, Innu and Naskapi groups could not realize the vision of government agents and utilize their paid employment at the mine as a mechanism to "climb the ladder" of industrial society; nor were they in a position to lay the foundations of a multi-sectorial, diversified regional economy that would outlast the life of the mine and bring about growth and development. This situation would have serious consequences for the viability of these two communities in the post-mining phase.

People's memories of the abandonment phase reveal the shock and the traumatic nature of the mine closure at Schefferville. As the Innu and the Naskapi express their disapproval of government and company actions, which created "a lot of troubles" for them,[44] they also recount, often with nostalgia, how people looked forward and continued to adapt and maintain their own life practices amidst the upheavals that characterized the failure of the modernization project. In view of these diverse interpretations of the past—inextricably linked to the contingency, complexity, and challenges of current everyday life—an intricate picture of local history, of indigenous homelands, and of the legacy of industrial development emerges. At Schefferville, local life stories generally reveal a deep sense of loss underlying Innu and Naskapi perspectives about the mine closure and the mining experience more generally, an experience which, in the final analysis, continues to defy any singular interpretation or meaning.

Several Innu and Naskapi individuals remember the closure and deindustrialization period in terms of severe disruption and the abandonment of their community by IOC and the government. These authorities, in their view, had operated illegitimately on a territory that did not belong to them in the first place. The sudden desertion by the mining company was especially disconcerting, according to people whom I interviewed at Schefferville, considering the immense wealth extracted during the operations years:

> We weren't even told and we didn't even know what they were using the ore for, why it was so precious to them. And years later, once they had distracted our way of life, we learned that the ore was in demand in foreign countries. They just came, took what they needed, and left.[45]

> Back then the company took out much iron from here, a lot of iron, an enormous amount of iron, because for them it was like gold. It brought a lot of money, back then, it was very profitable.[46]

Since the late interwar years, industry proponents had strived to first redefine and then exert control over the Québec-Labrador region in order to appropriate the riches of the subsurface. Now that the process was complete, Innu and Naskapi people were more or less left to pick up the pieces.

During the life of the mine, the somewhat marginal employment opportunities did provide, as some recall, non-negligible compensations for the local residents. This idea is articulated, for instance, by an Innu individual who explains that "according to my own knowledge, it did not bother the Innu to see the company dig holes in the ground and occupy our territory, because the company gave us work."[47] Yet as the company withdrew from Schefferville and wage labour was lost, it left few enduring benefits for the local communities, other than a small pension for a handful of long-time employees. "We got a certain amount from the company," a Naskapi elder recalls. "Certain people worked for so many years, and they received different amounts. But [the IOC managers] didn't care, they just left. They didn't care about us, and all the people, when they left from here. They didn't give any benefits to the people, or any other contributions."⁴⁸ People remember with some resentment that many indigenous workers were not in a position to even receive a pension after the company departed, because they were never employed—by choice or by constraint—as full-time labourers: "We got laid off and on, so some people like me didn't get the full benefits that the company offered after it left. There weren't a lot of benefits from the IOC. No other benefits were given to the people and the communities; it was only them and their money."⁴⁹

Some indigenous people maintain particularly vivid memories of the material removal of municipal buildings and houses in the years that followed that abandonment of the mine. As geographer John Bradbury describes, the relinquishment and destruction of infrastructure, considered by companies "as part of the production sector of the mine or mill," was a common strategy that served to sever a corporation's financial responsibility from the mining town or municipality in the closing phase of large industrial projects.⁵⁰ At Schefferville, IOC proceeded, in conjunction with the province, with a series of demolitions in order to dissolve its financial obligations (in addition to servicing all of Schefferville's debt, the company still contributed up to 80 per cent of municipal taxes in the post-closure years of 1984 and 1985).⁵¹ The provincial government led by the Ministère des Affaires municipales was also eager to shed its Schefferville responsibilities and recommended a $6.5-million compensation and demolition plan to shut down the municipality and remove a good portion of the infrastructure,⁵² including the local hospital, "for which Indians [were] by far the most frequent users."⁵³ This particularly dramatic event remains seared in the memory of many Innu and Naskapi individuals who have continued to reside in the vicinity of the once-cherished health facility.

In addition to the loss of town services and infrastructure, as well as shrunken employment and economic development opportunities, the communities near Schefferville have had to live, and continue to live, amidst a deeply scarred landscape. For the Innu and the Naskapi, the post-mining environment acts as an incessant material reminder of three decades of intensive land and resource exploitation by a company that naturally prioritized shareholder dividends and the needs of the American economy during the Cold War period,⁵⁴ but generated minimal

FIGURE 2: Abandoned mine with equipment near Schefferville. Photo by Jean-Sébastien Boutet.

returns for the local population. As a consequence of development, their homeland became littered with tailings piles, industrial wastes, leftover equipment, and pollutants that were abandoned in the wake of the mine closure (Fig. 2), but only marginally remediated many years later, when the IOC undertook a modest clean-up operation in the early 2000s.

At least some people in Schefferville are particularly moved by the former mining pits, which extend hundreds of metres deep over several hundred square kilometres across the subarctic landscape.[55] The removal of trees and other "overburden," the blasting of hilltops, the digging of holes, the large tailing piles and debris, and the oxidizing iron-rich earth have contributed to create a red-coloured, cratered, barren, and dangerous landscape (Figs. 3 and 4), which one Innu resident metaphorically depicts as extraterrestrial scenery. Another hunter and former chief describes the major changes to the Innu territory that resulted from open-pit mining, as he speaks of the unrecoverable beauty of the land near

FIGURE 3: Abandoned mine near Schefferville. Photo by Jean-Sébastien Boutet.

Schefferville: "We were already coming here before there was a mine. We came here to hunt. It was quite undulated here; there were beautiful mountains, beautiful rivers, beautiful lakes. It was nice here, before the mine."[56] In a similar vein, the same Innu who associates the mine with Martian landscape also emphasizes the profound changes to the territory compared with before industrialism, when people frequented the region for hunting and trapping activities. "When they came here ten years before [the mine], everything was pristine," he suggests. "Ten years later they made large holes. It's something to see the landscape where people hunted, and a decade later they see these big holes. It's something to see all that."[57]

For some individuals, then, the disfigured, deeply altered landscape evokes resentful memories of the company's actions. When probed about the overall record of IOC and its long-term legacy for Schefferville and the region, a few residents simply point to the impressive, scattered mining

FIGURE 4: Berms with warning sign bordering an abandoned mine near Schefferville. Photo by Jean-Sébastien Boutet.

pits as a kind of self-explanatory evidence for the disruption, the abandonment, and the reckless attitude of the mining managers:

> I would love to go up and show you over there in the mountains, all the holes that are there. I would show you how they left it all like that, and then ran away. It's still like that today.[58]

> They left, and after that the pits were flooded and everything. They just left everything.[59]

By contrast, though not necessarily contradiction, other individuals at times emphasize their awe at this surreal landscape, and even their appreciation of the physical and human immensity encapsulated by such scenes of emptiness (Fig. 5). Driving a truck with his grandson around the old mine, an elder repeatedly evoked his enjoyment of the landscape, whether we were passing by the rock waste and tailings, the open pits, or the surrounding lakes and rivers meandering through the abandoned sites.[60] Initially, these comments can seriously challenge one's own preconceived notions of another people's relationship with what is often denigrated as a kind of post-apocalyptic industrial wasteland. As an Innu from Matimekush seems to indicate, the mining landscape is indeed largely indicative of the ravage, squander, and unequal appropriation of resources by outside interests. But interestingly, these memories do not lead him to abandon or disdain the Innu territory around Schefferville. Rather, he makes use of powerful imagery and unusual metaphors in relation to the mining panorama to explain his own difficulty of living, as he understands it, "between two cultures"—or, of the struggle in sharing the land and livelihoods with strangers who, in the past, were mostly if not uniquely interested in reaping financial benefits from it.[61] As he set up his white canvas tent in the middle of the mined-out and desolate barrens, he seeks by this occasion to reaffirm his own Innu identity and demonstrate his people's enduring presence on their homeland: "It makes me feel good, to hang out here on the barren mountain. It doesn't discourage me. At least it tells me one thing: that even if I'm on a mountain without vegetation, I am on my territory."[62] Even as some areas of Innu and Naskapi homeland were pretty much obliterated by the mining activities, it does not follow that the territory has become insignificant or worthless for the people who continue to inhabit it and who frequently visit it under various circumstances.[63] Thus, reflecting and sharing stories about Schefferville, for this Innu man there is even a favourite time of the day to revisit the old mining sites. Right before sunset, the light reveals most effectively the scale of destruction and yet the strangely poetic beauty of this continuously modified, tirelessly engineered environment transformed by mine workers over more than a quarter century.

The decrepit industrial landscape, the various townsites (some of them now entirely derelict), and the corresponding diversity of life stories that are shared and kept alive thus interact in complex ways with people's present activities near and around Schefferville to create a fluid repository of interpretations of the longer past that include not only the closure but all phases of historical mining in the region. These place-based understandings do not seem to be strictly reducible to stories of

FIGURE 5: Abandoned mine near Schefferville. Photo by Jean-Sébastien Boutet.

appropriation, dispossession, destruction, and abandonment by an indifferent, perhaps egotistical, cast of mining characters. Several indigenous residents who continue to move extensively over the degraded territory, for example, do speak regretfully of the hazards associated with winter travels across a territory that was mined out and left virtually unremediated. At the same time, these very individuals may also continue to make use of the old mining roads to access the family cabins for hunting, fishing, or spending time on the land, while some also yearn—in the arduous context of a post-development economy—for the relatively plentiful jobs and active community life that the company once provided. If the dilapidation of the landscape and town continues to inflame people's discontent as they reminisce about the IOC company today, and if feelings of bitterness associated with past exploitation clearly linger (the mining holes are, in many ways, the physical embodiment of the "hit-and-run" model of resource exploitation that struck Schefferville at midcentury), these attitudes also coexist with subtler affirmations of personal and collective agency and, in some cases, of a vanished mining history that is in a sense still missed today.

The bare territory that typifies the former mine—repulsive, stunning, majestic, or mysterious—as well as certain abandoned areas of the municipality in fact continue to evoke nostalgic sentiments of a more active or entertaining past. When people revisit or come across former workspaces (an old shed, a rotten wooden bridge, a water-filled pit, and an overgrown road) they sometimes are led to share memories of a time when Schefferville was teeming with people and bustling with noise and activities. Such evolving human-material relations, grounded in memories of the changing landscape, do not only refer back to the harsh and unjust working conditions at the mine, the complicated coexistence of mining labour and activities on the land, or the difficult and fluctuating ecological conditions that coincided, according to local perceptions, with a disappearing caribou herd and the overfishing of surrounding lakes by workers brought in from the south. They also speak to an employment period that enabled working families to support their relatives and be involved in various ways with wage employment, and through which Innu and Naskapi employees sometimes nurtured close friendships with their fellow indigenous and even non-indigenous workers. One Innu evokes this sense of loss regarding the lively old days with friends that have either passed on or moved back to southern regions: "I knew many white people with whom I worked. I knew a whole bunch of them. Many of them have probably died. I also knew some bosses; at the end they were almost sixty-five years old. They must be dead today. It's troubling when I think about those I knew. They're almost all dead now."[64]

The abandonment of the town and some of the infrastructure near or on the reserves, as well as the emptiness of the sites that people of a younger generation formerly used for leisure activities, also call to mind sentiments of regret about a time when people actively participated in social and athletic activities. As he encounters the material leftovers dating back to the mining years, an Innu vividly reflects on his more youthful days when the company was in operation:

> When I think about the desolation here in the village, the desolation following the exploitation of the mine, when before the cultural and social life was so lively. It was lively here. My playground was everywhere, all the sidewalks that you see there; that was my playground. The hospital, for me, because it was asphalted, it was nice, there were some nice hills for someone who wanted to bicycle there. I could circulate freely when I lived in Matimekush.[65]

It is not uncommon for people in Schefferville, especially elders who used to work at the mine and befriended other workers, as well as adults who benefited from leisure and organized sport infrastructure in their youth, to remember elements of the social past with melancholy, given that many of these relationships and spaces died out soon after the IOC departure.

Overall, people's interpretations of the closure phase point to the failure of modernization through industrial mining at Schefferville. They also reveal the importance of considering long-term perspectives on the history of resource exploitation and the uneven development that characterized large-scale development in Québec's iron belt. Several individuals who live in the area today recall with much offence and distress the multiple rounds of infrastructure removal and the gradual disappearance of significant social spaces. They remain especially puzzled by the fact that a great number of houses, the cultural centre, and even a hospital were torn down during a period when there were outstanding needs in the domains of health, housing, infrastructure, social life, and youth support in Schefferville. As a former Naskapi leader explained to me, "they demolished the hospital, at a time when people really needed essential services, like health services. . . . Us, the owners of this territory, we were asking the government for more housing while the company that made money was demolishing houses. That scene was not respectful. It was a destructive scene."[66] The people of Schefferville painfully reminisce about the removal or abandonment of the swimming pool, the movie theatre, the bank, a few restaurants and bars, churches, the bowling alley, the town gymnasium, the ski hill infrastructure, and the asphalted roads and sidewalks (practically only the hockey arena was left standing). These mixed emotions of "anger, betrayal, resentment, exasperation and anxiety"[67] are quite a contrast with the official discourse that accompanied the closure of the mine, at the time largely justified to employees and the population at large in quite narrow technical and economic demonstrations.

For one Innu individual, the demolition exemplified the authorities' familiar paternalism toward indigenous populations, in this case fuelled by the belief that Innu and Naskapi communities could not possibly maintain and administer the buildings and town services without the backing of IOC, the municipality of Schefferville, or other government bureaucracies.[68] But he, as well as many others, believes that at a bare minimum, some of the installations could have been left for the local communities as compensation, and even new infrastructure might have been built for them upon IOC's departure. In the demanding context of the reconstruction years, when local residents attempted to carry on with their lives independently of an industrial economy, many maintain that the company should have been compelled to implement a mitigation program and leave behind some sort of positive legacy in order to offset the heavy social and environmental costs, perhaps in the form of roads to facilitate access to the territory and support hunting activities, financial donations to help out with the expensive airplane outings and especially the caribou hunt, or at the very least a proper rehabilitation and restoration of the mining sites.

DISCUSSION

> Et voilà que, trente ans plus tard, ces gens-là reviennent! C'est drôle, il y a seulement un an ou deux, ma communauté et moi n'existions pas dans le nord du Québec!
> —Réal McKenzie[69]

In light of these perspectives on mining, development, and history pertaining to the initial phase of extraction in the Labrador Trough, several concerns arise regarding the legitimacy and the soundness of the mode of industrial expansion favoured by *Plan Nord* and other such bureaucratic plans for the north. Nowadays, a new generation of miners actively looking to revive the historical deposits and bank on new discoveries cannot completely silence this problematic past,⁷⁰ especially the unfortunate period that followed the closing of mining pits and the abandonment of villages and communities like Schefferville in the 1980s. This history is, in a very real, everyday sense, etched into the land, remembered in a manner that, as I have argued, is deeply textured. "Mines," in the words of anthropologist Jamon Alex Halvaksz II, "are not merely extracting minerals, but are also marking time and space with their appearances." They "transform the landscape, but these transformations remain subject to multiple interpretations."⁷¹ But, ready as they may be to turn the page on this story, industrialists of today want to be reassuring about the future, confidently asserting that previous development errors will not be reproduced. As one mining executive explicitly defended, "we cannot underestimate this history, which belongs to the region. We certainly took it into account, to ensure that we do not repeat the same mistakes."⁷²

Much as was the case on the eve of the iron ore rush that took place at the midcentury, when federal bureaucrats encouraged Innu and Naskapi societies to adopt industrial livelihoods, nowadays mining-related employment is perceived and marketed—despite of all of its known historical shortcomings—as the device par excellence to engineer social, economic, and ecological progress for the region. According to this view, a new "'home grown' generation of people who will regenerate the mining industry in Quebec" is about to emerge: indigenous peoples inhabiting the region "will now see a brighter future thanks to Plan Nord; especially the grade schoolers who will learn more and more about mining as they continue their education," as well as "all Quebecois [who] will be given a chance to cash in on some of the province's fortunes."⁷³ While affirming that "women, the Aboriginal peoples and young people living in the territory that the Plan Nord covers are among the target populations to develop qualified local workers who take their place in sectors that are often non-traditional or little known,"⁷⁴ government planners nevertheless give little consideration as to how these communities' involvement with other sectors of the economy referred to as "traditional"—typically outside of formal, monetized networks of production and exchange—will be encouraged and supported. This approach in effect disregards the historical strategies used by indigenous groups such as the Schefferville Innu and Naskapi who, as I have explained elsewhere, often adjusted their practices to harmonize as best as possible their work at the mine with the crucially important life on the land.[75]

As for the contemporary mining industry, it generally sees "the development of Plan Nord [as] a very proactive initiative at the right time,"[76] in particular because of the considerable government resources assigned to the implantation of modern transportation and power networks. In this regard, the province's industrial expansion into the subarctic will be largely financed by Hydro-Québec and the public treasury ($60 billion is expected to be allocated to northern development in the next quarter century). With at least $1.2 billion reserved specifically for infrastructure upgrades and new construction during the initial five-year period (2011–16), "the government will first invest in projects that afford access to areas with the greatest economic development potential" in the domain of mining and energy.[77]

Given that miners never exhausted its ore reserves, it is not surprising that the Labrador Trough still holds much potential provided the economic conditions are favourable. As the region stands on the cusp of becoming "the gateway to northeastern Quebec, helping to play a vital role in Plan Nord,"[78] mining and steel conglomerates are knocking on the door, hoping to redeploy in the area with renewed intensity. Three large mining companies currently control roughly 40 per cent of the iron ore production worldwide and almost 80 per cent of the seaborne export trade; these statistics alone suffice to show the rapid consolidation of the sector, given that this triumvirate, Vale (headquartered in Rio de Janeiro), BHP Billiton (Melbourne–London), and Rio Tinto (London–Melbourne), was responsible for less than half of the seaborne trade in 1997.[79] From this veritable iron cartel emanates a strict control over ore prices, which have surged ninefold since 2000.[80] This phenomenon has created "major cost-inflation pressures"[81] and is a source of irritation for governments and steelmakers, whose leverage dwindles with growing demand for base metals in China—a country that now produces nearly half of the global crude steel output.[82]

From the perspective of the steelmaking industry—never entirely powerless thanks to enormous capital resources, a high level of corporate concentration, and the fact that, in the final analysis, the dependency relationship with the mining industry is reciprocal—a clever solution can serve to counter this hegemony: the vertical integration of enterprises. For the Labrador Trough, this ownership of upstream supplier firms by downstream producers represents an interesting return to a bygone era, since the period that preceded the iron crisis of the 1980s in Nouveau-Québec was in effect characterized, as the first section indicated, by the systematic acquisition of captive mining sites by large American steel interests. As the mining and steel industries formalize their integration, a restructuring of the industry is clearly looming on the horizon (if in fact it hasn't already begun, in particular among state-owned Chinese steelmakers[83]). It cannot be ruled out that, as a result of these companies consolidating their hold on the territory and the ore reserves, the Labrador Trough may once again be closely tied to the progress of distant urbanization and global consumer society.

Certainly the most prominent actors have changed over the course of sixty years. Bethlehem Steel and National Steel closed their doors at the turn of the twenty-first century and have been replaced by new multinational producers. The arrival of ArcelorMittal (Luxembourg), one of the leading steel-mining conglomerates in the world,[84] in the region of Fermont is symptomatic of this evolution. Already in command of the largest mine in Québec, with operations set up at the historical deposits of Mont-Wright and Fire Lake, the company recently divulged a massive expansion program evaluated at more than $2 billion. Forecasting a "breakneck" pace of development, ArcelorMittal seeks to increase substantially the capacity of its processing plant and magnify by a factor of two its ability to move ore and waste, notably with the support of an impressive fleet of 400-ton Caterpillar 797 trucks, the largest available on the market.[85] According to a company manager, "the commitment of the Quebec government" to the reindustrialization of Nouveau-Québec was a decisive factor in ArcelorMittal's decision to allocate this capital to the Mont-Wright expansion, making it clear that without Plan Nord, "the money might have gone to the United States, Mexico, Brazil or any of a number of African and European operations."[86]

At the earlier stages of both embryonic production and advanced exploration, Indian multinational Tata Steel (Mumbai) is a second potentially important player on the Québec portion of the iron ore trough. This Asian steelmaker, which figures among the ten largest producers in the world, is looking to secure "strategic captive iron ore" to supply its transformation operations in Europe.[87] In partnership with Calgary-based New Millennium Iron (NML), Tata Steel owns 80 per cent of a direct shipping ore project already in production near Schefferville,[88] a development that proposes to reactivate existing historical deposits, notably in the Timmins area which straddles both sides of the Québec–Labrador border.[89] Pursuing a long tradition of appropriating Innu and Naskapi territory in the name of corporate interests, NML recently renamed the geological formation stretching over more than two hundred kilometres west of Schefferville the *Millennium Iron Range*, "a huge iron ore district" over which the company now claims "control."[90] Tata is also studying the possibility of formalizing other investment partnerships with NML, with the goal of exploiting the much more imposing taconite deposits—these veritable company builders that rank among the largest iron deposits in the world, according to NML[91]—of Lac Harris (KéMag) and Howells River (LabMag). Pending the outcome of the ongoing feasibility study, the two partners hope to make use of the same railway installations constructed by QNS&L in the early 1950s, in addition to envisaging the laying of a 600- to 700-kilometre-long slurry "ferroduct" designed to transport fine grained concentrate to the port of Sept-Îles.[92] Necessary upgrades intended to transform the port installations into state-of-the-art shipping facilities are expected to be financed by a public-private partnership (several companies, including NML and Tata, are involved as part of an investment consortium) supported by the recently launched Atlantic Gateway and Trade Corridor Strategy, a federal initiative meant to "provide a quick, reliable, and secure transportation network between North American markets and markets in Europe, the Caribbean, Latin America, and Asia."[93]

A third steel giant is presently deploying its operations in the region. Already active as a minority partner on the Newfoundland side of the southern Labrador Trough, at the Bloom Lake mine and concentrator, WISCO International Resources Development & Investment (WISCO), a subsidiary of Chinese steelmaker Wuhan Iron & Steel Corporation, is at the helm of a longer-term, more ambitious project, also at the advanced exploration stage, which "centres on a huge iron deposit."[94] With the objective of exploiting the Lac Otelnuk formation 170 kilometres north of

Schefferville, the steelmaker entered a joint venture with Toronto-based junior miner Adriana Resources, and together they expect "to make the Otelnuk project nothing less than the biggest mine in Canadian history."[95] To justify the enormous investments required, Adriana's president and chief economic officer has suggested that "if we're right in our estimates, the mine life will be in excess of 100 years,"[96] an expectation echoed by the Québec government.[97]

Alerted by these clear echoes of past mining discourses in the region, obvious questions come to mind: To what extent have things effectively changed? Are observers and local communities confident that similar development mistakes will not be repeated? After all, Québec's *Plan Nord* assures that, in contrast with the previous phase of northern development, benefits will materialize "for all Quebecers" this time around, thanks in part to the creation of many new employment opportunities, particularly for indigenous labourers.[98] In order to fill these positions locally, "the objective is to ensure that the workers are ready to work when the projects are launched," which means that "Aboriginal and local communities [must] participate rapidly in the process that leads to the acquisition of the desired skills."[99] This government strategy mirrors the approach favoured by mining companies, who strive not only to hire Aboriginal individuals already settled near industrial sites, since this allows them to save "a lot of money"; but also, in light of the unfolding "cultural revolution" of corporate responsibility that is taking place in the industry, to secure a social licence to develop and operate mines in the region.[100] As to distant Asian and European steelmakers, they find themselves in need of "Canadian management" and administrators who "are used to dealing with Aboriginals, [who are] used to working in the north,"[101] in order to help them navigate the complexities inherent to the establishment of a mine in remote and isolated indigenous territories.

Apart from technical accounting discussions regarding modest alterations to royalty regimes, public officials, industry representatives, and Québec civil society have generally not engaged, however, in a fundamental rethinking of the business model—the common "dig-and-sell" paradigm[102]—that has guided, and continues to guide, mining development in the provincial north. If the means to carry out this overhaul remain to be invented and implemented, some industry observers, authors, and critics have theorized and in some cases empirically examined possible alternatives.[103] In economic terms, they have proposed:

- the creation of a resource rent tax and an associated sovereign wealth fund;
- revenue sharing with First Nations, Métis, and Inuit groups on whose land mineral development is occurring;
- regional economic diversification, in particular through the establishment and support of a balanced industrial base and smaller scale, revenue-generating ventures;
- the development of local and/or state-owned manufacture industries;
- a more careful examination and potential veto over foreign mergers and takeovers;
- government and/or Aboriginal control over the number, size, and scale of concurrent mining projects, and the staggering of operations over longer time horizons;
- transparency and systematic publishing of statistical information pertaining to industrial mining operations, including financial agreements and corporate taxes paid;
- support and development of the land-based and social economies of the north.

In the environmental realm, analysts have argued for:

- supervision of the weakly regulated mineral exploration industry and in particular, replacement of the free entry licensing system;
- stricter environmental permitting and the implementation of cumulative impact assessments and follow-up monitoring;

- close evaluation of rehabilitation and restoration plans, posting of steeper financial assurances, and more stringent auditing and certification processes related to mine closure;
- greater government oversight over land permits, implementation of integrated land management, and creation of zones of exclusion from development;
- comprehensive recycling, energy efficiency, and consumption reduction programs, including the fight against planned obsolescence.

Finally, in the social and political domains, experts have highlighted the potential for adapted work and training programs such as job-sharing arrangements for indigenous workers and, perhaps most importantly, active and meaningful participation of communities not only at the notification and assessment stages but also through the prospection, exploration, design, construction, implementation, monitoring, follow-up, and remediation, and also the partial ownership of industrial projects that have secured local support. These policy proposals have not really garnered serious considerations from regulatory authorities or the industry, which are generally intent on exploiting the region as rapidly and efficiently as possible so long as the social licence has been obtained.

CONCLUSION

By the mid-twentieth century, the "exploration" and gradual abrogation of indigenous homelands situated north of the forty-ninth parallel was already a long-standing phenomenon in the Québec-Labrador peninsula. These territories were travelled as early as 1578 by European crews looking for the Northwest Passage, before being visited, between the seventeenth and nineteenth century, by a series of missionaries, English and French traders, and, toward the end of that period, exploration parties assigned to diverse scientific and geological duties. As we have seen, the colonial process accelerated through the mining boom of the postwar period, when powerful American mining and steel interests formed corporate alliances, under the auspices of a generally proactive, interventionist state, to lay their hands onto the strategic iron ore reserves in the Labrador Trough. The end result turned out to be deleterious for the regional economy at large and for the well-being of local indigenous populations.

The initial search for valuable iron deposits in the interwar years led to the opening of the Ungava region to intensive mineral development, in part thanks to the provincial government's espousal of a liberal economic program that limited the ability or willingness of the state to sustain an endogenous manufacturing sector linked to primary extractive activities.[104] But, contradicting their own laissez-faire principles, government administrators adopted a hands-on approach to produce a mineral policy largely favourable to corporate interests. This state-sanctioned support of industry facilitated the incursion of massive foreign capital into the North, with the launching of modern infrastructure projects, notably in the domains of energy, mechanical transportation, and urban planning, and the consequent appropriation of indigenous homelands.

In the early 1980s, the development of the resource-dependent, mature mining municipality of Schefferville, located at the heart of the Labrador Trough, went tumbling to a sudden yet brutal crash. Despite the challenges thrown at the local communities to survive the death of their unique industry, many people, including most members of Schefferville's two indigenous groups, resolved to pursue their lives in this economically frail region. As authorities worked rather ineptly to promote alternatives to mineral production that could rescue the regional economy, Innu and Naskapi residents were forced to reorganize their livelihoods through the region's deindustrialization phase.

The post-closure phase did not come without immense obstacles and uncertainties, as one of the main sources of employment for the Innu and the Naskapi in Schefferville vanished virtually overnight. Indigenous residents remember with particular bitterness the difficulties associated with mine closure, holding both IOC and government authorities responsible for the failure to deliver on their promises of long-term prosperity for their communities. Notwithstanding their very real grievances and misfortunes, the dismantling of a wage economy at Schefferville implied that people had to find, much as during the preceding development and production years, alternate ways to make a living—only this time around, literally outside of an evanescent mining world that, for better or for worse, stopped exerting its overwhelming influence on the local people, their economies, and their environments.

In this chapter, I suggest that the mine's influence never fully disappeared from the memories, the imagination, and even the lived experiences of Innu and Naskapi residents. At Schefferville the past is certainly made, as Halvaksz illustrates in a different context, of the more constructive "aesthetic qualities of mineral extraction (the attractive constructions of town life, mining equipment, roads, etc.)."[105] Yet the mine also exhibits, to borrow an evocative metaphor from Sandlos and Keeling, a zombie-like character, where the industrial sites "continue to exert some sort of malevolent effect during their afterlife."[106] I contend that the undead nature of the mine is a powerful legacy of industrial extraction in the region.

Indeed, the past figured prominently in the minds of the Innu individuals who erected mine and rail barricades in Schefferville in the summer of 2010, contesting the redevelopment of the nearby iron deposits. For Innu leader Réal McKenzie, the lessons and details of this history were a central motivation to this action, as people engaged in the protest asserted that they did not "want to live the IOC story again."[107] In addition to physical blockades, the Innu communities of Matimekush–Lac John (Schefferville) and Uashat mak Mani-utenam (Sept-Îles) have launched a $900 million lawsuit against IOC/Rio Tinto to seek financial compensation for historical and ongoing damages resulting from the "colonization and dispossession" of their ancestral territories.[108] Through this judicial action, the Innu are hoping to recoup some of the profits earned by the IOC since 1954, thanks in part to company infrastructure such as the QNS&L railway, which, according to the Innu groups who filed the proceedings, continue to "violate their ancestral rights."[109] This prosecution should serve as an unambiguous reminder that, at the dawn of a new mining cycle and amidst the resource boom, these past mining developments continue to represent a deep historical wound for some societies that are about to experience the revival of the iron ore industry in Québec. In contrast with the dominant optimism that has seized the region, one Innu from Schefferville anticipates the future in his community in rather cautious terms, guided by his long experience with mining activities: "What do I see in twenty-five years? . . . I see the closure of this mine that is opening today. Then we will live through a second closure. They will say: 'Schefferville, goodbye.'"[110] By failing to engage the public and indigenous communities in a comprehensive consultation process and mineral policy overhaul, it is precisely this multifarious story (of colonization and dispossession) that public officials and corporate executives have not heard while planning to reindustrialize subarctic Québec.

ACKNOWLEDGMENTS

I would like to thank the Innu and Naskapi individuals from the regions of Sept-Îles and Schefferville who agreed to participate in the interviews and helped out with the project; as well as the Social Economy Research Network of Northern Canada, the Social Sciences and Humanities Research Council, and the Abandoned Mines Project led by Arn Keeling and John Sandlos.

NOTES

1. "Your ancestors have led you here to recount the images of your dreams." Québec, "Plan Nord: Building Northern Québec Together, The Project of a Generation" (Ministère des Ressources naturelles et de la Faune, 2011), xv. All translations from French to English are mine.

2. See, for example, Caroline Desbiens, "Défricher l'espace de la nation: lieu, culture et développement économique à la baie James," *Géographie et cultures* 49 (2004): 87–104.

3. Ibid., 6.

4. Alisha Hiyate, "The New Normal," *Mining Markets* 4, no. 3 (September 2011): 5.

5. In this chapter, I am concerned primarily with mineral developments occurring on the Québec portion of the Labrador Iron Trough.

6. Ocean Equities, "Iron Ore: The Labrador Trough" (London: Ocean Equities, January 18, 2013), 1.

7. David Ebner and Brenda Bouw, "Tata Joins Race for Canada's Iron Ore," *Globe and Mail*, March 8, 2011.

8   According to Markey, Halseth, and Manson, the concept of "resource bank" refers "to the practice of using the vast resource wealth of the hinterland for the purposes of either province building or supporting infrastructure and service spending within the metropolitan core" ["Challenging the Inevitability of Rural Decline: Advancing the Policy of Place in Northern British Columbia," *Journal of Rural Studies* 24 (2008): 412]. This situation "rarely results in significant investments in those regions from which resources were extracted" [Matthew Tonts, Kirsten Martinus, and Paul Plummer, "Regional Development, Redistribution and the Extraction of Mineral Resources: The Western Australian Goldfields as a Resource Bank," *Applied Geography* 45 (2013): 366].

9   "Americans own the capital that we lack. They enhance the value of our natural resources, awakening the sleeping wealth." Cited in Pierre Paquette, *Les mines du Québec, 1867–1975: une évaluation critique d'un mode historique d'industrialisation nationale* (Outremont: Carte blanche, 2000), 270.

10 Richard Geren and Blake McCullogh, *Cain's Legacy: The Building of Iron Ore Company of Canada* (Sept-Îles: Iron Ore Company of Canada, 1990), 31.

11 Ibid., 39.

12 Donna Yoshimatsu, "The Legacy of the Rail Lives On, But Could It Be Built Today?" *Canadian Mining Journal* 130, no. 5 (June/July 2009): 8.

13 Paquette, *Les mines du Québec*, 133, 143; Henri Dorion and Jean-Paul Lacasse, *Le Québec: territoire incertain* (Sillery: Les éditions du Septentrion, 2011), 326.

14 Paquette, *Les mines du Québec*, 133. In metric terms, the exploitation lease was equivalent to close to 777 square kilometres.

15 Up to 1958, IOC shareholders consisted of mining companies Hanna Mining (17.48%), Hollinger North Shore Exploration (12.62%), Hollinger (8.09%), M. A. Hanna (8.09%), and Labrador Mining & Exploration (6.47%); and steel companies Republic (16.18%), National (12.95%), Armco (6.47%), Youngstown (6.47%), and Wheeling-Pittsburgh (5.18%) (Geren and McCullogh, *Cain's Legacy*, 339).

16 See, for example, "Dépense initiale de 200 millions: 'Fortune' dit que l'on devra dépenser cette somme avant de retirer une tonne de minerai," *Action Catholique* (November 26, 1948), Fonds Ministère des Travaux publics et de l'Approvisionnement, file 52.1 (1947–65), E25, S105, SS1, SSS2, box 1960-01-41/704, Bibliothèque et Archives nationales du Québec.

17 Québec deed of sale number 5851 executed before John P. Rowat, Notary (Montréal, March 22, 1951), Fonds Ministère des Travaux publics et de l'Approvisionnement, Bibliothèque et Archives nationales du Québec.

18   Daniel Vachon, *L'histoire montagnaise de Sept-Îles* (Québec: Éditions Innu, 1985), 42–43.

19   Personal communication (Kawawachikamach, 2009: N-23a). Each Innu and Naskapi interviewee is anonymously identified with a unique numerical value that is consistent with other related works. The author conducted the interviews in the Schefferville region in the fall of 2009.

20   Personal communication (Kawawachikamach, 2009: N-25b).

21   Maurice Duplessis, cited in Évènement Journal, "Dès 1953 un train pour l'Ungava!," October 25, 1952, Fonds Ministère des Travaux publics et de l'Approvisionnement, Bibliothèque et Archives nationales du Québec.

22   Québec, *Arrêté en Conseil numéro 222 concernant le développement du Nouveau-Québec, et le progrès de la province*, Chambre du Conseil exécutif (March 7, 1951), Fonds Ministère des Travaux publics et de l'Approvisionnement, Bibliothèque et Archives nationales du Québec.

23   This percentage averaged 1.60 per cent for the pre-Schefferville decade 1945–1955 and 1.26 per cent for the following ten-year period, 1955–1965 (Paquette, *Les mines du Québec*, 177).

24   Ibid., 173, 190.

25   Ibid., 190.

26   In 1970, the value of iron ore production accounted for roughly one-sixth of the total value of production in Québec; by that measure it was over five times more important than gold production and about two-thirds as valuable as copper, the leading sector in the provincial mining economy. By 1975, iron ore was the leading sector in terms of value of output (Ibid., 14).

27   Ibid., 191.

28   Ibid., 66.

29   J. D. Jorgenson, "Challenges Facing the North American Iron Ore Industry" (Reston, VA: US Geological Survey, 2006), Open-File Report 2006-1061; Canada, "Iron Ore in Canada, 1886–1986," Mineral Policy Sector, Internal Report, MRI 88/2, ca. 1988, 3. I am indebted to John Thistle (Memorial University) for the latter reference.

30   John H. Bradbury, "Some Geographical Implications of the Restructuring of the Iron Ore Industry, 1950–1980," *Tijdschrift voor Economische en Sociale Geografie* 73, no. 5 (1982): 295.

31   Paul Sukagawa, "Is Iron Ore Priced as a Commodity? Past and Current Practice," *Resources Policy* 35, no. 1 (March 2010): 54. Sukagawa makes this characterization for the period that led to the late 1960s and the development of the Pilbara iron mines in Australia.

32   Canada, "Iron Ore in Canada, 1886–1986," 3.

33 Paquette, *Les mines du Québec*, 204.

34 In terms of this commodity, between the years 1955 and 1975 Québec exported "the totality of its production outside of its borders," 70 per cent of which shipped to the United States (ibid., 137–38).

35 Ibid., 258.

36 Ibid., 250. In 1944, the added value in the iron ore and steel industries accounted for 20 per cent of the country's total, with 9,760 jobs linked to the sector; by 1975, it had dropped to 11 per cent and 9,517 jobs.

37 In association with British Steel Corporation and Quebec Cartier Mining (an entity held by US Steel), Sidbec formed the company Sidbec-Normines, a half-state entity whose goal was to exploit and process the rich iron deposits of Fire Lake located near the municipality of Gagnon.

38 Bradbury, "Some Geographical Implications," 301–2.

39 *La ville de Gagnon*, directed by Christian Sénéchal and Hélène Brown (Plate-Formes-Prods, 2008), film, 19 min. 11 s.

40 Michel Nadeau, "Iron Ore ferme à Schefferville," *Le Devoir*, November 2, 1982, Fonds Cercle de presse de Sept-Îles, Le Soleil-Ouellet-Fessou (1960–85), P15, S2, dossiers 138 à 150.6.2, contenant 1982-11-002/3, Bibliothèque et Archives nationales du Québec.

41 Marie Tison, "Dossier de la fermeture de Schefferville: les Montagnais blâment Québec et Ottawa," *Le Soleil*, April 13, 1988, Fonds Cercle de presse de Sept-Îles, Le Soleil-Ouellet-Fessou (1960–85), P15, S2, dossiers 138 à 150.6.2, contenant 1982-11-002/3, Bibliothèque et Archives nationales du Québec.

42 Natural Resources Canada, Legal Surveys Division, *Historical Review–Matimekosh* (undated), 86–87, accessed December 1, 2014, http://clss-satc.nrcan-rncan.gc.ca/data-donnees/publications/indlanhisque-hisfonterindque/matimekosh_ang.pdf.

43 Personal communication (Kawawachikamach, 2009: N-23a).

44 Personal communication (Matimekush, 2009: I-17b).

45 Personal communication (Kawawachikamach, 2009: N-23a).

46 Personal communication (Matimekush, 2009: I-17a).

47 Personal communication (Schefferville, 2009: 1-29b).

48 Personal communication (Kawawachikamach, 2009: N-25a).

49 Ibid.

50 John H. Bradbury, "Towards an Alternative Theory of Resource-Based Town Development in Canada," *Economic Geography* 55, no. 2 (April 1979): 155.

51 André Bourbeau, "Projet de mémoire portant sur l'opportunité de la fermeture de la ville de Schefferville," report to Conseil des ministres (Québec,

March 4, 1986), Fonds Cercle de presse de Sept-Îles, Le Soleil-Ouellet-Fessou (1960–85), P15, S2, files 138 to 150.6.2, box 1982-11-002/3, Bibliothèque et Archives nationales du Québec.

52 Ibid.

53 Marc St-Pierre, "Fermer Schefferville: une question d'argent," May 14, 1986, Fonds Cercle de presse de Sept-Îles, Le Soleil-Ouellet-Fessou (1960–85), P15, S2, dossiers 138 à 150.6.2, contenant 1982-11-002/3, Bibliothèque et Archives nationales du Québec.

54 See Paquette, *Les mines du Québec*, 127.

55 Richard Laforest, Jacques Frenette, Robert Comtois, and Michel Mongeon, "Occupation et utilisation du territoire par les Montagnais de Schefferville" (Rapport au Conseil Attikamek-Montagnais, Village des Hurons, 1983), B74.

56 Personal communication (Schefferville, 2009: I-28a).

57 Personal communication (Matimekush, 2009: I-16a).

58 Personal communication (Matimekush, 2009: I-17b).

59 Personal communication (Kawawachikamach, 2009: N-24c).

60 Personal communication (Matimekush, 2009: I-18a).

61 Personal communucation (Matimekush, 2009: I-16a).

62 Cited in *Une tente sur mars*, directed by Martin Bureau and Luc Renaud (Productions Thalie and Les Films, March 3, 2009), film, 58 min. 06 s.

63 See Laura Cameron, *Openings: A Meditation on History, Method, and Sumas Lake* (Montréal: McGill-Queen's University Press, 1997), 18. In the context of large industrial projects that contributed to the destruction of a socially and culturally significant lake in British Columbia, Cameron seeks to grapple with the complicated array of local interpretations related to these developments: "Yes, Sumas Lake had been drained, but that hardly proved that it and the surrounding floodlands were valueless to all the people who lived there."

64 Personal communication (Matimekush, 2009: I-28c).

65 Personal communication (Matimekush, 2009: I-15a).

66 Personal communication (Kawawachikamach, 2009: N-23a).

67 I borrow this expression from Pini, Mayes, and McDonald, who use it in the context of the January 2009 closure of the Ravensthorpe nickel mine in Western Australia. The authors argue that "for the participants of our study the mine's end was a highly emotional event," which was "marked by a palpable sense of loss," despite the fact that it "was primarily represented as an economic and industrial issue" ["The Emotional Geography of a Mine Closure: A Study of the Ravensthorpe Nickel Mine in Western Australia," *Social & Cultural Geography* 11, no. 6 (September 2010): 570].

68 Personal communication (Matimekush, 2009: I-15a).

69 "And now, 30 years later, these people are coming back! It's funny, only one year or two ago, my community and I did not exist in northern Québec!," Réal McKenzie, "En finir avec la discrimination," *Recherches amérindiennes au Québec* 41, no. 1 (2011): 73.

70 "Ahead of the Curve: Working with Aboriginal Partners in the Race for Canada's Iron Ore," *Canadian Mining Journal* 133, no. 3 (April 2012): 26–27.

71 Jamon Alex Halvaksz II, "Whose Closure? Appearances, Temporality, and Mineral Extraction in Papua New Guinea," *Journal of the Royal Anthropological Institute* 14 (2008): 21–22.

72 Cited in Raymond St-Pierre and Dominique Landry, "Schefferville: les défis humains du Plan Nord," *Radio-Canada*, television, April 17, 2012.

73 Russel B. Noble, "La hommage [sic] au 'Plan Nord,'" *Canadian Mining Journal* 133, no. 3 (April 2012): 5.

74 Québec, "Plan Nord," 38.

75 See, for example, Jean-Sébastien Boutet, "Développement ferrifère et mondes autochtones au Québec subarctique, 1954–1983," *Recherches amérindiennes au Québec* 40, no. 3 (2010): 35–52.

76 Vice president of an exploration company, cited in Fred McMahon and Miguel Cervantes, "Fraser Institute Annual Survey of Mining Companies, 2011/2012" (Vancouver: The Fraser Institute, February 2012), 43.

77 Québec, "Plan Nord," 111.

78 Correy Baldwin, "Future Growth Built on Iron Ore Legacy," *CIM Magazine* 6, no. 2 (2011): 42.

79 Baffinland Iron Mines, "Iron Ore Industry Trends and Analysis" (August 31, 2009), 7, 18.

80 Jeffrey D. Wilson, "Chinese Resource Security Policies and the Restructuring of the Asia-Pacific Iron Ore Market," *Resources Policy* 37, no. 3 (September 2012): 331. Depending on the conditions, these dominant companies are also able, through overproduction and market flooding, to depress prices enough to stamp out competition from small iron producers.

81 Ibid.

82 World Steel Association, "Crude Steel Production 2013," accessed December 1, 2014, http://worldsteel.org/statistics/crude-steel-production.html.

83 See Wilson, "Chinese Resource Security Policies."

84 ArcelorMittal, "Core Strengths, Sustainable Returns: Annual Report 2011" (Luxembourg, April 2012), 7, 15.

85 Marilyn Scales, "Big Money: Mont-Wright Expansion to Pump $2.1 Billion into Quebec," *Canadian Mining Journal* 133, no. 3 (April 2012): 14–15.

86 Ibid., 13–14.

87   Dean Journeaux, "Breaking New Ground in the Labrador Trough," *Northern Exposure* presentation (St. John's, NL, January 22–24, 2013), 8.

88   In October 2010, Tata Steel (80%) and New Millennium Iron (20%) formed the Tata Steel Minerals Canada (TSMC) joint venture. The "Canadian" company "is part of Tata Steel Group of companies" headquartered in Mumbai (Tata Steel Minerals Canada, "Welcome to Tata Steel Minerals Canada Limited," accessed December 1, 2014, http://www.tatasteelcanada.com).

89   This site should not be confused with the Timmins mining district of northeastern Ontario; new reserves are also being identified in the area.

90   Journeaux, "Breaking New Ground," 6.

91   "New Millennium: Advancing the New Millennium Iron Range," *The International Resource Journal* 6, no. 3 (June 2011): 123.

92   New Millennium Iron, "On the Path to Production," corporate presentation (September 2011), 68.

93   Canada, "Speaking Notes for the Honourable Denis Lebel, Minister of Transport, Infrastructure and Communities" (annual conference of the Association of Canadian Port Authorities, Sept-Îles, QC, August 8, 2011).

94   Québec, "Plan Nord," 63.

95   Ibid. In 2011, the project was listed as the most expansive mining endeavor on the planet, all commodities considered [Magnus Ericsson and Viktoriya Larsson, "E&MJ's Annual Survey of Global Mining Investment," *Engineering and Mining Journal* (January 2012): 6].

96   Cited in D'Arcy Jenish, "Destined for Grandeur: Chinese Steel Giant Invests Heavily in Quebec Ore Property," *Canadian Mining Journal* 133, no. 3 (April 2012): 16.

97   Québec, "Plan Nord," 63.

98   Ibid., 117.

99   Ibid., 36.

100   St-Pierre and Landry, "Schefferville."

101   Robert Martin, chairman of the Strategic Advisory Committee of the Board of Directors, NML, cited in Ebner and Bouw, "Tata Joins Race."

102   See Timothy Prior, Damien Giurco, Gavin Mudd, Leah Mason, and Johannes Behrisch, "Resource Depletion, Peak Minerals and the Implications for Sustainable Resource Management," *Global Environmental Change* 22, no. 3 (2012): 585.

103   Ugo Lapointe, "L'héritage du principe de *free mining* au Québec et au Canada," *Recherches amérindiennes au Québec* 40, no. 3 (2010): 9–25; Ciaran O'Faircheallaigh and Ginger Gibson, "Economic Risk and Mineral Taxation on Indigenous Lands," *Resources Policy* 37, no. 1 (2012): 10–18; Prior et al.,

"Resource Depletion, Peak Minerals," 577–87; Vérificateur général du Québec, "Report of the Auditor General of Québec to the National Assembly for 2008–2009," Volume 2, April 1, 2009. This list is non-exhaustive.

104   See, for example, Paquette, *Les mines du Québec*, 271.

105   Halvaksz II, "Whose Closure?," 26.

106   John Sandlos and Arn Keeling, "Zombie Mines and the (Over)burden of History," *Solutions Journal* 4, no. 3 (June 2013): 81.

107   McKenzie, "En finir avec la discrimination," 72.

108   Radio-Canada, "Côte-Nord: des Innus poursuivent IOC pour 900 millions," March 20, 2013, accessed December 1, 2014, http://www.radio-canada.ca/regions/est-quebec/2013/03/20/010-ioc-innus-poursuite.shtml. See also the IOC/Rio Tinto "Pay the rent" campaign website (http://www.paytherent.info/). In 2000, Rio Tinto became the principal shareholder of IOC.

109   Sylvain Laroque, "Des Innus poursuivent IOC pour 900 millions," *La Presse*, March 20, 2013, accessed December 1, 2014, http://affaires.lapresse.ca/economie-et-ressources/201303/20/01-4633042-des-innus-poursuivent-ioc-pour-900-millions.php.

110   Essimeu Tite McKenzie, cited in Bureau and Renaud, *Une tente sur Mars*.

| CHAPTER 7

Indigenous Battles for Environmental Protection and Economic Benefits during the Commercialization of the Alberta Oil Sands, 1967–1986

Hereward Longley

INTRODUCTION

Since the late 1990s, the Alberta oil sands industry has become an economic powerhouse that employs thousands of indigenous and non-indigenous people, generates billions of dollars of economic activity, and produces over two million barrels of oil per day. However, it has also become the source of controversy and disputes over environmental impacts that include large-scale landscape disturbance and wildlife habitat destruction due primarily to open pit mining, atmospheric pollution, carbon dioxide emissions, and watershed pollution that may be related to high cancer rates in downstream communities.[1] Many indigenous people in the region have viewed government regulators as negligent in considering the impact of oil sands development on their traditional lands, treaty rights, and lives.[2] Though there is a substantial historical literature on hydrocarbon development in Alberta, there remains a shortage of research into social and environmental impacts and conflicts and the consequences of the initial development phase of the oil sands industry for indigenous communities.[3] This chapter demonstrates that while the environmental impacts of oil sands activities on indigenous communities are often understood to be recent controversies, they are contemporary manifestations of issues that first emerged during the initial commercial development phase of the oil sands industry from the late-1960s to mid-1980s.

The early impacts of the Great Canadian Oil Sands Ltd. (Suncor) and Syncrude oil sands operations were most acutely felt in the closest community, Fort McKay, located approximately twenty kilometres downstream on the west bank of the Athabasca River. Strip mining, atmospheric emissions, watershed contamination, and population increases from incoming workers and the industries that support large-scale synthetic oil production caused an array of adverse impacts on proximate ecosystems and undermined the capacity of the Fort McKay community to continue their hunting, trapping, and food gathering practices. In addition to these impacts on resources, Fort McKay was also left out of the employment opportunities and other economic benefits of industrialization. The Alberta government and the oil sands industry had minimal regard for indigenous peoples in the 1970s and 1980s, focused as they were on the rapid production of oil and dismissive of indigenous concerns as a federal responsibility. By the 1980s, the Fort McKay community was forced to respond to the environmental issues associated with oil sands development and their economic exclusion from the new industrial economy. To assess these issues, I use an approach drawn from several works on the history of resource development and indigenous people in Northern Canada, particularly those which call for a critical examination of the agency of indigenous peoples to shape and influence the colonizing forces of industrialization and the encroachments of western institutions.[4] As with many northern indigenous communities, Fort McKay representatives attempted to respond to the impacts of the oil sands industry in the 1960s to the 1980s through the various legal and political channels that were available to them. In spite of extensive efforts, however, the community was unable to extend any influence over developers or regulators to better protect their environment. Yet in forming the Athabasca Tribal Council (ATC) with other First Nations governments in the region, the community was ultimately able to make progress in the areas of employment and participation that increased the economic benefits of oil sands development for indigenous communities in northern Alberta.

DEVELOPMENT AND ENVIRONMENTAL IMPACTS

The development of the oil sands industry emerged as part of a larger twentieth-century process of industrialization in Northern Canada that exploited indigenous lands for resources and economic gain.[5] Change in northeastern Alberta began with the establishment of Fort McMurray as a major transport site during the 1930s and World War II. By the 1960s, the Lake Athabasca region had been affected by the uranium-mining boom at Uranium City, Saskatchewan, the construction of the Bennett Dam on the Peace River in British Columbia (which affected the Peace–Athabasca Delta in Alberta), and the establishment of commercial fisheries on Lake Athabasca.[6] Wider developments also prompted the commercialization of the Athabasca oil sands industry, including events in the oil-producing countries of the Middle East and the increasing volatility of the Cold War between the mid-1950s and the 1960s. In 1956, in response to the Suez crisis, the Sun Oil Company of Philadelphia took a majority position in Great Canadian Oil Sands Ltd. (GCOS). In 1966, Cities Service, Imperial Oil, Royalite, and Atlantic-Richfield formed the Syncrude consortium.[7] In 1967, GCOS opened as the first commercial synthetic oil production operation, followed by the Syncrude project in 1978.[8]

In the 1970s, the Organization of the Petroleum Exporting Countries (OPEC) imposed supply restrictions, and subsequently, the price of oil increased, creating an energy crisis in Canada and throughout the Western world. The supply crunch impelled the Alberta and federal governments to expand synthetic oil production irrespective of environmental consequences. In response to a federal report on the deficiencies of the 1973 Syncrude environmental impact assessment, Alberta Environment Minister William J. Yurko wrote to federal Environment Minister Jeanne Sauvé stating that a secure oil supply outweighed environmental risks:

> We know that major information gaps exist in respect to the baseline environmental data in the entire area. Nevertheless, in light of Canada's critical energy balance, it did not and does not appear prudent to delay oil sands development until all needed information is available.[9]

Energy security concerns prompted the federal, Alberta, and Ontario governments to take a combined 30 per cent equity in the Syncrude project following the withdrawal of Atlantic Richfield Canada (ARCAN) in December 1974.[10] With the price of oil and inflation rising through the 1970s and the early 1980s, the successful development of the oil sands industry was a top priority for the federal and Alberta governments until the collapse of oil prices in 1982.

The rush toward development caused dramatic environmental change in communities such as Fort McKay. By the 1980s, the Fort McKay community reported that they were seeing far fewer birds, squirrels, muskrats, and moose that had once been abundant and important sources of food and fur.[11] The community also reported that the influx of people to the region was compounding pressures on wildlife due to recreational hunting, particularly of moose populations.[12] They also noted a huge increase in waste dumping and garbage. The most significant and controversial impacts of oil sands development, however, stemmed from the pollution of the Athabasca River by tailings pond effluent and oil spills, and from atmospheric emissions from the upgrading process.[13]

The complex and energy-intensive process of removing surface soils and vegetation and extracting and processing bitumen produced huge quantities of toxic liquid tailings containing significant concentrations of ammonia and heavy metals, including copper, nickel, chromium, and zinc, as well as unextracted hydrocarbons that had to be stored.[14] From the mid-1960s, one of the most significant polluters of the Athabasca

River was the Great Canadian Oil Sands Ltd. tailings pond. Designed in 1964 as temporary storage on Tar Island pending the availability of an inland mined-out area for a permanent site, the GCOS tailings dyke, constructed of compacted earth, was initially twelve metres tall.[15] Because of unanticipated processing difficulties, more tailings storage was required than initially anticipated, and by 1974 the dyke grew to over 67 metres tall and 3.5 kilometres long. By 1976, effluent seeped from the tailings dyke into the Athabasca River at a rate between 1.5 and 1.6 million litres per day.[16] However, scientists from the Alberta Department of Environment thought that this seepage accounted for only 55 to 70 per cent of total seepage because of unknown quantities of groundwater contamination.[17] The Alberta government had no regulatory framework in place to control the effluent seepage from the GCOS tailings dyke. The company's 1973 Clean Water Act licence regulated effluents entering the tailings ponds but did not address seepage rates or quality.[18]

In the late 1970s, Dr. W. C. Mackay of Alberta Environment concluded that tailings pond water seeping from the tailings dyke was more toxic in composition than the organic carbon that naturally leached from exposed bitumen deposits.[19] Bioassay testing of the toxicity of tailings pond water conducted in 1974 found the heavy metal content to be lethal to rainbow trout.[20] The Athabasca River tended to dilute effluent flows by as much as 400 times in winter and 1,200 times in summer, one mile downstream of the dyke.[21] While this amount of dilution reduced the toxicity of contaminants to a non-lethal level for fish, Mackay maintained that sub-lethal concentrations of tailings water toxicants would impair various body functions and cause significant health problems in fish.[22] However, D. N. Gallup from Alberta Environment asserted that existing research had not yet assessed the long-term fish and human health implications of diluted chemical and organic contaminants in the Athabasca River.[23] The Athabasca River was also contaminated by biological pathogens from sewage produced by the rapidly expanding town of Fort McMurray.[24]

In addition to water problems, oil sands activities produced local air pollution affecting Fort McKay and the surrounding environment, as atmospheric emissions from the two operations are naturally funnelled northward along the Athabasca River.[25] In 1986, Fort McKay officials commissioned an environmental impact assessment (EIA) that concluded, "There has been a definite and statistically significant deterioration in long-term air quality of the region."[26] The Syncrude stack produced particulate emissions at a rate of 3,060 kilograms per day. Syncrude's analysis of these emissions revealed twenty-six toxic trace elements and metals emitted at seventy kilograms per day.[27] Of the trace element emissions, 95 per cent consisted of sodium, vanadium, magnesium, titanium, and manganese. The 1986 EIA report pointed out that vanadium, a transition metal emitted at three kilograms per day, was not monitored but had potential to cause deleterious effects on the human respiratory system.[28] The remaining 2,090 kg/day of emissions consisted of sulphur dioxide (a well-documented cause of acid rain, damage to vegetation, and respiratory issues among vulnerable individuals) and significant amounts of hydrocarbon particulates (a possible explanation for the presence of oily residue in water melted from snow in Fort McKay). The particulate emissions from oil sands operations could have adverse and long-term effects on terrestrial environments, including altering the mineral nutrient cycle in the region. Recent research has found that major increases in the atmospheric deposition of polycyclic aromatic hydrocarbons (PAHs) and dibenzothiophenes from oil sands operations over the last fifty years have had significant impacts on the surrounding watershed, causing oil sands lake ecosystems to enter "new ecological states completely distinct from those of previous centuries." These ecological changes may in turn be related to public health and environmental problems downstream from the oil sands industry.[29]

Biological and chemical atmospheric and water-borne pollution increased through the 1970s and had profound consequences for the Fort McKay community. Fort McKay had dealt with water quality issues since the late 1960s, when the community began to notice that drinking water from the Athabasca River induced nausea and vomiting and other illnesses, possibly due to industrial effluent or municipal sewage from Fort McMurray.[30] Between 1967 and 1975, the Alberta Department of Health warned the people of Fort McKay to stop drinking water from the river.[31] Two water storage tanks were installed at either end of the town. The tanks were meant to eliminate the problem of water supply, but were not cleaned as they should have been by government officials and quickly became contaminated. During the winter, the tanks had to be constantly heated by propane burners to prevent them from freezing.[32] By 1980, residents of Fort McKay reported that they could no longer wash clothes with river water because they would stink and cause skin irritation and rashes.

Before the Athabasca River became polluted, fishing was a significant food source for Fort McKay. Each family would catch over 2,000 fish each fall to dry and store for winter months. Members of the community reported that pike and pickerel caught from the Athabasca River tasted bad and induced vomiting. By the early 1980s, dead fish were regularly seen floating in the Athabasca River; fish from the Muskeg River began to taste like oil and were subsequently abandoned as a food source by the community.[33] In 1985, an EIA of Fort McKay, commissioned by the Energy Resources Conservation Board (ERCB), found that everyone in the community relied on the river, ice, snow, and rain for water, but that all of these sources were contaminated. The community reported that rainwater developed a "yellow scum" when collected and allowed to settle.[34] Fort McKay residents associated atmospheric emissions from oil sands operations with a decline in the health of regional vegetation.[35] They reported that the tops of birch trees were dying, and that those that were still alive had yellowing leaves and were unhealthy. All trees had generally declined in health and produced less foliage. They reported that Jack-pine needles were dying and falling off and that all coniferous trees were producing fewer cones and nuts, which had been a significant food supplement for Fort McKay. Soon after the GCOS plant began operations, Fort McKay residents observed that berries had become less abundant. Edible plants, herbs, and medicinal plants became more difficult to harvest, and the community trusted the safety of what could be collected less and less.[36]

The Fort McKay community participated in the regulatory process and opposed the environmental impacts of the oil sands industry through whatever channels they could. The community's intervention at the Energy Resources Conservation Board hearing on the expansion of GCOS operations in January 1979 highlighted the massive environmental, social, and economic effects that GCOS had had on the community:

> Before 1960, Fort McKay was a relatively isolated settlement having little contact with the "outside world." The building of the Great Canadian Oil Sands plant in the 1960s marked the beginning of the encroachment of major resource development upon the settlement. The plant was constructed on the summer residence for many families from Fort McKay. The construction of the plant provided the first major conflict between the traditional lifestyle of the community and an industrialized way of life. In such a conflict, the "old way" can not win [sic]. A giant like the GCOS has not changed its way because of Fort McKay. But certainly our community has had to turn "upside down" for GCOS and other specific resource developments.[37]

Additional research suggested that the GCOS plant on Tar Island destroyed a prime hunting area, including important summer camps and traplines. These sudden changes compromised the community's ability to subsist from hunting and trapping.[38] The community's intervention also expressed concerns about water quality in the Athabasca River, which residents perceived had "deteriorated significantly since the construction of the GCOS plant."[39]

In spite of the environmental and health concerns raised by Fort McKay, GCOS was granted approval to expand by Minister of Renewable Resources F. W. MacDougall on March 8, 1979.[40] In 1978, GCOS merged with Sun Oil, becoming Suncor shortly after the approval of the expansion project in 1979. In 1980, Suncor claimed $259 million in profit and continued to pollute the Athabasca River Valley on an even bigger scale.[41] Within two years, Suncor was responsible for another significant pollution spill in the Athabasca River. December 1981 was an unusually cold winter, affecting equipment throughout the region. In Fort McKay, the propane heater on the south water tank malfunctioned, and the entire structure burned down. The heater on the north tank failed and the tank froze, turning the remainder of the town's water into ice, which cracked and destroyed the tank as it expanded. The failure of the water system caused a crisis, and residents were forced to take water from the contaminated river. At the Suncor plant, cold temperatures caused significant equipment failures in late December 1981, which were compounded by fires in January 1982, causing major spills of oil, grease, and phenols into the Athabasca River that continued until the end of February. In the course of a few days, more than forty tons of toxic waste and chemicals were spilled into the river.[42] Suncor did not inform Fort McKay that a spill had occurred until February 23, despite having been told to do so by Alberta Environment on January 26. As news of the Suncor spill became widely known, an emergency water delivery system was established that was used into the mid-1980s.[43]

Environment Minister John Cookson told *Fort McMurray Today* that there would be an investigation into the spill: "Both the ERCB and my department are concerned [about] why this happened. The company has to tell us why machines failed, what staff was on duty to manage, and submit recommendations."[44] Fort McKay Chief Dorothy MacDonald was furious about the spill and how the situation was being handled. In March, she wondered in a press conference, "Where the hell was the government when all this was going on? Why didn't the Department of the Environment tell us what was going on and why didn't they conduct testing themselves? How foolish can you be to allow a company like Suncor to conduct its own monitoring? Do bank robbers turn themselves in after they've done the job?"[45] Commenting on Cookson's announcement of the investigation, Member of the Legislative Assembly Grant Notley told *Fort McMurray Today*, "It's a whitewash when they don't include an investigation of the department's performance. I think one thing that now is quite common throughout the province is we've got a Department of the Environment that is badly managed and incompetently led."[46] He drew attention to the Alberta Department of Environment pollution control division's "Summary of Suncor Inc. Wastewater Treatment System Performance, June 1978 to Date," which stated that Suncor had exceeded its water pollution limits in thirty-six of the preceding forty-three months. The minister for Workers' Health, Safety and Compensation reported that testing of Suncor effluents revealed an abundance of polychlorinated biphenyls (PCBs), which are toxic aromatic compounds.[47] The following week, *Fort McMurray Today* reported that samples of pickerel taken from Lake Athabasca near Fort Chipewyan had an oily taste and that the lake had high levels of PCBs. The government warned people downstream of

Fort McMurray not to eat fish from the lake or the river and delayed the opening of the commercial fishing season to June pending test results.[48]

In 1983, Suncor was charged with seven violations of the federal Fisheries Act and two violations of the Alberta Clean Water Act.[49] The initial trial was on the Clean Water Act violations, to which Suncor plead not guilty. The court found that the company had exercised due diligence in attempting to prevent the flow of oil into the Athabasca River, and Suncor was acquitted. The Crown then simultaneously appealed the judgment and pursued new charges of unlawful deposit of a deleterious substance in water frequented by fish under section 33(2) of the Fisheries Act.[50] The court ruled that the Crown had drawn a defective case that was further weakened by an unconventional appeal process and by the repetition of evidence that resulted from the Crown charging Suncor separately for each set of violations. The Alberta Court of Appeal acquitted Suncor and dismissed the appeals. For the residents of Fort McKay and Fort Chipewyan who suffered every day from the widely observed contamination of drinking water and the declining quality and quantity of fish, the Crown's failure to secure a conviction was a failure of justice for indigenous peoples in the Athabasca region. In spite of the collection of evidence from the trial of this case, no regional study of the potential impacts of industrial effluents on water quality and fisheries was conducted in the region such as had been previously conducted on the Mackenzie River below Norman Wells by Fisheries and Oceans Canada.[51]

Concurrent to the expansion of Suncor, Fort MacKay fought hard to be involved in the planning and development of the proposed $14 billion Alsands project on the east side of the Athabasca River near Fort McKay.[52] The community filed an intervention at the ERCB hearing calling for direct consultation and hearings in Fort McKay, but their concerns were largely disregarded. ERCB chairman Vern Millard wrote to Chief MacDonald stating that Fort McKay's claims did not justify further hearings, asserting that "the alleged long-term environmental and health impacts from oil sands development are, in the board's view, not substantiated. If they should be proven, the board and Alberta Environment would undoubtedly take the appropriate action." He stated that research into the ability of the new plant to deal with possible chemical and oil spills would not "serve any useful purpose." He also wrote off compensation and housing issues as not part of the ERCB's jurisdiction.[53] Chief MacDonald told *Fort McMurray Today* that "the response of the board is an absolute outrage." She criticized the review process, stating:

> The board says it won't act until there is evidence but it refuses to re-open the hearings to hear the evidence. They never considered health impacts at the hearings in 1979. It's fairly obvious that the ERCB is just a political body with absolutely no interest in human health.[54]

She continued, "the only acceptable evidence to them is if we rolled in with a wheel barrow with someone dead in it. The province is so intent on resource development that they don't care what impact it has on people. They just don't care what the public health cost is."[55] On June 5, 1979, Alberta Energy Minister Mervin Leitch announced that there would be no public ERCB hearings in Fort McKay, and that he was unaware of any significant local concerns about the plant.[56] He stated that the major consideration in building the Alsands project was economic viability. Other than the opportunity to intervene at the ERCB hearings, the Fort McKay community was largely excluded from the environmental review of the Alsands project. Under the "one window concept" introduced by Environment Minister William Yurko in 1973, environmental assessment was done by the company or the Department of Environment and was factored in as a component of the ERCB approval process but was not a separate decision-making criteria.[57] A review of the Alsands EIA by the Department of Indian Affairs and Northern Development stated, "It appears no effort has been taken to include or obtain the oral history of Indian elders in the area. It also appears that the Indian Association of Alberta and the individual Indian Bands were not consulted."[58] Though the community took significant steps to participate in the planning and regulation of the oil sands industry, Fort McKay never achieved the power to influence government or industry in a meaningful way.

EMPLOYMENT OPPORTUNITIES OR EXCLUSION?

While indigenous communities suffered from the crippling impacts of environmental degradation in the 1970s and 1980s, they were also largely excluded from the economic benefits of employment and participation. Indeed, the federal and provincial governments took only tentative steps to ensure indigenous employment in the oil sands industry.[59] The Alberta Conservation and Utilization Committee's 1972 "Tar Sands Development Strategy" advocated, for example, that the Alberta government create a "multi-purpose public awareness program which would emphasize the prospective developments and condition of the local population, and place special attention on the native people in order to encourage assimilation into the work force and overcome alienation."[60] Peter Lougheed, speaking in the legislature in 1973, suggested that this process would be slow:

> We have to keep in mind in this area that we, as a provincial government, cannot interfere, unless there are ways in which we are asked to, with the treaty rights of our Native people. We are all well aware that trapping and fishing is a phasing-out situation to some extent, and we are faced with skilled jobs in areas such as tar sands plants—and there is great transition going to be required in that, considerable patience and not too much false expectation. The progress will be slow and let no one pretend otherwise.[61]

In keeping with these comments, the Alberta government remained only minimally concerned with issues of indigenous employment.[62] When the federal government, Syncrude, and the Indian Association of Alberta (IAA) reached an agreement on the hiring of indigenous people in 1976, the provincial government refused to sign. The agreement contained plans for recruiting indigenous workers, establishing training programs, and forming institutional alliances to better the employment potential of indigenous peoples.[63] Even though the Syncrude agreement therefore focused mostly on training rather than setting definite quotas for indigenous employment, the Alberta government continued to maintain a hands-off approach to the issue. Indeed, a policy paper from the federal Department of Energy, Mines and Resources written in September 1980 stated that Alberta had "generally taken the position that special programs which operate in favour of status Indians (as proposed by the federal government) discriminate against non-status Indians and Métis."[64]

Despite such a weak commitment to indigenous employment, many within the provincial government publicly trumpeted the so-called triumph of indigenous employment in the oil sands. Local people and the federal government recognized that plans to hire indigenous peoples had more or less failed, but MLAs in the Alberta government argued as late as the 1980s that indigenous hiring had been a success. Norm Weiss, MLA for Lac La Biche–McMurray, championed the efforts of the private sector, stating that "the employment of natives by Syncrude and Great Canadian Oil Sands has shown a dedication to equality and human rights that our government can be proud of."[65] In response to a question from NDP MLA Grant Notley about the Alberta government's inadequate indigenous hiring policy in 1981, Dr. Don McCrimmon, Conservative Minister without portfolio responsible for native affairs, replied that "the history of Syncrude disproves what the Hon. Member is saying. When these megaprojects go ahead, I think the companies have been pretty conscientious and pretty good about trying to get the native people working in them as much as possible."[66]

In 1979, two consultants for the Cold Lake band, Roger Justus and Joanne Simonetta, produced a report that painted a very different picture of indigenous employment in the oil sands industry in the 1970s. Only thirty indigenous people in total, including twenty-four from Fort McKay, had ever been employed in the oil sands, and of these only seven people were still employed. Of those no longer employed, 33.3 per cent had been laid off, 16.7 per cent had left to go trapping, and 16.7 per cent had left because of illness. In terms of duration, 41.7 per cent had worked for less than six months, and only 23.6 per cent had worked for more than eighteen months. The majority of jobs were in menial labour. Respondents reported that there were only minimal salary increases, and only 13.3 per cent of respondents ever received a promotion.[67] The report further suggested that the Syncrude hiring agreement had meant that "Syncrude has made some effort to employ Indian people in all job categories. However, the number of Indian employees, particularly from the immediate local area, has remained relatively low."[68] The consultants also took a dim view of the pilot training program:

> The Syncrude Agreement represents a well-intentioned attempt by all parties to ensure Indian participation in employment training and business opportunities in the oil sands area. However, exploratory research in the communities and an analysis of the available documentation reveals a gap between the original intents of the Agreement and the results of implementation efforts, by all parties, to date.[69]

If the environmental impacts of this early period of development were justified in part through the promise of enhanced local wage labour opportunities, this pledge proved hollow for the people of Fort McKay.

Indigenous people were unable to find work in the oil sands industry for numerous reasons. Most of the jobs were in skilled labour and required training and education that most indigenous people in the region lacked. Another problem was that employment infrastructure was planned around work camps and busing workers in and out of Fort McMurray. Work was often not advertised in indigenous communities, and there were no indigenous-specific hiring and training programs. Also, full-time employment was incompatible with the hunting and trapping lifestyle of indigenous people. It was difficult to work a full-time, year-round schedule and pursue seasonal hunting and trapping opportunities. As was the case with other case studies in this volume (with Rankin Inlet being the notable exception), the unwillingness of many indigenous employees to commit to full-time employment was not acceptable in the oil sands industry.[70]

Nonetheless, the failure to employ significant numbers of Fort McKay residents in the oil sands was not an indication of local people's antipathy toward wage labour. The Justus-Simonetta report found, for example, that 60.5 per cent of the indigenous people surveyed, and 74 per cent of Fort McKay respondents, expected to get jobs in the oil sands industry. Indeed, over 76 per cent of respondents highly desired jobs and had applied for them despite most people not hearing of potential jobs before construction.[71] While Chief MacDonald primarily sought environmental protection in the construction of the Alsands project, the young secretary-treasurer of the band council, Jim Boucher, focused on the provision of local jobs. Age twenty-three in 1979, Boucher represented the generation that had grown up in a settled community and had been educated in residential schools. Though members of his generation continued to be highly dependent on the land for subsistence, they also had a greater connection to the industrial world. In an interview with the *Edmonton Journal*, Boucher stated that resource development in the area had made it impossible for community members to maintain a traditional way of life, and that within less than two decades the once-isolated community had been completely upset.[72] From Boucher's perspective, there was no choice but to work with government and industry to seek participation in the oil sands. He told the *Edmonton Journal* that Fort McKay supported the Alsands project and the proposal to build a new town. Boucher despised handouts, and sought autonomy, a guarantee of the town's survival, infrastructure improvements, land tenure, reduced pollution, and affirmative action hiring programs.[73]

As a further response to the disappointing hiring situation, the five indigenous communities of the Athabasca oil sands region, Fort McMurray First Nation, Chipewyan Prairie First Nation, Mikisew Cree First Nation, Fort McKay First Nation, and Athabasca Chipewyan First Nation, formed the Athabasca Tribal Council (ATC) to unify their voice on oil sands industry matters, especially employment and participation. As intervenors in the Alsands ERCB hearings, the ATC sought the implementation of an affirmative action hiring program as a condition of approval for the Alsands project.[74] The program would have legally bound Alsands to hiring indigenous workers. The ERCB concluded that it did not have power under section 43 of the Oil and Gas Conservation Act to mandate such a program.[75] The ATC appealed the decision to the Alberta Court of Appeal, which dismissed the case, ruling that the affirmative action program was out of the ERCB's jurisdiction, and that such a program might, as a form of reverse discrimination, be in breach of the Individual Rights Protection Act. The Supreme Court of Canada dismissed a further appeal, but ruled that affirmative action programs did not breach the Individual Rights Protection Act.[76] The ruling was a disappointment for the ATC, but the case established an important legal precedent that developers could not cite the Individual Rights Protection Act to prevent the tabling of future affirmative action programs.[77]

In 1980, both the ATC and the Indian Association of Alberta appealed to the highest levels of the federal government to seek improved participation in the oil sands industry. Joe Dion, president of the IAA, wrote to Prime Minister Pierre Elliot Trudeau:

> Development of Canada's resources has not been in partnership with Canada's Native people. Rather, it has occurred to the detriment of the traditional economies and lifestyles of Indian peoples. Being isolated from participation has caused no significant rise in income of Indian communities, and, as a result, Indian people do not have the capacity to finance their future developments. It is fundamental in our view, that the need for aid should eventually subside and this can only be accomplished with the growth in the capacity of Indians to help themselves.[78]

Dion advocated affirmative action and equity participation in the Alsands project. ATC Chairman Lawrence Courteorielle wrote to Marc Lalonde, Lloyd Axworthy, John Munro, and Jean Chrétien seeking greater participation in the Alsands project, specifically the establishment of affirmative action hiring programs, infrastructure spending, housing, and greater efforts to minimize the social impacts of industrialization.[79] The IAA and the ATC were finally able to influence the federal government to aid their interests by helping to encourage affirmative action programs. The National Energy Program explicitly required that Alsands implement a preferential hiring program for indigenous people as a condition of preferential oil pricing.[80]

Following the collapse of global oil prices in 1982, the Alsands project was cancelled and the people of Fort McKay were spared from further environmental destruction from a third oil sands plant designed to produce 125,000 barrels per day. But they also lost employment opportunities. In 1985, Fort McKay reported that the oil sands industry still had not delivered jobs. For example, Alsands had promised that during construction of the bridge, all who sought work could have it, but only one man was hired.[81] Although the Alsands project failed, the efforts of the IAA and the ATC were not a complete loss. In 1986, the Fort McKay First Nation established the Fort McKay Group of Companies, which provided basic services to the oil industry and evolved into a major business enterprise valued in the hundreds of millions of dollars.[82] In response to the efforts of indigenous communities, the oil sands industry, especially Syncrude, began to include communities in development planning and economic opportunities. In 1986 and 1987, Syncrude formed the Syncrude Application Review Group (SARG) and the Syncrude Expansion Review Group (SERG) to bring together communities, industry, and government in an alternative dispute resolution process to examine development and expansion issues.[83] Under the tenure of Eric Newell as CEO and chairman from 1989 to 2003, Syncrude became a more proactive employer of indigenous peoples. In a 2012 interview, Newell told the *Calgary Herald*, regarding the hiring of indigenous peoples in the 1980s, that Syncrude

> made every mistake in the book . . . We thought we were in a hiring program, but as fast as we could hire young aboriginal workers, we would let them go. We realized that taking some person from a little community of 250 people and throwing them into an industrial complex like Syncrude was not a formula for success.

Syncrude pursued indigenous education and development programs that eventually led the company to become a significant Canadian employer of indigenous peoples. Newell was later made an Officer of the Order of Canada for his indigenous employment initiatives, and has received the Award for Excellence in Aboriginal Relations from the Canadian Council For Aboriginal Business.[84] Although the efforts of indigenous people to challenge the environmental impacts of the oil sands industry in the 1980s failed, their efforts to gain increased participation represented the slow and painful beginning of what would, three decades after the first period of oil sands expansion, become a success story.[85]

CONCLUSION

The commercial development of the oil sands industry from the 1960s to 1980s, combined with the postwar industrialization of northeastern Alberta, radically transformed the environmental and economic landscapes of the Athabasca region. The environmental impacts of oil sands operations on nearby Fort McKay undermined the community's traditional economy, while mostly excluding Fort McKay residents from the jobs and economic benefits associated with industrialization. Although politicians and developers intended that economic benefits to indigenous people would offset the impacts of development, the oil sands industry in effect socialized the environmental costs of development by allowing adverse impacts, rather than the benefits of participation, to accrue to indigenous communities.

In the face of their new industrial neighbours' encroachments, indigenous peoples in the Athabasca region fought for environmental justice and participation by making interventions at ERCB hearings, voicing their concerns to politicians, contributing to published reports, and taking legal action. Despite extensive efforts, the indigenous organizations were unable to make industry or government take any meaningful action to protect their environment. In spite of the repeatedly documented and widely observed adverse environmental impacts reported by the Fort McKay community, neither industry nor government acknowledged the severity of or acted upon these environmental concerns. Perhaps indicative of a weakness of environmental rights in the Canadian legal system, government and industry disregard for indigenous peoples' environmental concerns has continued during recent booms in oil sands development.[86]

Indigenous peoples suffered from underemployment and inadequate economic participation in the oil sands industry from the 1960s to 1980s, but the principle of indigenous employment was eventually acknowledged by government and industry. Although the 1976 Syncrude hiring agreement was objectively an almost complete failure, it represented an intention by industry and government to include indigenous people in the benefits of the new industrial economy. By forming the Athabasca Tribal Council, indigenous communities demonstrated a powerful capacity to act independently and effectively within the Canadian legal and political systems. Efforts of the ATC to secure an affirmative action program were defeated in the Supreme Court but were successful in persuading the federal government to make indigenous hiring a requirement for the Alsands project in exchange for international preferential oil pricing. In spite of the inaction of the Alberta government, the ATC's hiring agreement with the federal government for the Alsands project and the earlier Syncrude agreement, alongside the efforts of progressive leaders like Dorothy MacDonald and Jim Boucher, were important early steps toward the economic development of First Nations communities in the oil sands region.

For indigenous communities like Fort McKay, environmental protection and economic development have never been mutually exclusive objectives.[87] While indigenous-owned businesses in the oil sands industry have thrived, major questions about the adverse environmental impacts of bitumen extraction remain unresolved and are seen as being inadequately addressed by the Alberta government. In the past year, the Fort McKay First Nation, the Athabasca Chipewyan First Nation, the Mikisew Cree First Nation, and the Fort McMurray First Nation have all abandoned Alberta's new Joint Oil Sands Monitoring Program, and have boycotted the new consultation policy, which they view as co-opting initiatives that were developed without indigenous input.[88] Communities including Fort McKay and the Athabasca Chipewyan First Nation are now funding their own health studies to investigate abnormal cancer rates that they believe are caused by the expanding oil sands industry, as the Alberta government has refused to do so.[89] Indigenous communities needed regional unification in the ATC, extensive advocacy, and legal action in order to secure employment and economic benefits in the 1980s. For indigenous communities seeking meaningful environmental regulation of the oil sands industry, regional indigenous unification, advocacy, and legal action may be the only path forward.

ACKNOWLEDGMENTS

John Sandlos and Arn Keeling provided fantastic supervision for the MA thesis that formed the basis for this chapter, as well as valuable comments on earlier drafts. The SSHRC-funded Abandoned Mines in Northern Canada research program was an exciting environment to conduct research, which greatly contributed to my MA research.

NOTES

1. Examples of this debate include: "Joint Community Update 2008 Reporting Our Environmental Activities to the Community" [Fort McMurray, AB: Regional Aquatics Monitoring Program (RAMP), Wood Buffalo Environmental Association (WBEA), Cumulative Environmental Management Association (CEMA), 2008]; "Wood Buffalo Environmental Association Human Exposure Monitoring Program (HEMP) Methods Report and 2005 Monitoring Year Results" (Fort McMurray, AB: Wood Buffalo Environmental Monitoring Association, 2007); "Alberta Oil Sands Community Exposure and Health Effects Assessment Program (HEAP) Summary report" (Edmonton: Health Surveillance, Alberta Health and Wellness, Government of Alberta, 2000); Y. Chen, "Cancer Incidence in Fort Chipewyan, Alberta, 1995–2006" (Edmonton: Alberta Cancer Board, Division of Population Health and Information Surveillance, Alberta Health Services, 2009); and Erin N. Kelly et al., "Oil Sands Development Contributes Elements Toxic at Low Concentrations to the Athabasca River and Its Tributaries," *PNAS Environmental Sciences* 107, no. 37 (2010): 16178–83.

2. Bob Weber, "Court Denies Aboriginal Bid to Block Ruling on Jackpine Expansion," *Ottawa Citizen*, November 26, 2012.

3. David H. Breen, *Alberta's Petroleum Industry and the Conservation Board* (Edmonton: University of Alberta Press, 1993); Paul Chastko, *Developing Alberta's Oil Sands: From Karl Clark to Kyoto* (Calgary: University of Calgary Press, 2004); Larry Pratt, *The Tar Sands: Syncrude and the Politics of Oil* (Edmonton: Hurtig Publishers, 1976); John Richards and Larry Pratt, *Prairie Capitalism: Power and Influence in the New West* (Toronto: McClelland and Stewart Limited, 1979); Arn Keeling, "The Rancher and the Regulator: Public Challenges to Sour-Gas Industry Regulation in Alberta 1970–1994," in *Writing Off the Rural West: Globalization, Governments, and the Transformation of Rural Communities*, eds. Roger Epp and Dave Whitson (Edmonton: University of Alberta Press, 2001), 279–300; Andrew Nikiforuk, *Saboteurs: Wiebo Ludwig's War against Big Oil* (Toronto: Macfarlane Walter and Ross, 2001);

Andrew Nikiforuk, *Tar Sands: Dirty Oil and the Future of a Continent* (Vancouver: Greystone Books, 2010).

4   Robin Brownlie and Mary-Ellen Kelm, "Desperately Seeking Absolution: Native Agency as Colonialist Alibi?," *Canadian Historical Review* 24, no. 4 (December 1994): 543–56; Frank Tough, *"As Their Natural Resources Fail": Native People and the Economic History of Northern Manitoba, 1870–1930* (Vancouver: UBC Press, 1997), 305; Lianne Leddy, "Cold War Colonialism: The Serpent River First Nation and Uranium Mining, 1953–1988" (PhD diss., Wilfrid Laurier University, 2011); Hans Carlson, *Home Is the Hunter: The James Bay Cree and Their Land* (Vancouver: UBC Press, 2008).

5   Kerry Abel and Ken S. Coates, "The North and the Nation," in *Northern Visions: New Perspectives on the North in Canadian History*, eds. Kerry Abel and Ken S. Coates (Peterborough: Broadview Press, 2001), 7–21; Ken Coates and William Morrison, *Forgotten North: A History of Canada's Provincial Norths* (Toronto: James Lorimer and Company, 1992); Kerry Abel, "History and the Provincial Norths: An Ontario Example," in *Northern Visions*, 127–140; David Quiring, *CCF Colonialism in Northern Saskatchewan* (Vancouver: UBC Press, 2004); and Jim Mochoruk, *Formidable Heritage: Manitoba's North and the Cost of Development, 1870 to 1930* (Winnipeg: University of Manitoba Press, 2004), among others.

6   Liza Piper, *The Industrial Transformation of Subarctic Canada* (Vancouver: UBC Press, 2009); Arn Keeling, "'Born in an Atomic Test Tube': Landscapes of Cyclonic Development at Uranium City, Saskatchewan," *The Canadian Geographer* 54, no. 2 (2010): 228–52; Patricia A. McCormack, *Fort Chipewyan and the Shaping of Canadian History, 1788–1920s: "We like to be free in this country"* (Vancouver: UBC Press, 2010); Tina Loo, "Disturbing the Peace: Environmental Change and the Scales of Justice on a Northern River," *Environmental History* 12 (October 2007): 895–919.

7   Chastko, *Developing Alberta's Oil Sands*, 90.

8   GCOS became Suncor in 1979.

9   W. J. Yurko to Jeanne Sauvé, 15 October 1974, in RG108 vol. 284 file 4833-3, Library and Archives Canada (hereafter LAC).

10 Don R. Getty, Alberta Minister of Energy and Natural Resources, to W. A. Posehn, 30 May 1975, 82.165, file. 49, Provincial Archives of Alberta (hereafter PAA).

11 Graeme Bethell, "Preliminary Inventory of the Environmental Issues and Concerns Affecting the People of Fort MacKay, Alberta," (Brentwood Bay, BC: Bethell Management Ltd., May 1985), 23.

12 Ibid., 25.

13 Bethell, "Preliminary Inventory of the Environmental Issues"; and Roger Justus and Joanne Simonetta, "Major Resource Impact Evaluation, Prepared for the Cold Lake Band and the Indian and Inuit Affairs Program" (Vancouver, BC: Justus-Simonetta Development Consultants Limited, December 1979); "From Where We Stand" (Fort McMurray, AB: Fort McKay Indian Band, 1983).

14 Bethell, "Preliminary Inventory of the Environmental Issues."

15 W. Solodzuk et al., "Report on Great Canadian Oil Sands Tar Island Tailings Dyke" (Design Review Panel, Alberta Environment, February 1977), 1.

16 P. H. Bouthillier, "A Review of the GCOS Dyke Discharge Water," in *Great Canadian Oil Sands Dyke Discharge Water* (Edmonton: Alberta Department of the Environment, August 1977), 1.

17 D. N. Gallup, "Impact Assessment of Discharge," in *Great Canadian Oil Sands Dyke Discharge Water*.

18 GCOS CWA Licence No. 73-WL-041 (1973) in Bouthillier, "A Review of the GCOS Dyke Discharge Water."

19 W. C. Mackay, "Toxicity of GCOS Tailings Pond Dyke Discharge," in *Great Canadian Oil Sands Dyke Discharge Water*.

20 S. E. Hrudey, "Characterization of Wastewaters from the Great Canadian Oil Sands Bitumen Extraction and Upgrading Plant" (Ottawa: Water Pollution Control Section, Environmental Protection Service, Northwest Region, Environment Canada, 1975).

21 Gallup, "Impact Assessment of Discharge."

22 Mackay, "Toxicity of GCOS Tailings Pond Dyke Discharge."

23 Gallup, "Impact Assessment of Discharge."

24 Bethell, "Preliminary Inventory of the Environmental Issues," 16.

25 Fort McKay Indian Band, "An Issues Assessment for Concerns Regarding Ongoing Oil Sands Developments and the Community of Fort McKay" (Fort McKay, AB: Fort McKay Indian Band, 1986), 16.

26 Ibid., 6.

27 Syncrude, "Biophysical Impact Assessment for the New Facilities at the Syncrude Canada Ltd. Mildred Lake Plant" (Calgary: Syncrude Canada Ltd., 1984), in ibid.

28 Fort McKay Indian Band, "An Issues Assessment," 20.

29 Joshua Kurek et al., "Legacy of a Half Century of Athabasca Oil Sands Development Recorded by Lake Ecosystems," *Proceedings of the National Academy of Sciences* (2013); Kelly et al., "Oil Sands Development Contributes Elements Toxic at Low Concentrations."

30 Bethell, "Preliminary Inventory of the Environmental Issues," 16.

31   Ibid., 38.
32   Ibid., 39.
33   Ibid., 16.
34   Ibid.
35   Ibid.
36   Ibid., 27.
37   "Intervention filed with the Energy Resources Conservation Board by the Fort McKay Community Committee in relation to the Proposed GCOS Expansion Application 780318." Energy Resources Conservation Board, Application No. 780318, 19 January 1979, ERCB Archives.
38   Michael G. Fox, "The Impact of Oil Sands Development on Trapping with Management Implications" (MA thesis, University of Calgary, 1977), 136.
39   "Intervention filed with the Energy Resources Conservation Board by the Fort McKay Community Committee in relation to the Proposed GCOS Expansion Application 780318."
40   G. B. Mellon to Don Getty, 3 May 1978, 82.165 file 466, PAA.
41   "Suncor Profit," *Ft. McMurray Express*, April 8, 1980, Alsands Press Clippings, Glenbow Archives (hereafter GA).
42   Bethell, "Preliminary Inventory of the Environmental Issues," 40.
43   Ibid., 39.
44   GA, Alsands Press Clippings, M-6328 box 2, Ken Nelson, "Charges Probable against Suncor during Waste-Water Probe," *Fort McMurray Today*, March 17, 1982, Alsands Press Clippings, M-6328 Box 2, GA.
45   "Suncor Faces Spill Inquiry," *Fort McMurray Today*, March 18, 1982, Alsands Press Clippings, M-6328 Box 2, GA.
46   Ibid.
47   Ken Nelson, "PCBs found in Suncor Fluid," *Fort McMurray Today*, May 5, 1982, Alsands Press Clippings, M-6328 box 5, GA.
48   Ken Nelson, "More Foul Fish Taken from River," *Fort McMurray Today*, May 14, 1982, Alsands Press Clippings, M-6328 box 5, GA.
49   Jackie MacDonald, "Fish from Athabasca Polluted," *Fort McMurray Today*, May 6, 1982, Alsands Press Clippings, M-6328 box 5, GA.
50   R. v. Suncor Inc., 1983, Alberta Court of Appeal, 219, Appeal #16352, September 15, 1983.
51   Fort McKay Indian Band, "An Issues Assessment," 10.
52   Ken Nelson, "Tiny McKay Battles a Mega-project," *Fort McMurray Today*, February 11, 1982, Alsands Press Clippings, M-6328 box 2, GA.

53 ERCB Chairman Vern Millard, quoted in Ed Struzik, "Indians' Demand Rejected," *Edmonton Journal*, February 18, 1982, Alsands Press Clippings, M-6328 box 2, GA.

54 Jackie MacDonald, "Indian Demand for Alsands Talks Nixed," *Fort McMurray Today*, February 19, 1982, Alsands Press Clippings, M-6328 box 2, GA.

55 Struzik, "Indians' Demand Rejected."

56 "No Public Hearings Being Planned on Fort McKay Oil Sands Plant," *Edmonton Journal*, June 5, 1979, Alsands Press Clippings, M-6328 box 1, GA.

57 William Yurko, *Alberta Hansard*, April 10, 1973, vol. 40, p. 2014, PAA.

58 Department of Indian Affairs and Northern Development, "General and Specific Comments on Alsands EIA," May 1979, RG131 vol.164 file 4300-12 (vol. 5) EMR – ALSANDS 4, LAC.

59 Albert Hohol and Albert Ludwig, *Alberta Hansard*, May 10, 1974, p. 1968, PAA.

60 Conservation and Utilization Committee, "Fort McMurray Athabasca Tar Sands Development Strategy," Policy Paper Prepared for the Executive Council, Government of Alberta, Edmonton, August 1972, 2, RG19 vol. 5238 file 9628-15-1 pt. 1, LAC.

61 Peter Lougheed, *Alberta Hansard*, April 18, 1973, vol. 45, p. 2410, PAA.

62 Ex. Bob Bogle's response to Grant Notley, *Alberta Hansard*, May 3, 1976, p. 1014.

63 Her Majesty The Queen in Right of Canada, Syncrude Canada Ltd., and the Indian Association of Alberta, "Syncrude Indian Employment Agreement," July 3, 1976, 82.165 file 273 pt. 1, PAA.

64 "Alsands Project Policy Paper," federal Department of Energy, Mines and Resources, September 1980, RG131 vol. 164 file 4300-12 (vol.1) EMR – ALSANDS, LAC.

65 Norm Weiss, *Alberta Hansard*, May 28, 1979.

66 Don McCrimmon reply to Grant Notley, *Alberta Hansard*, April 6, 1981.

67 Justus and Simonetta, "Major Resource Impact Evaluation," 40.

68 Ibid., 73.

69 Ibid., 76.

70 Ibid., 73.

71 Ibid., 41.

72 Tom Campbell, "Union Word Needed in Native Hiring," *Edmonton Journal*, July 5, 1979, Alsands Press Clippings, M-6328 box 5, GA.

73 Bobbi Lambright, "Fort McKay Residents Seek Assurances from Government," *Fort McMurray Today*, July 5, 1979, Alsands Press Clippings, M-6328 box 5, GA.

74 Athabasca Tribal Council, "Presentation to the Energy Resources Conservation Board," ERCB Hearings on the Alsands Project Group – Oil Sands Mining Project – Application #780724, June 1979, RG131 vol.164 file 4300-12 (vol. 5) EMR – ALSANDS 4, LAC.

75 Athabasca Tribal Council v. Amoco Petroleum Co., Supreme Court of Canada, December 4–5, 1980, and June 22, 1981.

76 Ibid.

77 Farrell Crook, "Alberta Indians Win a Big One – By Losing: A High Court Ruling Means Special Programs to Help Indians Are Not Legally Reverse Discrimination," *Toronto Star*, July 4, 1981, Alsands Press Clippings, M-6328 box 1, GA.

78 Joe Dion to Pierre Elliot Trudeau, February 6, 1980, RG131 vol.164 file 4300-12 (vol. 7) EMR ALSANDS, LAC.

79 Lawrence Courteorielle to Marc Lalonde, John Munro, Jean Chrétien, Lloyd Axworthy, April 25, 1980, RG131 vol.164 file 4300-12 (vol. 3) EMR – ALSANDS, LAC.

80 Marc Lalonde, "The National Energy Program," Department of Energy and Natural Resources (Ottawa: Minister of Supply and Services, 1980).

81 Bethell, "Preliminary Inventory of the Environmental Issues," 44.

82 Fort McKay Group of Companies LP, "Corporate Information," accessed June 15, 2013, www.fortmckaygroup.com.

83 Syncrude Application Review Group, "Report of the Syncrude Application Review Group to the Energy Resources Conservation Board on Application No. 851024 for New Mining and Discard Areas at Mildred Lake Plant" (1986), Alberta Energy Regulator Library.

84 Governor General of Canada, Order of Canada: Eric P. Newell, O.C., A.O.E., M.Sc., LL.D., F.C.A.E., P.Eng., October 21, 1999, accessed November 15, 2014, http://archive.gg.ca/honours/search-recherche/honours-desc.asp?lang=e&TypeID=orc&id=5572; Canadian Council For Aboriginal Business, "Eric P. Newell, O.C. to receive the Award for Excellence in Aboriginal Relations," September 17, 2012, accessed November 15, 2014, https://www.ccab.com/uploads/File/Excellence_Award-_2012.pdf.

85 Robert Remington, "Remington: 'Syncrude Solution' May Tap Potential of Aboriginals," *Calgary Herald*, June 8, 2012.

86 Weber, "Court Denies Aboriginal Bid to Block Ruling on Jackpine Expansion"; David R. Boyd, *Unnatural Law: Rethinking Canadian Environmental Law and Policy* (Vancouver: UBC Press, 2003).

87 Meagan Wohlberg, "Fort McKay First Nation Objects to Oilsands Project, Company Says Buffer Zone Will Hurt Economics of Project," *Northern Journal,* April 29, 2013.

88 Vincent McDermott, "First Nations Chiefs Boycott Consultation Meetings with Alberta Government," *Fort McMurray Today*, August 29, 2014, accessed November 15, 2014, http://www.fortmcmurraytoday.com/2014/08/29/first-nations-chiefs-boycott-consultation-meetings-with-alberta-government.

89 Vincent McDermott, "Fort McKay Launching Health Study," *Fort McMurray Today*, August 30, 2014, accessed November 15, 2014, http://www.fortmcmurraytoday.com/2014/08/30/fort-mckay-launching-health-study.

| CHAPTER 8

Uranium, Inuit Rights, and Emergent Neoliberalism in Labrador, 1956–2012

Andrea Procter

The struggle for Aboriginal rights in Canada is closely connected with resource conflicts. The threat of mining projects, hydroelectric dams, pipeline construction, and other industrial developments has often pushed Aboriginal groups to mobilize in order to reclaim control over traditional lands. As a number of chapters in this book demonstrate, many Aboriginal peoples have fought to prevent mining developments or have struggled against the harmful effects of mines on their lives and lands. The mining industry and nation-states have dispossessed Aboriginal groups of their territories and sovereignty on so many occasions that scholars often automatically situate the relationship between Aboriginal peoples and mining as a conflict. Yet increasingly, some Aboriginal governments are choosing to negotiate with mining companies and are entering into impact and benefit agreements and development proposals. Some analysts have connected the shift from protest to negotiation with

FIGURE 1: Labrador, showing territory of Nunatsiavut in shaded areas and key settlements and locations mentioned in the chapter. Map by Charlie Conway.

hegemonic neoliberal values and practices (see Levitan and Cameron in this volume).[1] The current global economy encourages states to facilitate the unimpeded exploitation of resources, and neoliberal governance tends to prioritize decentralization, a sense of responsibility for self-improvement, and decreased dependency on the state.[2] As Gabrielle Slowey argues, within this context, Aboriginal "self-determination is consistent with normative and neoliberal goals of economic, political, and cultural self-reliance."[3] How does the willingness of Aboriginal governments to negotiate with mining interests, therefore, correlate with Aboriginal goals of self-determination? What are the implications of this engagement with neoliberal projects?

This chapter explores these questions by examining how Inuit in Nunatsiavut, Labrador, have dealt with the prospect of a uranium mine on their territory since the 1950s, and how the relationship between neoliberal and Inuit goals has become complex and entangled. Based on a year of ethnographic fieldwork in Labrador in 2007–8, as well as on archival and media analysis, this chapter examines the historical articulation between changing and contradictory ideas about Aboriginal rights, economic development, and regional autonomy. By attending to the diverse perspectives of individuals and organizations, I illustrate the lived experience of regional and global processes such as modernization, Aboriginal self-government, and emerging neoliberalism, and explore how Inuit self-government and mining have become interconnected in unexpected ways. Over a sixty-year timeframe, Labrador Inuit, state authorities, and industry have come to align Inuit sovereignty, resource development, and regional self-sufficiency in order to further their own political and economic ambitions. As Charles Hale and others have argued, the neoliberal combination of autonomy and constraint can offer some space for Aboriginal rights claims, but, as I demonstrate here, a neoliberal framework can also limit economic and conceptual possibilities and can exacerbate and obscure inequalities.[4]

URANIUM, MODERNIZATION, AND INUIT RELOCATIONS IN THE 1950s

When uranium was found in northern Labrador in 1954, the discovery set off a string of events that profoundly transformed the political, social, and economic circumstances of the region. In the 1950s, the majority of Inuit in Labrador lived in dispersed settlements, homesteads, and camps on the northern coast and in central regions. In the winter months, many chose to stay in the villages established in the late 1700s by Moravian missionaries, in order to trade with the mission store and to attend church events. Most Inuit relied on a combination of commercial fishing for cod and char, fur trapping, and subsistence harvesting for their livelihood. Since World War II and the arrival of the American and Canadian militaries in the region, some also worked on the construction of various military sites, including those near Hopedale and Goose Bay.[5]

In provincial affairs at the time, Labrador was an afterthought. The region only gained political representation in 1946, and responsibility for education and health care had largely been left to the churches and the privately funded International Grenfell Association.[6] The provincial government in St. John's viewed Labrador as predominantly a place for Newfoundlanders to fish and, more recently, to find employment at military bases. Labrador was both a source of resources and a burden; its people were of little consequence to the government. In fact, Newfoundland had tried to sell Labrador to Quebec in 1925 and again in 1933 in order to deal with its financial difficulties.[7] Aboriginal people in the province, including Inuit, Innu, and those of mixed ancestry in Labrador, were "pencilled out" of the Terms of Union agreement when Newfoundland joined Canada in 1949, as Premier Smallwood argued that there were only Canadians in the province.[8]

It is within this context of political and economic marginalization that the uranium story begins. In 1953, the Newfoundland government leased vast resource rights over much of Labrador to the British Investment Company (or "Brinco"), for very little in return.[9] The one-sided Brinco concession was an example of the government's approach to economic development at the time: natural resources were the property of the Crown and were to be developed for the public good; no special rights to lands or resources based on Aboriginal rights or historical ownership existed.[10] Brinco prospectors found some uranium deposits near Makkovik and Postville in 1954 and even more in 1956. Although the company tried to keep the discovery secret until the timing better suited it, Premier Smallwood leaked the news during the Labrador Conference of 1956, which had been convened by the provincial government in order to discuss future plans for Labrador and its people.[11]

One major theme at the conference was the social and economic future of the Labrador Inuit. Conference participants from the government, the Moravian mission, and the International Grenfell Association discussed their problems with administering health and social services to such a dispersed Inuit population and their concerns about the Inuit harvesting economy, which some felt should be replaced by wage labour. One option that was considered, in the absence of any Inuit participants, was relocating Inuit in the northernmost communities of Nutak and Hebron to places farther south, where they might find paid employment. In the midst of these discussions, and much to the dismay of Brinco officials, Smallwood announced to the press, "It is quite likely that mining of uranium ore and processing of uranium concentrates could commence in 1957."[12] The possibility of a mine near Makkovik created great optimism in government circles about development potential, especially given the bleak economic and health conditions on the coast that the conference participants had described.

Propelled in part by the possibility of wage labour opportunities, authorities decided to relocate more than four hundred Inuit from Nutak and Hebron to more southern communities in 1956 and 1959, and almost two hundred went to Makkovik. The main rationale was the provision of improved and more efficient services in centralized locations, but the discovery of uranium near Makkovik and the potential for jobs for the relocated Inuit did play a role in the decision.[13] The prospect of a uranium mine influenced the idea of Makkovik as a "growth centre" and therefore as a suitable location for the relocatees. Officials at the time generally assumed that Inuit could adapt easily, both to wage labour jobs and to new environments. Some Inuit who were relocated recount promises made to their families about jobs at the potential mine. One woman recalled, "They said that we're moving to a place where there is lots of things, lots of seals everywhere, lots of animals and fish. That's what they said. That there were jobs available also. We had to go to Makkovik because they said Makkovik had everything."[14] Another said, "We were told on July 12 that we had to leave Okak and we left on July 25. My father was told there would be lots of work with Brinex. They took dad away from his fishing in Silutalik. My father could not get work so he ended up fishing there."[15]

The Inuit themselves were not consulted about the relocations, and they expressed their opposition to the plans, but paternalistic government authorities argued that the transition to an industrial, modern economy would be beneficial for them. Wage labour jobs would keep Inuit in settlements where they could also be provided with education and health services, and would encourage them to move away from what authorities felt were non-modern harvesting practices. "Both the Eskimos and the Indians [sic] have been encouraged and assisted in hunting and fishing," wrote the minister of public welfare in 1959, "but we regard these activities as 'holding operations' until the economy in the area becomes more diversified. However, with the development of the mineral resources of Labrador, there is hope that some progress in this direction will be possible."[16] Modernization would provide the government with the dual benefits of creating "productive" and sedentary citizens, as well as removing people from the land, thus making it available for development.[17]

Despite the promises of houses and jobs, the almost two hundred Inuit from Nutak and Hebron arrived in Makkovik to find nothing of the sort. Most spent many cold months living in tents before houses were eventually built on the outskirts of town in an area called "Hebron village." The relocatees were therefore spatially segregated from the rest of Makkovik and were also socially segregated by the original residents, who, although mainly of Inuit ancestry, considered themselves to be "Settlers," and not Inuit. The ethnic tensions and sometimes outright racism between the two groups added considerably to the already stressful and disruptive relocation.[18] Rose Pamack-Jeddore of Nain describes the relocations from an Inuit perspective:

> Resettlement is one of the gems of Confederation. Hebron and Nutak became non-existent in the rush for centralization. The rationale for resettlement was improvement of services. To

the Inuit, resettlement meant living in tents while waiting for accommodation, leaving behind personal belongings, adjusting to a different hunting environment, the inconvenience of returning to fishing grounds in open boats and living in tents in the summer, and it led to an increase in community conflicts. The ensuing insecurity of relocation and the futility of attempting to adjust to depleted hunting, fishing, and wooding grounds drove the Inuit to the escape mechanism of drunkenness. Their powerlessness and insignificance in the dominant society had been made all too clear to them.[19]

The Hebron and Nutak relocations caused widespread social upheaval in Labrador, and the ongoing social devastation and inequalities caused by the resettlement policy are still evident today.[20]

The enthusiasm caused by the discovery of uranium near Makkovik was relatively short-lived. By 1958, Brinco's subsidiary exploration company Brinex had built tunnels and developed plans for a mine at the Kitts site, but by the end of 1959, uranium prices dropped, and the company realized that the mine would be too late to qualify for Atomic Energy of Canada contracts. Brinex exploration stopped soon afterwards. No jobs materialized for the almost two hundred relocated Inuit in Makkovik, who were now far from home and without a meaningful or sufficient livelihood.

NEW LIVELIHOODS: BRINEX IN THE 1970S

Brinex's interest in uranium surfaced again in the early 1970s with the international energy crisis and an improved uranium market. Company employees returned to the Kitts-Michelin site and built a camp fifteen kilometres from the communities of Postville and Makkovik. Labradorians viewed the company's renewed interest in the region with some skepticism. Other Smallwood-initiated developments in the province had failed to create many local or provincial benefits, and people were beginning to call for more local participation in development decision making.[21]

The company's resurgence also coincided with the rise of the Aboriginal rights movement in Labrador. Influenced by social movements in the United States—and driven by the controversy over the 1969 White Paper[22] and large-scale developments such as the Mackenzie Valley pipeline and the James Bay hydroelectric dam—Aboriginal peoples across Canada increasingly challenged state incursions into their lives and lands. In Labrador, the injustice of the 1950s northern relocations prompted many Inuit to voice their grievances and to call for restitution. After the federal government announced in 1973 that it would agree to negotiate comprehensive land claims with Aboriginal groups who had never signed treaties, Inuit in Nain[23] formed the Labrador Inuit Association (LIA). The LIA granted full membership to Kablunângajuit (the Inuttitut word for those of mixed Inuit-settler ancestry) in 1975 and filed a Labrador Inuit land claim in 1977.[24]

The Brinex project proposals of 1976 and 1979 were some of the first development issues tackled by the LIA. Within the context of the global anti-nuclear movement, the uranium development in Labrador was controversial in itself, and the LIA connected environmental and ethical concerns with calls for the recognition of Inuit authority over their homeland. Because they framed their opposition to the project in terms of Inuit rights, identity, and culture, however, the LIA leaders faced some difficulty in finding widespread support for their claims among those who lived closest to the mine site. Many Settler/Kablunângajuit residents from Makkovik and Postville did not self-identify as Inuit at that point, and they expressed some ambivalence toward political claims based on Inuit cultural difference.[25] The Labrador Inuit Association tried to use the sense of crisis created by the Brinex project to convince its potential members to recognize their shared interests and to foster a sense of Inuit identity and solidarity: "Development from big multi-national companies threaten[s] our traditional way of life, not to mention our resources and our land and waters . . . LIA encourages its members to unite and have one voice speaking for the rights of all its members. Let's begin seriously discussing our future and get the most of LIA's land claim for the good of all northern Labrador."[26]

Many residents of Makkovik, especially, framed their opposition to the project and their concerns about the dangers of uranium mining in terms of the government's obligation to protect "a way of life that is vibrant and strong and where the traditional lifestyle makes for a fiercely independent people."[27] Residents stressed the importance of their "way of life," which encompassed shared values, activities, self-sufficiency, and circumstances,[28] but few labelled it as "Inuit." Evelyn Plaice and Lawrence Dunn have argued elsewhere[29] that this focus on "lifestyle" or "way of life" was an approach that North Coast residents and the Labrador Inuit Association used to avoid issues concerning ancestry.[30] In the 1970s, ethnic tensions and differentiation between Inuit and Settlers/Kablunângajuit were prominent features of social life in the region; the northern relocations had created persistent social and economic inequalities among groups in the communities.[31]

In the 1970s, the provincial government was receptive to the "way of life" argument, but it firmly denied any possibility of Aboriginal rights. It issued an official development policy for Labrador in 1979 that suggested local lifestyles should be taken into consideration, but only to the extent that these lifestyles were deemed desirable. Premier Brian Peckford expanded on how this approach would affect Labrador:

> The special relationship of the people to the land must be accounted for. The traditional lifestyle of Labrador, based on the harvesting of renewable resources, fishing, hunting, trapping, etc., requires a sensitive and symbiotic relationship between man and his delicate northern environment. That relationship permeates almost every aspect of the society and culture of Labrador and has to be accounted for in future development. However, we must also recognize the challenges, opportunities, and rewards of new lifestyles which can be ours through a rational program of resource development.[32]

Although government policy statements allowed that the way of life "must be accounted for," it was framed as something that could (and ideally should) be changed into "rewarding new lifestyles," and not as something that was inherently valuable or vital. Aboriginal rights, on the other hand, as championed by the LIA, had a fundamentally political and anticolonial basis that was much less malleable, more challenging, and therefore less appealing for the government.

Brinex's official response to calls for the recognition of Aboriginal rights was to defer to the provincial government's handling of the matter. In practice, however, the company criticized and denied such claims, especially in 1979, when public support for the recognition of Aboriginal rights was increasing,[33] and the issue became prominent in discussions about the potential mining development. In response to this pressure, Brinex worked to minimize the effect of Aboriginal rights on the project, first by arguing that the mine site would not interfere with any current Inuit land use. "I believe the project in question does not directly involve the LIA," argued Brinex's vice president of mining. "I would think they cannot be considered to be residents of the land in the project area south of Kappokok Bay [sic]."[34] The company also responded to local concerns by arguing that the mine could be designed on a very local scale and in isolation from places and people who did not wish to be "involved." A Brinex official stated, "The communities of Postville, Makkovik and North West River have special concerns relating to impacts such as lifestyles. I am confident that we can work closely with these communities to resolve concerns and design our systems to have them participate only to the extent they wish."[35] At a public meeting in 1976, when residents commented that local people would likely only get the lowest-paid jobs, Brinex's Murray Poloski replied that they "have to start somewhere—have to choose a lifestyle . . . The communities can choose their amount of involvement."[36]

Brinex thus acknowledged the "special concerns" of residents by appropriating the claim for local rights through isolating the residents geographically, socially, and economically from any development. Any economic benefit that may accrue to local people from the mine would be a result of their own choice to "change their lifestyle"—they would not be forced to participate. The company thus attempted to defuse and neutralize the residents' claims of local rights by conflating "local" with "isolated," and by relying on the liberal championing of the individual right to choose.

Despite the fierce denial by industry and government of local arguments against the Brinex proposal, moral and political support for the Inuit claims grew. In 1980, the Environmental Assessment Board for the Kitts-Michelin uranium project concluded that the Brinex mining proposal did not prove that the project was environmentally, socially, and economically acceptable, and that serious concerns about Aboriginal rights must be addressed.[37] By the time of the board's report, however, internal company politics and poor global markets for uranium after the Three Mile Island accident in 1979 had prompted the company to shelve the project for the final time.

NEOLIBERAL CONTEXTS: NUNATSIAVUT IN THE TWENTY-FIRST CENTURY

In the years following the Brinex proposals, Inuit land claims negotiations between the Labrador Inuit Association and the provincial and federal governments were often stalled.[38] In 1980, the provincial government reluctantly acknowledged Aboriginal rights, but it limited its definition of Aboriginal rights to culture, heritage, and the traditional use of renewable resources, and refused to consider any rights to subsurface or "non-traditional" resources.[39] In the 1980s and early 1990s, the province framed its approach to the land claims negotiations as affirmative action for a disadvantaged minority, who, once lifted to the economic status of others in the province, would presumably then become socially indiscernible.[40] This perspective on Inuit concerns contrasted greatly with the LIA's position that the Inuit were a political community with inherent rights to resources and autonomy.

Even in 1989, when LIA President William Andersen III articulated a possible alignment between Inuit and provincial goals, the Newfoundland and Labrador government was slow in following the LIA's lead. Andersen argued that Labrador Inuit must be "guaranteed their own lands and resources in sufficient quantities to be as self-sufficient a people as possible ... Of fundamental importance to us, as Labrador Inuit, is our future as a distinct and viable people. We are looking to the next 200 years—not the next 20."[41] The LIA's approach changed slightly from demanding recognition of Inuit rights to demanding the means by which Inuit could sustain themselves in a self-contained territory. Inuit were looking to land claims to help create political and economic independence, he argued: "To us, land claims is not a threat to non-aboriginal people, it's a way to self-sufficiency."⁴² By situating the Labrador Inuit as a political community in search of self-sufficiency, the LIA framed its claim in the dominant society's terms by appropriating concepts from the state's neoliberal project of producing self-reliant subjects.⁴³ Given the government's insistence on limiting Inuit rights to those based on an image of Inuit as self-sustaining subsistence harvesters, aligning the goals of self-sufficiency and resource development offered some middle ground to the provincial government.

Yet it was only when a massive nickel deposit was found in 1994 at Voisey's Bay, south of Nain, that land claims negotiations heated up. With the increasingly mobile nature of capital markets, investors jumped at the prospect of a mining development, but they were also increasingly sensitive to uncertainties about land tenure, and the provincial government felt new pressure to settle Aboriginal land claims in Labrador. Its first step was to take the Voisey's Bay area completely off the negotiation table, but the government also fast-tracked the Labrador Inuit land claim talks. An agreement in principle was signed in 1999, and the final agreement was signed in 2004, at which point the provincial government also issued a formal apology and financial compensation for its role in the 1950s relocations. The Labrador Inuit Land Claims Agreement created the Labrador Inuit Settlement Area, a region administered by the Nunatsiavut Government.

Within the Settlement Area, the Nunatsiavut Government has enhanced authority over a proportion of land (15,799 square kilometres of the total 72,520 square kilometres of land) specifically selected by Inuit and designated as Labrador Inuit Lands.⁴⁴

Shortly after its creation, the Nunatsiavut Government (NG) faced one of its first major crises. Uranium prospects had improved in the early 2000s, due to uncertain global geopolitics concerning oil resources and renewed interest in nuclear power as a possible response to climate change. Accordingly, the price of uranium had risen from US$7/lb in 2001 to US$136/lb in 2007.⁴⁵ Exploration companies returned in full force to Labrador, and Nunatsiavut beneficiaries were embroiled in an intense debate about how to approach the renewed possibility of uranium mining on their lands near Makkovik and Postville.

The Kitts-Michelin site, now under lease to Aurora Energy Resources Ltd., was called "the largest undeveloped uranium deposit in Canada," and Aurora spokespeople claimed that a mine would be many times the size of the nearby Voisey's Bay nickel mine.[46] However, this project would mine uranium—a substance that was much more controversial than nickel. The global anti-nuclear movement had quieted significantly since the Brinex proposals, due in no small part to the post-Chernobyl decline of the nuclear power industry, but uranium remained a contentious topic. The jurisdiction over the land and the allocation of potential economic benefits had, however, significantly changed since the 1970s. The Michelin deposit is on Labrador Inuit Land, which means that NG has jurisdiction over surface access and would receive a 25 per cent share in any provincial royalties of a future development.

In August 2007, Aurora notified the Nunatsiavut Government of its intention to register its project for an environmental assessment by the end of the year. NG had not yet developed a regional land use plan or legislation on environmental assessment, and so the NG minister of lands and environment tabled a motion in the Nunatsiavut Assembly to ban uranium development on Labrador Inuit Lands for three years.[47] Debate about "the moratorium," as it was called, was fierce and endured for many months until the Nunatsiavut Assembly voted to pass the bill in April 2008.

For those who supported the moratorium, the idea that the Inuit government was actively protecting its land and people from potentially harmful developments was a source of pride based in ideas about post-colonial control, environmentalism, and Inuit culture. Given the history of frustrated efforts to defend their lands from industrial incursions, many Inuit took great comfort in knowing that the NG has, in the words of one man, "complete sovereignty" over the land and is willing to exercise that authority.[48] The lengthy land claims negotiations had resulted in some concrete authority over land governance, and some people expressed their relief that the NG was not simply going to "give the land away" for quick economic gain, or capitulate to industrial and economic pressures.[49]

Some who supported the ban described aspects of the Nunatsiavut Government's approach as characteristically Inuit: "When people are hasty to encourage economic development for the sake of accessing jobs and revenues, important details get overlooked. Nunatsiavut Government does need time to be able to stand on solid ground before taking part in an environmental assessment for a proposed uranium development on Labrador Inuit Lands. One common virtue that Inuit culture is based on is patience. We are an Inuit Government."[50] Unlike the situation in the past, the Inuit now had the ability to control development within Nunatsiavut, and many argued that it should do so carefully, at its own pace, and based on its own priorities.

To many Inuit, the connection between the land and the well-being of future generations was fundamental and must be protected. As in the 1970s, the issue of people's "way of life" was prominent in the debate, but this phrase took on various meanings. For some, protecting the "way of life" entailed protecting the integrity of the land for future generations by prohibiting mining. For others, protecting the "way of life" entailed protecting the viability of living in Nunatsiavut by developing resources in order to support Inuit jobs, housing, and infrastructure.

Some argued that the potentially disastrous environmental effects of uranium tailings could destroy the land, water, and wildlife in the vicinity of the development, and could pose a health risk. Many felt that this potential harm was not worth the short-term economic benefits of a mine: "The employment benefits are not going to be there forever, so it's really not worth the environmental and the health (risks) and the loss of our traditional hunting areas," argued Terry Rice of Makkovik.[51] Others felt that the use of land for the sustenance of future generations did not preclude resource development, as long as the environment is not ruined:

> We're not against development—we see development as providing economic opportunities for beneficiaries that's greatly needed and we see need in communities for infrastructure, for housing, for other projects and maybe revenue from mining can allow us to deliver these programs. But if our environment is contaminated then these things don't really matter. You know, we need to ensure that, first and foremost our land is protected for future generations and the onus is on the Assembly to ensure that we do this.[52]

On the other hand, some Inuit argued that mining could "help to protect our people" by allowing them to remain in their home communities and "learn their culture by living near the land instead of moving away for work."[53] Many people on the coast were worried that young people would have to leave the community in order to make a living elsewhere if mining were not permitted. One woman connected this out-migration with the loss of culture, as young people would not have the opportunity to experience and learn their culture by living in Nunatsiavut and being taught by elders:

> It's a sin if our children have to move away because they can't get anything here. They're not going to work at the fish plant if there's something better somewhere else . . . If all the young people have to leave the communities for work, there'll be nothing left to govern. Everything will just die out! Yes, there will be land, but there will be nobody there.[54]

The idea of "protecting our people" included providing the means by which they can remain in Nunatsiavut. According to this perspective, people in Nunatsiavut should be "using the land to its fullest" and using Nunatsiavut's resources to "sustain ourselves in perpetuity."[55]

Some Inuit also argued that people should adopt a stronger sense of individual responsibility so that they gain the education and develop the skills needed to take advantage of potential mining jobs. "There's more to life than 420 hours of work to get EI. The young people need to get educated, and then come home to get work in the mine, if it goes ahead. We need to take control of our lives."[56] In some ways reminiscent of Brinex's claim that people could "choose their own level of participation" in the mine, these comments illustrate how some people have now internalized the desire for "self-improvement," and how modernization themes are reappearing as hegemonic neoliberal values. Arguments for increased responsibility over one's future were echoed by government and industry campaigns encouraging people to train for jobs and to seize mining-related opportunities. Although these perspectives were quite widespread among both Inuit and industry representatives, other Inuit countered these arguments by stating that the role of the Nunatsiavut Government should be to provide compensation or jobs to those who might be hurt by a moratorium.[57]

Neoliberal calls for individual responsibility and self-improvement are also evident at the governmental level. The land claim agreement's fiscal structures strongly encourage the Nunatsiavut Government to sustain itself and its programs through revenues that it raises itself. In general terms, the agreement states that the Nunatsiavut Government would gain a 25 per cent share of provincial royalties from development on Labrador Inuit Lands, while it would only see a 5 per cent share of royalties from development on other lands in the settlement area.[58] Because the Michelin site is on Labrador Inuit Lands, the prospect of a 25 per cent share of royalties from a mine was a serious consideration for many, especially given the huge financial pressure of funding a new government.

The land claim agreement therefore puts Inuit in a new relationship with the land, one that is determined by formal economic agreements. Inuit no longer stand in direct opposition to industry or to a government intent on facilitating resource development; as the "landlord," the Nunatsiavut Government now has a vested interest in development, as well as the responsibility of protecting Nunatsiavut beneficiaries.[59] The pressure on the NG to develop its resources is therefore quite high, and the decision to prevent any uranium mining for three years was made by some members of the Assembly only for administrative reasons—not because they opposed the development of a uranium mine in the future.[60] One politician described the pressure on the government to finance itself in this way: "We need economic development and in a government where we're all aware that the finances are not that great and we will be facing some hard times in the next few years, to delay any process that may give us a light at the end of the tunnel could be extremely detrimental to our success as a government."[61]

As a product of the alignment between Inuit goals of self-determination and neoliberal state goals of increased regional self-sufficiency, the land claim agreement promotes particular ways of conceptualizing Nunatsiavut land and resources.[62] One candidate for the position of NG president in 2008 illustrated the framing of resources as commodities:

> As our North Coast tax base is significantly smaller than what we need to run our self-government, the Nunatsiavut government will be dependent on extracting its natural resources in order to have the funds necessary to sustain our communities and the running of our government. The generation of own source revenue is essential, and right now, mining seems to be the short-term answer. So, even though I am not a mining advocate, I still have to consider it strategically as an income source for government, a source of employment and opportunity for Beneficiaries, and a way to utilize one of our most important assets.[63]

The Nunatsiavut Government should therefore keep this new economic relationship in mind, he continued, as "there is no sense in negatively affecting the reputation of our potential business partner [i.e., the exploration company] or the value of our asset" by establishing a moratorium.[64] Government and corporate interests are aligned, in this perspective, and actions that have a negative impact on industry will likewise have a negative impact on Nunatsiavut.

When the Nunatsiavut Assembly voted to ban temporarily uranium development in 2008, this prediction proved correct, at least initially. As uranium exploration company stocks plummeted in the days after the vote, many Inuit watched with either trepidation or quiet cynicism about industry's flair for melodrama. As one woman told me, "They're just riling people up and getting them to support mining. Give them a year or two years and they'll be back with more support—just before the three years is up."[65] For its part, Aurora Energy Ltd. offered a measured response, intended to quell environmental concerns about uranium mining and to minimize the perceived distance between the company and the Inuit communities: "Aurora shares the goal of careful stewardship of the land that Labrador Inuit have been a part of for over 5,000 years . . . In light of a growing world demand for clean, safe energy, Aurora is looking to the future benefits of moving forward in the spirit of co-operation with the people of North Coast Labrador."[66]

The political, economic, and social ramifications of Aboriginal self-government within a neoliberal context are thus multiple and often contradictory. While the provincial government has an increasing appetite for regional self-sufficiency and, therefore, the settlement of Inuit land claims, the lands and resources on which the Inuit are expected to subsist are extremely limited. In the midst of this debate, mining and exploration companies offer consistently positive information and images about the solution that uranium mining offers to individuals, families, and governments, and the role that Inuit can play in decision making.

The consistent focus of industry, government, and many individuals is on the ability of the Nunatsiavut Government and Labrador Inuit to control the outcome of the uranium issue. In this view, Inuit are now "empowered" to engage in decision-making processes and to decide what their own futures will hold. Yet, as Shore and Wright argue, this focus on new-found "empowerment" can obscure many underlying issues that the new governance structures have not resolved.[67] The ideal neoliberal citizen—self-managing, self-governing, and self-sufficient—is empowered to work as a partner in management and to take responsibility for his or her own success. In this "project of self-improvement . . . any discussion of poverty as inequality or disadvantage has been erased from the discourse."[68] Larger issues such as the lingering inequalities created by the 1950s relocations of Hebron and Nutak, for instance, are overshadowed by talk of individual choice and self-governance, responsibility to improve oneself, and job training.

While many embrace the opportunity to engage in "self-improvement" and development projects, many others feel further alienated by this emphasis on empowerment, as a woman from Nain told me:

> Most [Inuit] live in too much depression to really do anything, can't understand English most of the time and do not benefit from programs and services. It's sickening . . . The Inuit population always seems as though [they] are always in the position of a high price chip: worth a lot but never really benefiting from all what is happening. People say they want the money instead [of programs], they say the leaders don't listen and they only take care of their own family and many don't like it, including me.[69]

The forced relocations from Hebron and Nutak in the 1950s, as well as the more general social and economic impacts of centuries of colonialism, have caused lasting social divisions and inequalities among Inuit in

Nunatsiavut. Some have benefited from the land claims settlement and new economic opportunities, but others have felt further marginalized.

THE POWER OF THE PROSPECT

The relationship between uranium and Inuit self-government is still evolving. In 2011, Aurora Energy Ltd. was purchased by the giant Australian uranium producer Paladin Energy Ltd., and in March 2012, the Nunatsiavut Government lifted its moratorium on uranium mining, but the development of a mine in Labrador has not progressed to formal deliberation. As the global market for uranium continues to fluctuate at the time of writing (early 2013) with the impact of the 2011 Fukushima disaster, the development of a uranium mine in Nunatsiavut remains uncertain, and the ultimate influence of the deposit is yet to be known.

For over fifty years, the prospect of a uranium mine in Nunatsiavut has highlighted the complex and often contradictory correlation between Inuit self-governance and mining development. Speculation about the uranium deposit has fuelled many of the advances in Inuit self-government, but the relationship has not been straightforward. Over the years, the Inuit and provincial authorities have used the crises surrounding uranium both as a catalyst to solidify their political goals and as an opportunity to launch a different future. In many ways, however, the anticipation of a mine has also deepened social inequalities and constrained the range of economic possibilities.

The history of the debate about uranium in Labrador illustrates the entwined connections between Inuit and global dynamics. In the 1950s, following the worldwide propensity for modernization schemes, governmental authorities assumed that the Inuit harvesting economy should be converted into a wage-earning economy, and that Inuit would adapt to enforced (yet inevitable) social and economic transformation. The potential for jobs at a mine was therefore one of the justifications used for the massive relocation and dispossession of northern Inuit from their lands, an event that has caused pervasive social trauma and inequality in the region. In the 1970s, the Brinex proposal was a catalyst for Inuit and Kablunângajuit to join forces and to fight for their rights as Aboriginal people, as national and international debates about local and Aboriginal participation in resource governance were challenging state domination. The Brinex experience also illustrates the emerging influence of neoliberalism in community-industry relationships, both in the autonomy that Brinex granted individual citizens to "participate only to the extent they wish," and in the increasing autonomy that the new political context granted to local voices such as the Labrador Inuit Association to express their own concerns.

In the early twenty-first century, uranium plays a central role in highlighting the contradictory neoliberal engagement with Aboriginal self-governance. With the creation of Nunatsiavut, Inuit are now "empowered" to control their own government and region, but as such, as Linda Tuhiwai Smith says about another indigenous situation, the Inuit are perhaps "made responsible for their own oppression and freedom."[70] They have taken advantage of the space offered by neoliberal policies to advance their claims, and now have extraordinary authority over some aspects of governance in Nunatsiavut, including the freedom to make many of their own decisions about issues such as uranium mining. But hand in hand with this authority comes pressure to achieve economic self-sufficiency. The limited Inuit land base, diminished through historical dispossessions, relocations, policies that discounted the significance of the harvesting economy, and, finally, the land claims negotiations, must now support the needs of the Nunatsiavut Government and its beneficiaries. Inuit may therefore have the authority to ban uranium mining, but, within the neoliberal framework of the land claim agreement, they face significant pressure to develop their limited resources.

ACKNOWLEDGMENTS

Andrea Procter would like to thank the people of Nunatsiavut, the Labrador Institute, SSHRC, David Natcher, Jean Briggs, Adrian Tanner, August Carbonella, John Sandlos, and Kirk Dombrowski for their support of her research and thoughtful critiques of this chapter.

NOTES

1. Charles Hale, "Neoliberal Multiculturalism: The Remaking of Cultural Rights and Racial Dominance in Central America," *PoLAR* 28, no. 1 (2005): 10–28.

2. Noel Castree, "Neoliberalism and the Biophysical Environment 1: What 'Neoliberalism' Is, and What Difference Nature Makes to It," *Geography Compass* 2, no. 12 (2010): 1725–33; David Harvey, *A Brief History of Neoliberalism* (New York: Oxford, 2005).

3. Gabrielle Slowey, *Navigating Neoliberalism: Self-Determination and the Mikisew Cree First Nation* (Vancouver: UBC Press, 2008), xv; Castree, "Neoliberalism."

4. Hale, "Neoliberal Multiculturalism"; Gabriela Valdivia, "On Indigeneity, Change, and Representation in the Northeastern Ecuadorian Amazon," *Environment and Planning A* 37 (2005); Harvey Feit, "Neoliberal Governance and James Bay Cree Governance: Negotiated Agreements, Oppositional Struggles, and Co-governance," in *Indigenous Peoples and Autonomy: Insights for a Global Age*, eds. M. Blaser, R. de Costa, D. McGregor, and W. Coleman (Vancouver: UBC Press, 2010), 49–79; Elizabeth Povinelli, *The Cunning of Recognition: Indigenous Alterities and the Making of Australian Multiculturalism* (Durham, NC: Duke University Press, 2002).

5. Peter Evans, "Abandoned and Ousted by the State: The Relocations from Nutak and Hebron, 1956–1959," in *Settlement, Subsistence, and Change among the Labrador Inuit: The Nunatsiavummiut Experience*, eds. D. Natcher, L. Felt, and A. Procter (Winnipeg: University of Manitoba Press, 2012); Carol Brice-Bennett, "The Redistribution of the Northern Labrador Inuit Population: A Strategy for Integration and Formula for Conflict," *Zeitschrift fur Kanada-Studien* Nr. 2, Band 26 (1994).

6. John Kennedy, *People of the Bays and Headlands: Anthropological History and the Fate of Communities in the Unknown Labrador* (Toronto: University of Toronto Press, 1995).

7. William Rompkey, *The Story of Labrador* (Montreal: McGill-Queen's University Press, 2003).

8. Adrian Tanner, John C. Kennedy, Susan McCorquodale, and Gordon Inglis, "Aboriginal Peoples and Governance in Newfoundland and Labrador," a report for the Governance Project, Royal Commission on Aboriginal Peoples (St. John's, NL, 1994); Edward Tompkins, "Pencilled Out: Newfoundland and Labrador's Native People and Canadian Confederation, 1947–1954," report written for Jack Harris, MP (Ottawa: House of Commons, 1988); Evelyn Plaice, "Response to Kuper's Return of the Native," *Current Anthropology* 44, no. 3 (2003).

9   Philip Smith, *Brinco: The Story of Churchill Falls* (Toronto: McClelland and Stewart Ltd., 1975).

10 Plaice, "Response to Kuper's."

11 Smith, *Brinco*.

12 Ibid., 76.

13 John Kennedy, "Local Government and Ethnic Boundaries in Makkovik, 1972," in *The White Arctic: Anthropological Essays on Tutelage and Ethnicity*, ed. Robert Paine (St. John's, NL: ISER, 1977); Evans, "Abandoned and Ousted by the State."

14 Brice-Bennett, "Redistribution of the Northern Labrador Inuit Population," 78, 87.

15 Carol Brice-Bennett, *Reconciling with Memories: A Record of the Reunion at Hebron Forty Years after Relocation* (Nain, NL: Labrador Inuit Association, 2000), 84.

16 S. J. Hefferton (Minister of Public Welfare, Government of Newfoundland), "Letter to Minister of Citizenship and Immigration, Ottawa, 26 March 1959," in Government of Newfoundland (Department of Public Welfare), 1964, The Administration of Northern Labrador Affairs, Appendix 1.

17 Frank Tester and Peter Kulchyski, *Tammarniit (Mistakes): Inuit Relocation in the Eastern Arctic* (Vancouver: UBC Press, 1994); Harvey, *Brief History*; Michael Watts, "Development and Governmentality," *Singapore Journal of Tropical Geography* 24, no. 1 (2003).

18 S. Ben-Dor, *Makkovik: Eskimos and Settlers in a Labrador Community* (St. John's, NL: ISER, 1966); John Kennedy, *Holding the Line: Ethnic Boundaries in a Northern Labrador Community* (St. John's, NL: ISER, 1982); Brice-Bennett, *Reconciling*; Evans, "Abandoned and Ousted by the State."

19 Rosina Pamack-Jeddore, "What Confederation Has Meant to the Labrador Eskimo." *Decks Awash* 3, no. 5 (1974): 7.

20 Evans, "Abandoned and Ousted by the State."

21 James Overton, "Progressive Conservatism? A Critical Look at Politics, Culture, and Development in Newfoundland," in *Ethnicity in Atlantic Canada*, Social Science Monograph Series 5 (Saint John: University of New Brunswick, 1985).

22 The Trudeau government's 1969 White Paper suggested that the Indian Act be abolished, thereby severing all ties between the Aboriginal peoples under the Act and the federal government. Aboriginal peoples in Labrador were not under the Indian Act at the time, and so were not directly affected, but they were influenced by the Aboriginal rights movement that grew as a result of the White Paper (Tanner et al., "Aboriginal Peoples and Governance").

23 By the early 1970s, most of the Inuit who had been relocated to Makkovik from Nutak and Hebron in the 1950s had moved to other communities, predominantly Nain (Brice-Bennett, "Redistribution of the Northern Labrador Inuit Population").

24 Terje Brantenberg, "Ethnic Commitments and Local Government in Nain, 1969–76," in *The White Arctic*; John Kennedy, "Aboriginal Organizations and Their Claims: The Case of Newfoundland and Labrador," *Canadian Ethnic Studies* 19, no. 2 (1987): 13–25; Carol Brice-Bennett, ed., *Our Footprints Are Everywhere: Inuit Land Use and Occupancy in Labrador* (Nain, NL: Labrador Inuit Association, 1977).

25 Kennedy, "Aboriginal Organizations."

26 Enoch Obed, *Kinatuinamut Ilingajuk* (Nain, August 10, 1979), 1, 3; John Kennedy, "The Changing Significance of Labrador Settler Ethnicity," *Canadian Ethnic Studies* 20, no. 3 (1988); Brantenberg, "Ethnic Commitments"; V. Haysom, "Labrador Inuit Land Claims: Aboriginal Rights and Interests vs. Federal and Provincial Responsibilities and Authorities," *Northern Perspectives* 18, no. 2 (1990).

27 Environmental Assessment Board, "Report of the Environmental Assessment Board: Brinex Kitts-Michelin Uranium Project" (St. John's, NL: EAB, 1980), 42.

28 Linda Fong, "An Overview of Some of the Issues, Attitudes, and Events Concerning the Proposed Uranium Development in the Kitts-Michelin Area of Labrador" (Happy Valley, NL: Labrador Resources Advisory Council, 1977).

29 Evelyn Plaice, "'Making Indians': Debating Indigeneity in Canada and South Africa," in *Culture Wars: Context, Models, and Anthropologists' Accounts*, eds. Deborah James, Evelyn Plaice, and Christina Toren (New York: Berghahn Books, 2010); Lawrence Dunn, "Negotiating Cultural Identity: Conflict Transformation in Labrador" (PhD diss., Syracuse University, 2002).

30 On the other hand, the "way of life" concept was common in resource conflicts with Aboriginal peoples throughout Canada in the 1970s. The influential Berger Inquiry of the early 1970s concerning the Mackenzie Valley pipeline discussed in great detail the dynamics and the importance of maintaining Aboriginal ways of life. Anthropologist Michael Asch argues that negotiations concerning Aboriginal rights of the 1970s and early 1980s focused on Aboriginal "way of life" rights, as opposed to political rights, such as the right to self-government. This conceptualization of "way of life" rights, however, seems to be limited to socio-economic harvesting practices, and not the wider political aspect of self-determination that is inherent in Makkovik residents' use of the term in Labrador. Michael Asch, "From Calder to Van der Peet: Aboriginal Rights and Canadian Law, 1973–1996," in *Indigenous*

Peoples' Rights in Australia, Canada, and New Zealand, ed. P. Havemann (Auckland, NZ: Oxford University Press, 1999).

31   Kennedy, *Holding the Line*; Terje Brantenberg, "Ethnic Values and Ethnic Recruitment in Nain," in *The White Arctic*.

32   Government of Newfoundland and Labrador, "Statement by Premier A. Brian Peckford on the Question of Native Land Claims in the Province" (Telecommunication to Toby Andersen, President of Labrador Resources Advisory Council, Happy Valley, Labrador, October 10, 1980).

33   Many organizations stood behind the Aboriginal rights movement both nationally and in Labrador in 1979, including the Labrador Resources Advisory Council, town councils, and the United and Anglican churches.

34   J. C. O'Rourke, "Comments on Public Hearings," in *Summary of Kitts-Michelin Hearings* (St. John's, NL: Brinex Ltd., 1979), 2.

35   *Summary of Kitts-Michelin Hearings*, December 10, 1979, 4

36   Murray Poloski, in "Minutes of Public Meeting" (Makkovik, NL: Labrador Resources Advisory Council, July 27–28, 1976), 12.

37   Environmental Assessment Board, "Report," 4.

38   Haysom, "Labrador Inuit Land Claims."

39   Government of Newfoundland and Labrador, "Statement by Premier A. Brian Peckford"; Government of Newfoundland and Labrador, "Land Claims Policy" (St. John's, NL: Intergovernmental Affairs Secretariat, Native Policy Unit, 1987).

40   Louise Humpage, "Tackling Indigenous Disadvantage in the Twenty-First Century: 'Social Inclusion' and Maori in New Zealand," in *Indigenous Peoples and Poverty: An International Perspective*, eds. Robyn Eversole, John-Andrew McNeish, and Alberto Cimadamore (London: Zed Books, 2005); R. Maaka and A. Fleras, *The Politics of Indigeneity: Challenging the State in Canada and Aotearoa New Zealand* (Dunedin, NZ: University of Otago Press, 2005); Stephen Cornell, "Indigenous Peoples, Poverty, and Self-Determination in Australia, New Zealand, Canada, and the United States," in *Indigenous Peoples and Poverty*, 199–255; Nancy Fraser, *Justice Interruptus: Critical Reflections on the "Postsocialist" Condition* (New York: Routledge Press, 1997).

41   William Andersen III, "Address by William Andersen III at the Opening of Land Claims Negotiations, 22 January 1989," *Northern Perspectives* 18, no. 2 (1990): 5.

42   Labrador Institute of Northern Studies, "Labrador in the 90s" (Happy Valley–Goose Bay, Labrador, NL: Oct 29–Nov 1, 1990): 56.

43 Watts, "Development"; Michel Foucault and Michel Senellart, *The Birth of Biopolitics: Lectures at the College de France, 1978–79* (New York: Palgrave Macmillan, 2008); Castree, "Neoliberalism."

44 Indian and Northern Affairs Canada, Labrador Inuit Land Claims Agreement (Ottawa: 2005).

45 Andy Hoffman, "Labrador Goes Radioactive," *Globe and Mail*, June 28, 2007; Cameco spot price, accessed February 5, 2014, http://www.cameco.com/investors/markets/uranium_price/spot_price_5yr_history/.

46 Stephen Stakiw, "Aurora Builds Labrador Uranium Inventory," *Northern Miner*, March 5, 2007, 6–7.

47 Nunatsiavut Government, press release: "October 15, 2007 – Nunatsiavut Government Assembly begins fall session in Hopedale with the tabling of a motion on uranium development in Labrador Inuit Lands, the tabling of audited financial statements and reports related to the status of women and the announcement of a new pharmacy" (Hopedale, NL: Nunatsiavut Government, 2007).

48 Fieldnotes, February 2008.

49 Fieldnotes, March 2008.

50 Todd Broomfield, *Nunatsiavut Government Hansard*, April 2008.

51 Alisha Hiyate, "Nunatsiavut Government Mulls Uranium Moratorium: Motion May Ice Aurora's Michelin," *Northern Miner*, November 26, 2007.

52 Todd Broomfield, *Nunatsiavut Government Hansard*, October 2007.

53 Fieldnotes, March 2008.

54 Fieldnotes, March 2008.

55 Fieldnotes, June 2009.

56 Fieldnotes, February 2008.

57 Fieldnotes, February 2008.

58 Indian and Northern Affairs Canada, "Labrador Inuit Land Claims Agreement."

59 Fieldnotes, April 2008.

60 Fieldnotes, February 2008.

61 Keith Russell, *Nunatsiavut Government Hansard*, April 2008.

62 Foucault and Senellart, *The Birth of Biopolitics*; Tania Li, "Governmentality," *Anthropologica* 49, no. 2 (2007): 275–81.

63 Fieldnotes, April 2008.

64 Ibid.

65 Fieldnotes, April 19, 2008.

66 Aurora Energy Ltd. "Aurora Community Newsletter," Vol. 4 (July 2008): 1.
67 Cris Shore and Susan Wright, "Policy: A New Field of Anthropology," in *Anthropology of Policy: Critical Perspectives on Governance and Power*, eds. Cris Shore and Susan Wright (London: Routledge, 1997).
68 Ibid., 231.
69 Fieldnotes, July 2008.
70 Linda Tuhiwai Smith, "The Native and the Neoliberal Down Under: Neoliberalism and 'Endangered Authenticities,'" in *Indigenous Experience Today*, eds. Marisol de la Cadena and Orin Starn (Oxford: Berg, 2007), 345.

| CHAPTER 9

Privatizing Consent? Impact and Benefit Agreements and the Neoliberalization of Mineral Development in the Canadian North

Tyler Levitan and Emilie Cameron

This chapter considers the extent to which impact and benefit agreements (IBAs)—agreements between indigenous communities and mining companies that seek to extract resources from their traditional territories—relate to broader processes of neoliberalization in Northern Canada.[1] Although IBAs have been a focus of scholarly inquiry for some time, scholars have not yet explicitly theorized IBAs in relation to the broader political-economic context within which they unfold, or in relation to the neoliberalization of resource and indigenous governance in Canada.[2] Drawing on interviews and participant observation in the Northwest Territories (NWT), analysis of federal and territorial policy, and a critical review of various academic literatures, we consider the ways in which IBAs relate to these broader formations.[3]

We begin with a brief overview of IBAs and the history of their development in the Canadian North. Next, we discuss key insights into processes of neoliberalization in Canada, with an emphasis on the ways in which relations between indigenous peoples, corporations, and the state are being reconfigured. With these insights in mind, and drawing on our research in the NWT, we outline some of the ways in which IBAs might relate to processes of neoliberalization, including their role in removing barriers to accumulation of capital, privatizing the federal duty to consult and to accommodate indigenous peoples in regard to development on their lands, and naturalizing market-based solutions to social suffering. We conclude with some thoughts on the implications and limitations of these findings.

IBAs: OVERVIEW

IBAs are a relatively new component of the northern resource governance regime, having emerged only in the late 1980s and early 1990s.[4] Since then, IBAs have become a standard component of mineral development in the Canadian North; they are a de facto requirement for corporations interested in developing mines within the traditional territories of northern indigenous groups and have been negotiated in relation to every major mine proposed or developed since the late 1990s.[5] For indigenous communities, IBAs typically include provisions for employment quotas, skills training and other educational benefits, contracting and joint venture opportunities, financial compensation, environmental mitigation-related measures, and even culture-related benefits.[6] Industry proponents, for their part, secure good working relations with indigenous communities and enhance their "social licence to operate"—but they also, it should be noted, frequently negotiate clauses prohibiting public critique of the company and any form of protest against the mine on the part of indigenous community members, as well as other measures to secure ongoing consent.[7] It is typically industry proponents, moreover, that insist on the confidentiality of IBAs, although indigenous signatories have also advocated for and defended the confidentiality of IBAs. As such, the details of the range of impact-benefit agreements signed in the

North are not in the public domain, and are sometimes not made public even to members of signatory communities, and thus much of the literature on IBAs comments on the agreements in general terms.

Indeed, what distinguishes IBAs from other components of resource governance in the North (such as mining codes, Crown land regulations, environmental impact assessment processes, comprehensive land claim agreements, various forms of licensing, and so on) is their bilateral and private nature: they are typically signed between an industrial proponent and an indigenous government, with no direct involvement by federal or territorial government representatives or agencies, and no public policy framework guiding their negotiation, terms of reference, or implementation. The first IBAs signed in the North, as part of the development of the Ekati diamond mine in the NWT, remain the first and only IBAs in which government[8] played a formal role, insofar as Indian and Northern Affairs (INAC[9]) mandated that "satisfactory progress"[10] be made in negotiations regarding the realization of benefits and the mitigation of impacts for indigenous communities affected by the Ekati mine within a sixty-day period in 1996. INAC itself did not participate in the negotiations. Since then, despite numerous calls for a comprehensive federal policy to guide IBAs in the region, INAC (now Aboriginal Affairs and Northern Development Canada, or AANDC) has issued no such guidelines, and appears not to be actively pursuing the matter.

There is a relatively robust literature on IBAs, some of which emphasizes the practical benefits of IBAs for Aboriginal communities[11] and some of which raises concerns about the broader power relations shaping IBA negotiation and implementation.[12] Caine and Krogman argue, for example, that IBA researchers have thus far been insufficiently attentive to the ways in which power infuses not only the negotiation and implementation of IBAs, but also the broader social, political, and institutional context within which they have come to make sense. Their findings echo and extend concerns articulated in the late 1990s, when scholars first began to comment on IBAs after the federal government mandated the negotiation of IBAs for the Ekati diamond mine.[13] Since then, as IBAs have become an established part of the northern resource governance regime, research has tended to focus more on the extent to which IBAs enhance the equitability, sustainability, and benefits of mining than on critiques of the political-economic conditions under which they have emerged.[14]

Perhaps because IBAs do not formally and directly involve any government ministries, the role of the state in the development and implementation of IBAs has not been a primary focus of the literature. IBAs are frequently referred to as "supraregulatory,"[15] for example, and their role in northern resource extraction is primarily considered as a supplement or alternative to state regulations, policies, and practices. While Isaac and Knox[16] note that, ultimately, IBAs are subject to contract law and may well be tested by the judicial system, in general it is the *absence* of the state from IBA negotiation and implementation that tends to be noted by scholars. The implications of that absence have been probed by a number of authors. Prno and Slocombe understand IBAs as evidence of the emergence of "local communities" as "particularly important governance actors" in northern resource extraction, arguing that "conventional approaches to mineral development no longer suffice for these communities, who have demanded a greater share of benefits and increased involvement in decision making."[17] In such formulations, less state is good for local communities, insofar as the state has been understood as a barrier to community involvement in governing extractive activity. Others have been much more critical of the lack of active state involvement in this dimension of resource governance, both in relation to the imposed sixty-day timeline with respect to the Ekati IBAs and more broadly around the question of federal fiduciary responsibilities.[18]

In the wake of the government's ad hoc position with respect to Ekati, for example, a comprehensive discussion paper was prepared by Steven Kennett for the Mineral and Resources Directorate of the Department of Indian Affairs and Northern Development (DIAND) outlining a range of policy options that might clarify the government's position in relation to IBAs.[19] The report found that IBAs were already seen "as a *de facto* regulatory requirement" in the North because of DIAND's role in the Ekati process, but also because DIAND provided informal advice to mining companies and funding to indigenous organizations that were negotiating IBAs. But, because the negotiation and implementation of IBAs "lacks the procedural and substantive parameters normally associated with regulation," Kennett argued, "the roles of mining companies, aboriginal organizations and government are not properly defined and some inappropriate off-loading of responsibilities by government is occurring." Kennett also noted issues with identifying communities that were eligible to negotiate IBAs and the emergence of "marked inequalities" and "tensions" between communities who secure varying benefits under bilateral, confidential negotiations. He urged DIAND to develop a more comprehensive, reasoned policy on IBAs, a recommendation that, almost fifteen years later, has yet to be taken up.

Although it has not been studied in-depth, there is a distinction made in the literature between IBAs signed by communities with settled land claims and those whose territorial claims remain unresolved. Many comprehensive land claim agreements (CLCAs) include language around IBAs, and all CLCAs formally clarify surface and subsurface land rights within a given territory. To varying extents, CLCAs also clarify and establish the broader regulatory regime through which mineral development will be assessed. In regions where land claims have been settled, IBAs are thus often framed as one among several mechanisms whereby an indigenous community can obtain benefits from extraction. In communities without settled claims, however, IBAs may be one of the only means through which an affected community can realize limited benefits from extraction on their lands. Thus, Galbraith notes that respondents in her study "cynically" argued that "IBAs are like historical treaties between aboriginal groups and the federal government . . . these agreements stem from government and developers' interest in clarifying the legal rights of aboriginal people and aim to limit these rights with respect to the diamond mine development."[20] But while IBAs may act as a kind of substitute for treaties and land claims in regions where CLCAs have yet to be finalized, the conditions under which indigenous communities negotiate IBAs are vastly different than the conditions under which comprehensive land claims are negotiated. Because the federal government has provided "no principled basis for determining eligibility to negotiate IBAs in areas of unsettled land claims or where project impacts cross settlement area boundaries," and because development projects can proceed whether or not an IBA has been signed, mining companies are to some extent free to decide which communities they will negotiate with, and can threaten to terminate negotiations at any time.[21] Although it is part of the Akaitcho

Dene Nation (which has not yet signed a CLCA), for example, Deninu Kue (Fort Resolution, NWT) was left out of the IBA process for the Snap Lake mine despite arguing that their traditional territory would be greatly impacted, even while other Akaitcho communities were consulted.[22]

To the extent that the discovery of promising mineral deposits on indigenous lands has historically motivated governments to settle comprehensive land claims, moreover, IBAs may actually slow progress on CLCA negotiations, insofar as corporate and state interests in securing indigenous consent for specific developments are satisfied by IBAs. Indeed, Gogal et al. advise developers negotiating IBAs with Aboriginal groups who have not signed comprehensive land claim agreements to include "a covenant to the effect that the Aboriginal group will not advance any land claim that will negatively impact or impede the project."[23] In such cases, IBAs would seem to function not merely as temporary measures to ensure consent to a proposed development while CLCAs are under negotiation, but rather as explicit deterrents to the pursuit of comprehensive claims.

O'Faircheallaigh has made the important observation that while IBAs can offer clear and substantial benefits for participating indigenous communities, "they also have implications beyond the contractual relationship they create between a developer and a community that must be addressed if their contribution to community development is to be maximized."[24] IBAs, he argues, effectively curtail the "two powerful weapons" indigenous groups have historically used to intervene in the assessment and development of extractive projects on their lands: a) formal engagement in various legal and regulatory processes (such as environmental impact assessment hearings, court challenges, and so on) and/or direct action aimed at halting a project; and b) "the ability to embarrass government politically by using the media to appeal to its constituents."[25] Both "weapons" are contractually limited or prohibited by various clauses typical of IBAs.

In sum, the literature on IBAs has thus far established that the lack of federal policy and regulation with respect to the negotiation and implementation of IBAs has a range of repercussions for indigenous communities, that IBAs are functioning as a kind of substitute for the discharge of duties that have typically been the responsibility of the federal Crown (and, in many cases, remain Crown responsibilities), and that IBAs contain clauses that are themselves changing the relationship between indigenous peoples and the state. Thus far these shifts have been noted and critiqued by IBA scholars but have not yet been theorized in relation to broader shifts in resource and indigenous governance in Northern Canada, and particularly in relation to processes of neoliberalization.

NEOLIBERALISM AND INDIGENOUS PEOPLES IN CANADA

Although neoliberalization is commonly understood to involve a retrenchment of the welfare state, deregulation, and privatization, and its proponents frequently appeal to the notion of an excessive and overextended state to rationalize the rollback of various programs and services, it is crucial to note that such shifts do not represent the disappearance of the state so much as its restructuring.[26] Thus, "neoliberalization" typically involves the "rolling back" of traditional state functions and the subsequent "rolling out" of different state functions, policies, and practices that serve to deepen institutional linkages with globalized capital. As Peck observes, through the "rolling back" of the state, the state per se is not being hollowed out; what is being hollowed out is "a historically and geographically specific institutionalization of the state, which in turn is being replaced, not by fresh air and free markets, but by a *reorganized* state apparatus."[27]

The "rolling out" of the state, Peck argues, is characterized, in part, by non-governmental entities or third parties taking on some of the responsibilities and assets previously held by the state. In practice, neoliberal adjustments tend to "purge the system of obstacles to the functioning of 'free markets'; restrain public expenditure and any form of collective initiative; celebrate the virtues of individualism, competitiveness, and economic self-sufficiency; abolish or weaken social transfer programs while actively fostering the 'inclusion' of the poor and marginalized into the labour market, on the market's terms."[28] Over the last several decades, such adjustments have been documented in a number of jurisdictions, including Canada where, according to Albo, although identifiable much earlier, neoliberal ideology and policies came to dominate by the 1990s, regardless of the party in power.[29] We pay particular attention here to the commentary on how processes of neoliberalization have reshaped indigenous governance in Canada.

If there is any realm where a reorganization of state practices and functions might be welcomed in Canada, including by many indigenous leaders and communities, it is in the realm of indigenous–state relations. For indigenous peoples across Canada, the welfare state has been associated with some of the most egregious and illegitimate interventions into their lives, including residential schooling, the apprehension of indigenous children through child welfare systems, the promotion of Band governments over traditional forms of governance, the exertion of various forms of control over wildlife, hunting, health, education, and justice, ongoing amendments to the Indian Act, extraction of natural resources from indigenous lands, and other colonial policies and practices.[30] Opposition to the state, and identification of the colonial dimensions of state intervention into the lives of indigenous peoples, has thus been central to anticolonial and self-determination movements over the past several decades.[31] It should come as no surprise, then, that the devolution or dismantling of certain state functions and policies has been welcomed by some indigenous communities, who see rejection of the colonial state and the assertion of indigenous sovereignty as a cornerstone of self-determination.

Perhaps because of the resonance between indigenous challenges to state jurisdiction over their lands and lives and neoliberal rhetoric about the merits of "less state" and freedom from state intervention, several authors have identified a kind of convergence between neoliberal ideologies and indigenous self-determination movements. Thus, Slowey and MacDonald argue that neoliberal and indigenous critiques of the Canadian welfare state share the same disdain for the paternalism that both claim are endemic to the welfare state and the ways in which excessive state oversight impedes autonomy and self-determination.[32] The devolution of federal and provincial authority over indigenous education and child welfare to indigenous governments in various jurisdictions in Canada, for example, can be understood as both an off-loading of state responsibilities and a form of redress for colonial dimensions of state control over indigenous children and families.[33] While Slowey acknowledges that "the assumptions guiding federal policy and activity may not reflect First Nations concepts of self-determination," federal interest in devolving responsibilities to indigenous governments, she argues, "still facilitates the realization of First Nation's goals of economic self-reliance and jurisdictional autonomy."[34]

We are skeptical of this apparent convergence of interests, and follow scholars who insist that indigenous and state interests in neoliberal times are informed by distinctly different histories, material circumstances, and political commitments. Although the notion that there should be "less state" meets the objectives of some indigenous communities under some circumstances, indigenous communities, organizations, leaders, scholars, and activists across the country (including Northern Canada) consistently argue that the state's failure to meet its obligations and commitments undermines their well-being. The Idle No More movement that arose, in part, in response to federal omnibus bills aiming to weaken environmental legislation and "streamline" government, citizen, indigenous, and corporate involvement in various dimensions of resource governance, called for the implementation and enforcement of treaty and non-treaty relationships, and aimed to hold the federal government to account for its failure to meet its constitutional, fiduciary, treaty, and land claims obligations.[35] The movement explicitly rejected the notion that "less state," or the retreat of the state, responds to the demands of indigenous peoples in Canada.

Furthermore, while understandings of indigenous self-determination vary across and within the diverse indigenous communities impacted by resource development in Canada, the United Nations Declaration on the Rights of Indigenous Peoples not only delineates a range of indigenous rights (including the right to self-determination; the right to free, prior, and informed consent about any activities that might impact indigenous territories; and the right to determine land and resource use), it also emphasizes the role of the state in ensuring such rights are upheld and respected.[36] Asserting and enacting self-determination, in other words, can include holding the state accountable for its activities and obligations. There is an important distinction to be made between reorganizing the state to meet these obligations and reorganizing the state to further the interests of mining corporations.

Indeed, a number of scholars have expressed skepticism about the potential for neoliberal policies and practices to support indigenous self-determination and indigenous rights. Dempsey et al. argue, for example, that the neoliberal push to expand private property rights on First Nations reserves is based on a shallow idea of "equality" that assumes that equality with white settler society can only be experienced when Western notions of property are institutionalized, a notion that "is made possible only by a complete bracketing of historical geographies of dispossession."[37] The language of self-reliance and responsibility "hijacks" notions of self-determination, they argue, leaving First Nations with the opportunity to "self-determine within the narrow confines that are stipulated by the white capitalist elite."[38] Papillon points to the contradictory effects of neoliberalization in indigenous communities, arguing that it represents a new form of cultural and economic colonization insofar as it promotes resource extraction in indigenous territories and restricts the sovereign control of communities over affected lands.[39] MacDonald similarly observes that "[n]eoliberal manifestations of indigenous autonomy or self-government are . . . vulnerable to criticisms launched against practices of privatization. These practices include a variety of policies that promote shifting contentious issues out of the public sphere, thereby limiting public debate and collective—that is, state—responsibility."[40] Finally, Kuokkanen challenges Slowey's contention that neoliberal shifts in indigenous–state relations reduce dependency. In the case of the Mikisew First Nation, she argues that while the nation's dependency on "the government might have decreased to some extent, it has only created a new dependency on corporations exploiting the natural resources in the Mikisew territory."[41] As Kuokkanen makes clear, the retreat of the state is often accompanied by an intensification and acceleration of corporate involvement in indigenous lands and governance.

It is by no means certain, then, that indigenous peoples are better placed to advance their interests through the forms of privatization, marketization, and individualization that underpin processes of neoliberalization in Canada, even while the retreat of the state from certain dimensions of indigenous peoples' lives has been welcomed by some. Howlett et al. thus call for research that engages the "not-so-positive implications for Indigenous peoples" of neoliberal reforms in order to develop a more comprehensive picture of the ways that neoliberalization exacerbates "the inequalities that already exist for Indigenous peoples."[42] But just as there is a danger of overemphasizing the benefits of neoliberal reforms, we share Feit's concern about simplistic analyses that cast indigenous peoples as either naive victims of neoliberalization or active opponents, and echo his emphasis on the importance of nuanced, contextualized accounts of contemporary shifts in indigenous, state, and corporate relations.[43]

IBAs AND THE NEOLIBERALIZATION OF RESOURCE AND INDIGENOUS GOVERNANCE

How might we understand IBAs, then, in relation to processes of neoliberalization in Canada? As noted, IBAs have not been explicitly theorized in relation to neoliberalism in Canada, although scholars have identified a number of issues associated with IBAs that might productively be read as manifestations of neoliberal processes and policies.[44] We consider, here, the ways in which IBAs relate to neoliberal processes of marketization, privatization, and individualization in the Canadian North, and to broader state interests in ensuring that large-scale, industrial resource extraction proceeds in the region. Drawing on interviews conducted with indigenous leaders, government representatives, consultants, and lawyers involved in the negotiation, implementation, and regulation of IBAs in the Northwest Territories, as well as on a review of key documents that have shaped the evolution (or lack thereof) of federal policy on IBAs over the past ten to fifteen years, we argue that the emergence of IBAs as a key dimension of resource governance in the North since the late 1990s has advanced neoliberal objectives in the region, which we understand to include the removal of barriers to accumulating capital; the privatization of state assets, functions, and services; and the promotion of market-based solutions for various social, economic, environmental, and political struggles.[45] These shifts are made to make sense through neoliberal discourses emphasizing entrepreneurial, individualized understandings of citizenship and social life.

Although it is possible to identify processes of neoliberalization operative both within and beyond the Canadian North, we follow Hayter and Barnes in refusing to assume "neoliberalism's hegemony" in this study.[46] We do not begin with the assumption that the Canadian North is wholly shaped by processes of neoliberalization, nor do we argue that IBAs are themselves a primarily neoliberal phenomenon. To make such claims would require much more attention to the specific geographies of neoliberalization in the region, and a more thorough study of the ways in which IBAs articulate with other institutional, historical, and political processes. Our analysis is both more preliminary and more schematic. We aim here to bring understandings of neoliberalism to bear on the study of IBAs in order to sharpen our analysis of precisely what IBAs *do*, and draw attention to the ways in which IBAs are both emblematic of and contribute to shifting relations between indigenous peoples, mining corporations, and the state.

Removal of barriers

If, in the 1970s and 1980s, northern indigenous peoples were understood by government and industry to be formidable and effective "barriers" to the development of industrial resource extraction in the region, to what extent do IBAs assist in the removal of such barriers? As discussed, IBAs contain features that act as explicit deterrents to community-level resistance, such as non-compliance provisions. Indeed, from an industry perspective, part of the purpose of IBAs is to ensure that there will be no interruptions to the project, and that there is certainty that the community (or at least its leadership) has consented to the development and will allow it to go forward without any hindrances. Non-compliance provisions—contained within a legally binding IBA—serve as what Caine and Krogman refer to as a "gag order,"[47] whereby indigenous groups cannot voice concerns in light of new information as the project proceeds. Along with confidentiality clauses that limit a community's capacity to seek allies by releasing details of the agreement that may assist in creating campaigns of public support and political pressure, non-compliance provisions are tools used by industry to mitigate the risk of indigenous resistance, which has historically been a significant barrier to the accumulation of capital through the extraction of natural resources in Canada.[48]

Securing indigenous consent for a given project through an IBA, whether or not it is accompanied by non-compliance and confidentiality clauses, and whether or not that consent is meaningful, informed, and legally binding,[49] is also an enormously efficient, cost-effective, and rapid means of addressing the single largest threat indigenous peoples have historically posed to resource development in Canada: their capacity to assert unceded claims to a given piece of land. Negotiations of comprehensive land claims, which themselves aim (from the federal perspective) to clarify surface and subsurface land rights, are vastly more expensive[50] and time consuming, and do not directly secure a developer's legal or social licence to develop a given project. As we detail in the following section, moreover, to the extent that IBAs are viewed as de facto (if not de jure) satisfaction of the Crown's duty to consult and accommodate indigenous peoples with respect to actual or potential infringement of their rights to a given territory, they act as particularly potent removals of barriers to capital accumulation in the region. In Nunavut, efforts to challenge the development of a proposed uranium mine have been thwarted, in part, by assertions that the land claims corporation (Nunavut Tunngavik Incorporated) has contractual obligations to proceed with its approval of the project, regardless of dissent among beneficiaries and the fact that environmental assessment processes are ongoing.[51] Here, the lack of clarity around what constitutes adequate, legitimate, and appropriate consultation for proposed resource development is acutely felt, and raises the question of whether bilateral agreements between developers and indigenous organizations can undermine public review and assessment processes.

Privatization

The privatization of state assets, functions, and services is considered a hallmark of neoliberal ideology and practice. IBAs, we argue, can be understood as a means by which the Crown's duty to consult and accommodate indigenous peoples is privatized. The duty to consult and accommodate rests solely with the Crown; as established through *Haida Nation vs. British Columbia*,⁵² the Crown is required to consult and accommodate indigenous peoples in Canada whenever their traditional rights may be infringed upon by resource development. From a legal standpoint, the Crown can delegate procedural aspects of this duty—third parties, for example, can work to determine real or potential infringement of Aboriginal rights and title with respect to a particular mineral development project.⁵³ Third parties are under no legal obligation to consult with and accommodate Aboriginal peoples, however, as that obligation is assumed entirely by the Crown. But as Fidler and Hitch observe, in practice, governments increasingly operate as though the Crown's duties have been effectively relieved through IBAs, insofar as a signed IBA is taken as evidence that the indigenous nation or community in question has been both consulted and accommodated in relation to a proposed resource extraction project.⁵⁴ This observation is widely shared by government, legal, and indigenous sources interviewed as part of this research, and corroborated by corporate sources.⁵⁵ As a consultant working in the indigenous governance sector put it, the Crown's approach is:

> instead of being active, being more passive but watching and monitoring. I think they hold their nose over what IBAs are, but also part of them likes the fact that they're being done because it sort of gets them off the hook for taking care of their consultation and accommodation responsibilities. The company's going to do it and get the First Nation to sign; then really the Crown can sort of wash its hands and say "consultation and accommodation accomplished. Problem solved. We don't have to get involved. Someone else has solved it for us."⁵⁶

Indeed, the process of consulting and accommodating indigenous governments for every potential infringement of Aboriginal rights and title is extremely cumbersome for the state, and there is certainly a financial incentive for the state to delegate this duty to the private sector. A lawyer who has been involved with IBA negotiations in British Columbia and in the North observed, moreover, that "savvy" mining companies endeavour to include language in their IBAs that specifically characterizes the IBA as satisfying the Crown's duty to consult and accommodate.⁵⁷ Such language would not likely hold up in court,⁵⁸ but it demonstrates the extent to which a historically defined relationship between the state and indigenous peoples has been delegated to the private sector. Similarly, Gogal et al. advise developers pursuing IBAs to keep the Crown informed "of the substance of the negotiations" and ensure the Crown "is satisfied with the level of consultation, and, if possible, signs off on the IBA" to protect against future challenges to the legitimacy of the IBA as satisfying the Crown's duty.[59] The Crown and the courts have justified the delegation of procedural aspects of consultation and accommodation to industrial proponents partly because the proponent is most familiar with the details of a proposed development project and is in a "financial position to offer mitigation and other benefits."[60] But as a high-ranking AANDC employee rightly observed, "if we want this to be an effective part of our duty—the Crown's duty to consult—then we need to know what's in them,"[61] and confidentiality provisions precisely delimit such knowledge.

The de facto satisfaction of the Crown's duty to consult and accommodate has been rationalized, in part, through neoliberal discourses of efficiency and enhanced individual agency. Crown–indigenous relations that have historically (and constitutionally) been framed in terms of rights and responsibilities are being shifted to the private sector in the name of cost effectiveness, timeliness, efficiency, and the enhancement of local or indigenous agency. Indeed, AANDC not only appears to see itself as appropriately absent from IBA negotiations, but when asked about the unequal playing field within which IBAs are negotiated, a high-ranking AANDC employee responded:

> You can't say to a company "you will have less lawyers than these guys. You will look out for their interests." This is a naturally evolving equilibrium here. It's a free economy. Like I said, if you were living in Russia or China or in a communist system—there has to be some sort of a free market that the system itself establishes its rule.[62]

Here, neoliberal discourses of a "free market" economy are invoked to justify the negotiation of agreements between multinational corporations and indigenous communities, "stakeholders" with vastly different legal and consulting resources at their disposal. Cognizant of the risks of

IBA negotiations being (or appearing to be) one-sided, Gogal et al. advise developers to include provisions indicating that the negotiations have not been coerced, and even to provide funding to an indigenous group to assist with securing independent counsel and consulting assistance. The authors go on to suggest that, faced with both the practical requirement to negotiate IBAs and the lack of certainty about whether IBAs can secure legally enforceable, ongoing access to a given territory (what they term "a problem that has no clear answer"), "building meaningful business partnerships" with indigenous groups "may be the best solution."[63] Here, business partnerships are not just ideologically sensible but also a profoundly practical and efficient solution to the "problem" of unresolved indigenous territorial claims.

Although AANDC sources frame IBA negotiations as appropriately regulated by a "free economy," not all stakeholders are satisfied with the government's lack of involvement in this dimension of resource governance in the North. As an indigenous negotiator in the BHP-Ekati IBA negotiations pointed out:

> Government shouldn't just wash their hands. They have a fiduciary obligation which clearly states that when Aboriginal peoples are in a situation that requires certain resources, to ensure that there's a fair treatment of their citizens in their own territory, and that they're there to ensure that things are done properly and that industry isn't just ramrodding whatever they want through with Aboriginal people in their own homeland. So it's a concern.[64]

Here, the ongoing failure of the federal government to develop a policy on IBAs becomes particularly significant. When asked in an interview whether he was aware of any follow-up to his 1999 report, Steven Kennett commented that he "handed it off to them; they thanked me very much. I think there were a few comments from people, but when I stopped tracking this issue a long time ago, there was no indication they would do anything."[65] A broad review of northern regulatory systems commissioned by DIAND in 2008 specifically recommended that the "federal government should give priority to developing an official policy on the purpose, scope and nature of Impact Benefit Agreements in the North,"[66] but this, too, appears not to have resulted in meaningful action. Interviews with both Government of the Northwest Territories (GNWT) and AANDC employees confirm that there has been no apparent movement on the IBA file.[67] Meanwhile, as Gogal et al. observe, although "once seen as the Crown's fiduciary obligation, the duty to consult has now shifted to the developer,"[68] even if in legal terms it remains wholly and clearly the responsibility of the Crown.

Market-based solutions to social problems

The past decade has seen a marked rise in discourses of indigenous entrepreneurialism and market-based understandings of indigenous–state relations and indigenous economic, social, and cultural well-being. From treatises (and, more recently, proposed federal legislation) advocating the privatization of reserve lands to celebrations of indigenous–corporate joint ventures and indigenous-run corporations, it has become increasingly common to associate the aims of increased well-being, independence, autonomy, and self-determination with involvement in capitalist labour, property, and investment markets.[69] As a number of scholars have made clear, these shifts are highly affiliated with neoliberalism.[70]

IBAs, we would argue, are consistent with such shifts. Not only do they promote market-based solutions to various social, economic, and political problems, but in a time of reduced federal spending on social transfer programs, as well as a lack of spending on infrastructure needed in northern, predominately indigenous communities, the benefits resulting from contracts with mining companies have in many ways replaced the traditional role of the state in providing for these vital services and infrastructural needs.[71] While the state continues to provide subsidies to the private sector,[72] private industry is becoming the primary point of contact for some indigenous communities regarding the impacts and benefits of extraction. Industry is increasingly involved in the provision of social, environmental, and cultural services and benefits in the North, as well as in the development of roads, airports, and other physical infrastructure, and in monitoring and assessment processes.

Indeed, as a number of scholars have observed, indigenous signatories aim to use IBAs to extract benefits that they have repeatedly failed to secure from federal and territorial governments.[73] But while government is often—and rightly—blamed for creating many of the conditions under which northern indigenous peoples experience social, economic, cultural, and political suffering,[74] and while *less* government intrusion into the lives of indigenous peoples is often demanded by indigenous communities, the obligation of the state to redress colonial policies and practices, to honour historical treaties and contemporary claims, and to provide adequate health, educational, judicial, and other services is also emphasized by northerners. To the extent that IBAs act as market-based proxies for satisfying obligations that would otherwise be the responsibility of the state, they are in line with a broader neoliberalization of indigenous–state relations in Canada, and they fortify the notion that jobs and cash will resolve the structural, systemic, historically informed struggles faced by northern indigenous peoples.

Consider, for example, the preponderance of employment quotas, job-training programs, and other educational and employment benefits in IBAs. The federal government has aimed to integrate northern indigenous peoples into the wage labour market for decades,[75] but thus far the educational system has not been intimately tied to particular industrial requirements. IBAs act as a mechanism for integrating public education with the employment needs of mining companies, which in communities with severely restricted employment and educational opportunities can significantly shape available options. Concerns have been raised by northerners that the explicit reorientation of educational and training programs to support industry needs will further restrict the development of a professional class in northern communities, and entrench the racialized division of labour already documented at northern mine sites, where indigenous peoples predominantly work as labourers, and non-indigenous, transient workers occupy professional, skilled managerial roles.[76] But within market-based assessments of indigenous well-being, an increase in waged work of any kind is seen as a measure of success, even though mine work is linked to increased rates of sexually transmitted infections (STIs), substance abuse, and family violence, and even though mines themselves pose risks to land-based economies.[77] Indeed, when asked about the role of IBAs in the contemporary North, a former NWT politician characterized IBAs as "business agreement[s] between one company and another"[78] and emphasized the capacity for IBAs to provide "tools" to facilitate the integration of indigenous peoples into the labour market. Here, the historical lack of indigenous involvement in waged work is framed as a matter of missed opportunity and lack of capacity, not as a conscious choice to engage in land-based and mixed economies, and not as a function of colonial histories of education and employment in the North.

Unlike Crown obligations to indigenous peoples, which predate the formation of the Canadian state and are grounded in a larger nation-to-nation framework (as well as being reaffirmed in the Constitution and clarified in case law), the obligations of developers are reduced to their delineation in the IBA itself and are ultimately subject to contract law. This distinction is most acutely felt in regard to designating responsibility for ongoing, unplanned, or significant impacts. In discussing the ways in which IBAs intersect with environmental assessment processes, Gogal et al. note that environmental assessment "deals with mitigation of planned or known environmental effects," not long-term, unforeseen, or otherwise complex impacts. While IBAs "often include mitigation or remedial measures over and above those commitments made during the environmental assessment process," the authors warn that "Aboriginal groups will often want to negotiate through an IBA additional compensation for unplanned events or effects that are more significant than planned. Negotiation of such provisions should be approached with caution so as to avoid a 'bottomless pit' of compensation."[79] If, in fact, developers are securing measures to ensure they are not liable for a "bottomless pit" of compensation, and if the Crown is functioning as though IBAs satisfy their duty to consult and accommodate, it remains unclear who will address these more complex dimensions of extraction. As an indigenous IBA negotiator observed, although he is satisfied with the IBAs that he has been involved in, "IBAs [are] a quick fix. It's not going to solve our problems. I think it's going to make more problems for us in the future."[80] These "future problems," he noted, will arise when mines close and the retreat of government from key services and programs—justified, in part, because of IBAs—will be most acutely felt.

Finally, while the services, quotas, training programs, contracting, partnerships, and joint venture opportunities negotiated through IBAs are frequently framed as novel opportunities that enable a kind of latent entrepreneurialism to flourish in indigenous communities, and while the federal government has explicitly supported such developments,[81] it would seem that such "flourishing" also becomes a justification for government clawbacks. According to an indigenous leader from the Northwest Territories, the federal and territorial governments have encouraged the proliferation of IBAs in order to reduce their spending in indigenous communities:

> There's no doubt that [the government] encourage[s IBAs] because it lessens the pressures on their social purse strings . . . If you totally just go on the social purse, which is what all governments look at—it's to ensure you have the basic requirements of survival—but IBAs are a little bit different because it becomes a benefit. Governments will try to claw back on those benefits any time, every chance they get.[82]

Indeed, following the signing of IBAs regarding the Ekati mine, the GNWT began clawing back income support payments to welfare recipients from communities that were signatories to these agreements.[83] After nearly ten years of protest, the territorial government finally decided to allow up to $1,200 in IBA-related "gifts" to be received without affecting income support payments.[84] The clawbacks themselves resulted in minimal cost-saving for the GNWT, and it would seem that they were as much about reframing the terms of indigenous–government relations as about the savings themselves. According to O'Faircheallaigh, fear of government clawbacks motivates some indigenous signatories to keep the content of IBAs confidential.[85]

In sum, viewed through the lens of neoliberalism, several dimensions of IBA policy, negotiation, and implementation can be theorized in terms of broader political-economic shifts in indigenous and resource governance in the North. IBAs contribute to the removal of barriers to accumulation in the region, effectively privatize the federal duty to consult and accommodate indigenous peoples, and both facilitate and validate the development of market-based solutions to the historically-rooted, structural, and systemic challenges that confront northern indigenous peoples.

DISCUSSION AND CONCLUSION

Studies of neoliberalism have been subject to several important critiques. As some critics observe, the term "neoliberalism" has come to be associated with any and all shifts in social, economic, political, and environmental relations. According to Barnett, neoliberalism has become a catch-all term, a "consolation" for the Left, which actually "compound[s] rather than aid[s] in the task of figuring out how the world works and how it changes."[86] Indeed, just as it has become standard to invoke processes of neoliberalization in a wide range of studies of social, political, and economic restructuring, it has also become standard to acknowledge that the term itself can lack both conceptual clarity and analytical purchase. Furthermore, as Feit argues, many studies of neoliberalism provide simplistic analyses that cast indigenous peoples as either naive victims or active opponents rather than as empowered subjects facing complex choices.[87]

We are sympathetic to these critiques, and are reluctant to attribute the historically and geographically specific shifts in resource governance and indigenous–state relations in the North entirely to processes of neoliberalization. We note, for example, that far from being passive victims of a retreating, neoliberalizing state, some northern indigenous leaders and governments have in many ways *pushed* the federal government out of various dimensions of resource governance, insisting that that the state has no jurisdiction over resource development on indigenous lands. The responsibility to negotiate impacts and benefits related to resource development has not so much been downloaded onto indigenous communities by the federal government, these leaders insist, but rather indigenous peoples have asserted jurisdiction over what was always their responsibility, and in this sense, IBAs must be understood as a hallmark of self-determination.[88] While some of the benefits, services, and infrastructure northern indigenous peoples are securing through IBAs have previously been provided by the state, moreover, many have not; the most successful and comprehensive IBAs are celebrated for their capacity to bring meaningful benefits to northerners, benefits that governments have repeatedly failed to provide.

Furthermore, the historic timeline taken up in so many studies of neoliberalization, in which neoliberal ideologies begin to emerge in the 1970s in the United States and United Kingdom, and come to flourish by the 1990s in these and other jurisdictions, is an awkward fit in Northern Canada. Although one can certainly identify forms of state retrenchment in Northern Canada over the last several decades, lack of federal involvement in the region is anything but novel. The rapid expansion of state services in the North in the post–WWII era was an historical anomaly, insofar as previous governments had aimed to minimize federal involvement in the lives of indigenous northerners and ensure that they maintained land-based livelihoods (except, of course, in areas where the state aimed to extract natural resources). Indeed, as indigenous northerners are keenly aware, the state has quite happily engaged in both "laissez faire" and "aidez faire" policies for centuries, from the days of the fur trade (when a corporation, the Hudson's Bay Company, was permitted to govern Rupert's Land as it saw fit, so as to maximize fur trade profit) through to the granting of oil, gas, and mineral leases to corporations without regard to indigenous rights and claims. A state that promotes the interests of capital, seeks to limit its involvement in the lives of its citizens, and works to enrol its citizens in market-based, individualized, and privatized forms of life is nothing new to indigenous northerners.

What is new, we suggest, is the ways in which a neoliberal reorganization of the state is being framed as consistent with northern indigenous self-determination, and the alleged consensus that "less state" represents a win-win-win situation for indigenous communities, corporations, and the state. According to such framings, it is precisely around the notion of "less state" that indigenous peoples, mineral developers, and neoliberal governments find agreement. But as Peck and others have made clear, neoliberalization involves not so much *less* state as a *different* state, one organized to more fully accommodate the needs of capital, and one that relies upon its citizens to enact entrepreneurial, individualized forms of self-governance and self-regulation. Although framed as a morally and politically appropriate shift away from paternalism, the neoliberalization of indigenous–state relations and state involvement in northern resource extraction risks ushering in new forms of dependency, even while it is rightly celebrated by indigenous northerners as a rejection of long-standing, deeply colonial relations. As Kuokkanen observes:

> In attempts to break the cycles of dependency, poverty and dire socioeconomic circumstances in their communities, many indigenous groups and institutions have, as a part of their self-governance efforts, embarked on the path of neoliberal economic development which has often meant further exploitation of the natural resources in their territories, now in the form of joint ventures and in partnerships with corporations. What is surprising in these contemporary political efforts is that very little attention has been paid to the economic processes that played a significant role in creating dependency historically or linking the historical dependency creation to the contemporary forms of dependency on corporations and their conditions for partnerships which may include restructuring key institutions in indigenous communities.[89]

Indeed, while a rejection of state paternalism and an assertion of jurisdiction over land and livelihoods underpins many northern indigenous peoples' support for resource governance mechanisms like IBAs, it is far from clear that these mechanisms will result in precisely what they aim at: meaningful, long-term, effective control over development on indigenous lands.

Our intention in raising questions about IBAs and their affiliation with processes of neoliberalization, then, is not to condemn IBAs as components of northern resource governance. Indigenous signatories are acutely aware of the vulnerabilities of their communities to the whims of capital, and many framed their negotiation and signing of IBAs not as a panacea, but as an effort to secure needed (if limited) benefits and to assert control over a portion of resource development. Many indigenous groups redirect IBA monies toward community initiatives in an effort to ensure that IBAs do not simply work to "get you into the wage economy, and turn you from an Indian to a Canadian," but also help to "maintain your identity, your language, your culture."[90] We concur with Feit that insufficient attention has been paid to the diversity of relations between indigenous peoples and various neoliberal practices and shifts; identifying the neoliberal dimensions of IBAs does not imply that indigenous signatories have uniform relations, experiences, or interests in these agreements.[91] Our concern, here, has been with the broader political-economic work that IBAs perform, and their role in both facilitating and naturalizing significant shifts in indigenous–state relations and resource governance in the Canadian North. To the extent that IBAs remove barriers to capital accumulation; privatize federal duties and responsibilities; naturalize market-based solutions to social, political, economic, cultural, and environmental struggles; and delimit the capacities for members of signatory communities to assess proposed developments or to express concern and dissent, the framing of IBAs as expressions of convergence between indigenous, corporate, and state interests must be challenged.

In conclusion, then, the neoliberalization of northern resource governance and indigenous–state relations, as manifested in IBAs, gives us reason to echo Caine and Krogman's "healthy suspicion" that IBAs do what they claim to do, and that they are, in fact, the best available tool for ensuring that indigenous northerners secure benefits and minimize the impacts of extraction on their lands, particularly in regions without settled comprehensive claims.[92] When compared to past practices, where indigenous peoples gained almost no benefits from extraction on their lands, IBAs surely represent an improvement. But this improvement must be weighed against concerns that IBAs are merely a "quick fix"[93] whose primary function is to secure consent for projects with significant long-term effects and that IBAs absolve the Crown of its responsibility to consult and accommodate indigenous peoples, silence potential critics of mining development, delay resolution of comprehensive land claims, and naturalize individualized and entrepreneurial forms of citizenship and community. Viewed in this light, Caine and Krogman's healthy suspicion is thoroughly warranted.

ACKNOWLEDGMENTS

We gratefully acknowledge the research participants in this study, as well as funding from ArcticNet and the Northern Scientific Training Program. Portions of this chapter were published in the journal *Studies in Political Economy* in 2014.

NOTES

1. This chapter is a based on a journal article recently published by Emilie Cameron and Tyler Levitan entitled "Impact and Benefit Agreements and the Neoliberalization of Resource Governance and Indigenous–State Relations in Northern Canada," *Studies in Political Economy* 93 (2014): 23–50.
2. But see Cathy Howlet et al., "Neoliberalism, Mineral Development, and Indigenous People: A Framework for Analysis," *Australian Geographer* 42, no. 3 (2011): 309–23, for an analysis in the Australian context.
3. Twenty-one interviews were carried out between August and December 2011. Although we cannot guarantee anonymity, we have attempted to maintain the anonymity of research participants, as stipulated by the *Carleton University Research Ethics Board*.
4. Steven Kennett, *Issues and Options for a Policy on Impact and Benefit Agreements for the Northern Territories* (Calgary: Canadian Institute of Resource Law, 1999), identifies the signing of the Strathcona Sound/Nanisivik Agreement in 1974 as the first impact and benefit agreement in the North. It was an agreement between the federal government and Mineral Resources International Ltd., however, and involved little to no consultation with Inuit [see Chapter 10, this volume, and Tee W. Lim, "Inuit Encounters with Colonial Capital: Nanisivik – Canada's First High Arctic Mine" (MA thesis, Institute of Resources and Environmental Sustainability, UBC, 2013)]. The agreements negotiated in relation to the Ekati diamond mine thus represent the first IBAs in the North that did not involve government negotiators or signatories and that were directly negotiated by indigenous signatories. Not surprisingly, indigenous leaders celebrated this assertion of jurisdiction over their own affairs (I-4), even while they were critical of the federal government's sixty-day timeline mandate and the lack of funds made available to support negotiations.
5. "IBA Research Network," accessed November 30, 2012, http://www.impactandbenefit.com/IBA_Database_List/; NRCan, "Agreements between Mining Companies and Aboriginal Communities or Governments," accessed February 4, 2014, https://www.nrcan.gc.ca/mining-materials/aboriginal/14694.
6. Ginger Gibson and Ciaran O'Faircheallaigh, *IBA Community Toolkit: Negotiation and Implementation of Impact and Benefit Agreements* (Toronto: Walter and Duncan Gordon Foundation, 2010).
7. Jason Prno and Scott D. Slocombe, "Exploring the Origins of 'Social License to Operate' in the Mining Sector: Perspectives from Governance and Sustainability Theories," *Resources Policy* 37, no. 3 (2012): 346–57; Ken Caine and Naomi Krogman, "Powerful or Just Plain Power-Full? A Power Analysis of

Impact and Benefit Agreements in Canada's North," *Organization & Environment* 23, no. 1 (2010): 76–98; Gibson and O'Faircheallaigh, *IBA Community Toolkit*.

8   We use "government" in this chapter to refer to the specific governmental ministries or institutions in question (e.g., the federal Ministry of Aboriginal Affairs and Northern Development) and "state" to refer to the broader political, legal, military, police, and administrative institutions though which the interests of capital are maintained. Following Rianne Mahon, we emphasize the relative autonomy of the administrative apparatus of government in the development of public policy and do not conceptualize government ministries or bureaucrats as directly and uniformly implementing policy that uniquely serves particular interests (Rianne Mahon, "Canadian Public Policy: The Unequal Structure of Representation," in *The Canadian State: Political Economy and Political Power*, ed. L. Panitch (Toronto: University of Toronto Press, 1977), 165–98. We do, however, begin with the assumption that the Canadian state has a long-standing and ongoing interest in large-scale resource extraction in the Canadian North in particular and in capitalist economic development in general.

9   The federal ministry charged with "Indian affairs" and northern development has changed names several times over its history. Interviewees and publications cited in this study refer, variably, to INAC, DIAND (Department of Indian Affairs and Northern Development, in use after 1966), and AANDC (Aboriginal Affairs and Northern Development Canada, the name in place since 2011).

10 CIRL, *Independent Review of the BHP Diamond Mine Process* (Calgary: 1997), 15.

11 Lindsay Galbraith, Ben Bradshaw, and Murray Rutherford, "Towards a New Supraregulatory Approach to Environmental Assessment in Northern Canada," *Impact Assessment and Project Appraisal* 25, no. 1 (2007): 27–41; Jason Prno, "Assessing the Effectiveness of Impact and Benefit Agreements from the Perspective of their Aboriginal Signatories" (MA thesis, University of Guelph, 2007).

12 Caine and Krogman, "Power Analysis of Impact and Benefit Agreements"; Gibson and O'Faircheallaigh, *IBA Community Toolkit*.

13 Caine and Krogman, "Power Analysis of Impact and Benefit Agreements"; Janet Keeping, "Local Benefits and Mineral Rights Disposition in the Northwest Territories: Law and Policy," in *Disposition of Natural Resources: Options and Issues for Northern Lands*, eds. Monique Ross and John Owen Saunders (Calgary: Canadian Institute of Resources Law, 1997); Kennett, *Issues and Options*; Kevin O'Reilly and Erin Eacott, "Aboriginal Peoples and Impact and

Benefit Agreements: Summary of a National Workshop," *Northern Perspectives* 25, no. 1 (1999).

14 Prno, "Assessing the Effectiveness"; Prno and Slocombe, "Exploring the Origins of Social License"; for a critique of this trend in the literature, see Caine and Krogman, "Power Analysis of Impact and Benefit Agreements."

15 See Galbraith, Bradshaw, and Rutherford, "Towards a New Supraregulatory Approach."

16 Thomas Isaac and Anthony Knox, "Canadian Aboriginal Law: Creating Certainty in Resource Development," *University of New Brunswick Law Journal* 53 (2004): 3–42.

17 Prno and Slocombe, "Exploring the Origins of Social License," 354.

18 Janet Keeping, *Thinking about Benefits Agreements: An Analytical Framework*, prepared for Canadian Arctic Resource Committee, 1998; O'Reilly and Eacott, "Aboriginal Peoples and Impact and Benefit Agreements"; Kennett, *Issues and Options*.

19 Kennett, *Issues and Options*.

20 Lindsay Galbraith, "Understanding the Need for Supraregulatory Agreements in Environmental Assessment: An Evaluation from the Northwest Territories, Canada" (MA thesis, Simon Fraser University, 2005), 75–76.

21 Kennett, *Issues and Options*, ix.

22 Ginger Gibson, "Negotiated Spaces: Work, Home, and Relationships in the Dene Diamond Economy" (PhD diss., University of British Columbia, 2008); Tyler Levitan, "Impact and Benefit Agreements in Relation to the Neoliberal State: The Case of Diamond Mines in the Northwest Territories" (MA thesis, Carleton University, 2012); see also Ellen Bielawski, *Rogue Diamonds: Northern Riches on Dene Land* (Seattle: University of Washington Press, 2003) for a discussion of similar issues in relation to the Ekati mine. Interviewees allege that, although the federal government is not formally involved in regulating IBA negotiations, there is "cooperation between industry and government to determine who should get an IBA . . . I think government had some influence in terms of telling industry, 'Ya OK. You wanna leave them out? We won't squawk about it. We're not going to protest'" (I-4).

23 Sandra Gogal, Richard Reigert, and JoAnn Jamieson, "Aboriginal Impact and Benefit Agreements: Practical Considerations," *Alberta Law Review* 43, no. 1 (2006): 156.

24 Ciaran O'Faircheallaigh, "Aboriginal – Mining Company Contractual Agreements in Australia and Canada: Implications for Political Autonomy and Community Development," *Canadian Journal of Development Studies* 3, nos. 1–2 (2010): 74.

25 Ibid., 76.

26 Jamie Peck and Adam Tickell, "Neoliberalizing Space," *Antipode* 34, no. 3 (2002): 380–404; Leo Panitch and Sam Gindin, *The Making of Global Capitalism: The Political Economy of American Empire* (London: Verso, 2012).

27 Jamie Peck, "Neoliberalizing States: Thin Policies/Hard Outcomes," *Progress in Human Geography* 25, no. 3 (2001): 447.

28 Ibid., 445.

29 Greg Albo, "Neoliberalism, the State, and the Left: A Canadian Perspective," *Monthly Review* 54, no. 1 (2002): 46–55.

30 See, for instance, T. Alfred, *Peace, Power, Righteousness: An Indigenous Manifesto* (Don Mills, ON: Oxford University Press, 1999); S. de Leeuw, M. Greenwood, and E. Cameron, "Deviant Constructions: How Governments Preserve Colonial Narratives of Violence and Mental Health to Intervene into the Lives of Indigenous Children and Families in Canada," *International Journal of Mental Health and Addiction* 8 (2009): 282–95; P. Kulchsyki and F. Tester, *Kiumajut (Talking Back): Game Management and Inuit Rights 1900–1970* (Vancouver: UBC Press, 2007); V. Satzewich and T. Wotherspoon, *First Nations: Race, Class, and Gender Relations* (Regina: Canadian Plains Research Centre, 2000).

31 Alfred, *Peace, Power, Righteousness*; Dene Nation, "Dene Declaration," in *Dene Nation: The Colony Within*, ed. M. Watkins (Toronto: University of Toronto Press, 1975); P. Nadasdy, *Hunters and Bureaucrats: Power, Knowledge, and Aboriginal–State Relations in the Southwest Yukon* (Vancouver: UBC Press, 2003).

32 Gabrielle Slowey, *Navigating Neoliberalism: Self-Determination and the Mikisew Cree First Nation* (Vancouver: UBC Press, 2008); F. MacDonald, "Indigenous Peoples and Neoliberal 'Privatization' in Canada: Opportunities, Cautions, and Constraints," *Canadian Journal of Political Science* 44 (2011): 257–73.

33 See MacDonald, "Indigenous Peoples."

34 Slowey, *Navigating Neoliberalism*, 53.

35 See, for example, "Idle No More (INM)'s Canadian Concerns," *Indigenous Policy Journal* 24 (2013).

36 United Nations Declaration on the Rights of Indigenous Peoples, G.A. Res. 61/295, Annex, U.N. Doc. AIRES/61/295 (Sept. 13, 2007).

37 J. Dempsey, K. Gould, and J. Sundberg, "Changing Land Tenure, Defining Subjects: Neoliberalism and Property Regimes on Native Reserves," in *Rethinking the Great White North: Race, Nature, and Whiteness in Canada*, eds. A. Baldwin, L. Cameron, and A. Kobayashi (Vancouver: UBC Press, 2011), 4.

38 Ibid., 26; see also S. Pasternak, "How Capitalism Will Save Colonialism: H. De Soto, the Settler Colony of Canada, and the Privatization of Reserve Lands"

(paper presented at the Association of American Geographers Conference, New York, NY, February 25, 2012).

39 M. Papillon, "Les peuples autochtones et la citoyenneté: quelques effets contradictoires de la gouvernance néolibérale," éthique publique 14 (2012).

40 MacDonald, "Indigenous Peoples," 258.

41 R. Kuokkanen, "From Indigenous Economies to Market-Based Self-Governance: A Feminist Political Economy Analysis," *Canadian Journal of Political Science* 44 (2011): 285.

42 C. Howlett, M. Seini, D. Mcallum, and N. Osborne, "Neoliberalism, Mineral Development, and Indigenous People: A Framework for Analysis," *Australian Geographer* 42 (2011): 320.

43 H. Feit, "Neoliberal Governance and James Bay Cree Governance: Negotiated Agreements, Oppositional Struggles, and Co-governance," in *Indigenous Peoples and Autonomy: Insights for a Global Age*, eds. M. Blaser, R. de Costa, D. McGregor, and W. Coleman (Vancouver: UBC Press, 2010), 49–79.

44 Keeping is skeptical of the breadth of IBAs, arguing that benefits such as adult education for basic literacy should be provided by the state rather than industry, and arguing that IBAs are consistent with government retrenchment at all levels ("Local Benefits," 205). Sosa and Keenan have hinted at the neoliberal nature of IBAs as well, pointing out that "[i]n the context of government cutbacks to social programs and environmental regulation, the wide scope of these agreements and the reduced government role in their negotiation and execution has led to criticism that IBAs are a form of government downloading that sees companies act as welfare providers and communities as environmental watchdogs" [Irene Sosa and Karyn Keenan, "Impact Benefit Agreements between Aboriginal Communities and Mining Companies: Their Use in Canada" (Calgary: Canadian Environmental Law Association, 2001), 9]. Caine and Krogman ("Power Analysis of Impact and Benefit Agreements") make similar observations, as does O'Faircheallaigh ("Aboriginal – Mining Company Contractual Agreements"). None, however, have explicitly conceptualized IBAs in relation to processes of neoliberalization.

45 See Noel Castree, "From Neoliberalism to Neoliberalisation: Consolations, Confusions, and Necessary Illusions," *Environment and Planning D* 38 (2006); Wendy Larner, "Neo-liberalism: Policy, Ideology, Governmentality," *Studies in Political Economy* 63, no. 1 (2000); Peck and Tickell, "Neoliberalizing Space."

46 Roger Hayter and Trevor Barnes, "Neoliberalization and Its Geographic Limits: Comparative Reflections from Forest Peripheries in the Global North," *Economic Geography* 88, no. 2 (2012): 198.

47 Caine and Krogman, "Power Analysis of Impact and Benefit Agreements," 86.

48 Gibson and O'Faircheallaigh, *IBA Community Toolkit*.

49 IBAs have not yet been tested in the courts, and although some companies go so far as to include language stating the IBA satisfies the Crown's duty to consult and accommodate, this is unlikely to be proven in a court challenge (I-10; Isaac and Knox, "Canadian Aboriginal Law").

50 As O'Reilly and Eacott note ("Aboriginal Peoples and Impact and Benefit Agreements"), the community of Lutselk'e, NWT, was provided by INAC with only $7,500 to support their IBA negotiations with BHP in relation to the development of the Ekati mine, a grossly insufficient amount to allow for the necessary background research, translation, community consultation processes, and various legal and other demands.

51 Warren Bernauer, "Uranium Mining, Primitive Accumulation, and Resistance in Baker Lake, Nunavut: Recent Changes in Community Perspectives" (MA thesis, University of Manitoba, 2011); Nunavummiut Makitagunaraningit, "Discussion Paper – Kiggavik Draft Socioeconomic Impact Statement," June 2012, accessed November 10, 2012, http://makitanunavut.files.wordpress.com/2012/06/makita-socioeconomic-discussion-paper.pdf.

52 Haida Nation v. British Columbia (Minister of Forests), 2004, S.C.J. No. 70.

53 Gibson and O'Faircheallaigh, *IBA Community Toolkit*, 30.

54 Courtney Fidler and Michael Hitch, "Used and Abused: Negotiated Agreements" (paper presented at the "Rethinking Extractive Industry: Regulation, Dispossession and Emerging Claims" conference, York University, Toronto, 2009).

55 See Northern Regulatory Improvement Initiative, Submission by the Northwest Territories and Nunavut Chamber of Mines, the Prospectors and Developers Association of Canada, and the Mining Association of Canada to Neil McCrank, 2008.

56 I-1.

57 I-10.

58 Isaac and Knox, "Canadian Aboriginal Law."

59 Gogal, Reigert, and Jamieson, "Practical Considerations," 156.

60 Ibid., 133.

61 I-2.

62 I-5.

63 Gogal, Reigert, and Jamieson, "Practical Considerations," 156, 157.

64 I-4.

65 Stephen Kennett, pers. comm., December 14, 2011.

66   Neil McCrank, *Road to Improvement: The Review of the Regulatory Systems across the North* (Ottawa: Minster of Public Works and Government Services Canada, 2008), 21.

67   I-7; I-2; I-5; see also Government of the Northwest Territories, *Approach to Regulatory Improvement*, March 2009.

68   Gogal, Reigert, and Jamieson, "Practical Considerations," 157.

69   Pasternak, "How Capitalism Will Save Colonialism."

70   Ibid.; Dempsey, Gould, and Sundberg, "Changing Land Tenure"; Kuokkanen, "From Indigenous Economies"; Slowey, *Navigating Neoliberalism*.

71   We note, however, that the state has often failed to provide the very services and infrastructure that indigenous peoples are attempting to secure through IBAs. To suggest that the federal government is "retreating" from various forms of service provision in the North would be to imply that the state did, at one point, provide adequate services and infrastructure, and many northerners would no doubt challenge such a characterization.

72   It is crucial to note that many of the services, programs, and infrastructure associated with IBAs remain heavily subsidized by government, but they are framed as either wholly private or as public-private partnerships. See Levitan, "Impact and Benefit Agreements in Relation to the Neoliberal State."

73   Courtney Fidler and Michael Hitch, "Impact and Benefit Agreements: A Contentious Issue for Environmental and Aboriginal Justice," *Environments Journal* 35, no. 2 (2007): 45–69; Keeping, "Local Benefits and Mineral Rights"; Kennett, *Issues and Options*.

74   Stephanie Irlbacher-Fox, *Finding Dashaa: Self-Government, Social Suffering, and Aboriginal Policy in Canada* (Vancouver: UBC Press, 2009).

75   Frank Tester and Chris Flannelly, "The Rankin File: Public Responsibility, Private Provision, and Health Care at the Edge of the Canadian Liberal Welfare State" (paper presented at the ArcticNet Annual Scientific Meeting, Ottawa, Ontario, December 14–17, 2010); John Sandlos and Arn Keeling, "Claiming the New North: Mining and Colonialism at the Pine Point Mine, Northwest Territories, Canada," *Environment and History* 18, no. 1 (2012): 5–34.

76   Emilie Cameron and Warren Bernauer, "Accumulation, Dispossession, and Self-Determination: Wage Labour and the Expansion of Industrial Resource Extraction in Nunavut" (paper presented at the Association of American Geographers conference, New York, NY, February 26, 2012); Lindsay A. Bell, "Economic Insecurity as Opportunity: Job Training and the Canadian Diamond Industry," in *Humanizing Security in the Arctic*, eds. Michelle Daveluy, Francis Lévesque, and Jenanne Ferguson (Edmonton: Canadian Circumpolar Institute, 2011), 293–304.

77   Nunavummiut Makitagunaraningit, "Discussion Paper"; Ginger Gibson and Jason Klinck, "Resilient North: The Impact of Mining on Aboriginal Communities," *Pimatisiwin: A Journal of Aboriginal and Indigenous Community Health* 3 no. 1 (2005): 115–39.
78   I-5.
79   Gogal, Reigert, and Jamieson, "Practical Considerations," 139.
80   I-6.
81   NWT and Nunavut Chamber of Mines, *Sustainable Economies: Aboriginal Participation in the NWT Mining Industry, 1990–2004* (Yellowknife, NWT: 2004).
82   I-12.
83   Nathan VanderKlippe, "GNWT Reconsiders Clawbacks," *Northern News Service*, October 23, 2002, accessed November 30, 2012, http://www.nnsl.com/frames/newspapers/2002-10/oct23_02claw.html.
84   CBC, "NWT Announces Income Support Changes," August 2, 2007, accessed November 30, 2012, http://www.cbc.ca/news/canada/north/story/2007/08/02/nwt-income.html?ref=rss.
85   O'Faircheallaigh, "Aboriginal – Mining Company Contractual Agreements."
86   C. Barnett, "Critical Review: The Consolidations of 'Neoliberalism,'" *Geoforum* 36 (2005): 7, 10. See also R. Hayter and T. Barnes, "Neoliberalization and Its Geographic Limits: Comparative Reflections from Forest Peripheries in the Global North," *Economic Geography* 88 (2012).
87   Feit, "Neoliberal Governance."
88   I-15.
89   Kuokkanen, "From Indigenous Economies," 291.
90   I-8. Interview with indigenous leader and former IBA negotiator (September 2011), taperecorded, Behchoko, NWT.
91   Feit, "Neoliberal Governance."
92   Caine and Krogman, "Power Analysis of Impact and Benefit Agreements," 78.
93   I-6.

Section 3
Navigating Mine Closure

| CHAPTER 10

Contesting Closure: Science, Politics, and Community Responses to Closing the Nanisivik Mine, Nunavut

Scott Midgley

Hit by a decline in the price of zinc and with a looming deadline to pay millions of dollars for an annual order of sealift supplies, the Nanisivik lead-zinc mine on north Baffin Island closed in 2002 after twenty-six years of successful operation. News of Nanisivik's closure was delivered by Nunavut's territorial newspaper *Nunatsiaq News* with the morose headline "Nanisivik Mine to Die Four Years Early."[1] Similarly, some community members viewed Nanisivik as a deceased entity, eulogizing the personified Nanisivik during public meetings:

> In some ways, it's sad for me because it was a town for a long time, and we were working there, and we were friends with the people that I worked with, and Inuit from our communities were there too. And when you, one of your family members dies, it looks like you're losing some of your family members even the

non-Inuit there were — they too were your friends . . . It was emotional for me that I could still feel the life in that building.[2]

These statements mirrored classic narratives of mine closure that often consider ruination and dereliction as inevitable, in what some scholars call the "mining imaginary": the idea that mining is a linear process that must naturally terminate in ecological destruction and economic devastation.[3] In its productive phase, the Nanisivik mine extracted ores as well as economic value. Once the mine's operation halted, it seemed, the mine had died. The mine was no longer productive; the ore deposit was no longer valuable.

While acknowledging that ore deposits are finite resources and mines must inevitably close, this chapter encourages a different reading of Nanisivik's closure that moves away from the "mining imaginary." Rather than suggesting that the Nanisivik mine became an unproductive, lifeless, and valueless site after its closure, this chapter builds on arguments developed by geographers and historians such as Ben Marsh, William Wyckoff, and David Robertson who suggest that mining communities, memories, and legacies persist long after mining activity formally ends.[4] While analyses of industrial development and resource exploitation by scholars in fields such as resource geography often investigate how natural resources are culturally produced within particular socio-technical arrangements and historical-geographical circumstances,[5] these literatures focus on resource production rather than post-production when these arrangements and circumstances shift. Yet focusing on the process of closure itself reveals ways in which former sites of commodity production—their communities, economies, and environments—continue to be negotiated and transformed by numerous actors and reclamation practices in ways that upset the mining imaginary. Mine closure is a time when the history of a mine resurfaces, and the landscape, already transformed by mining, is remade again by closure and reclamation activities.

This chapter examines the ways in which Nanisivik's closed mining landscape became an object of experimentation, subjected to scientific activities that extracted environmental data from the mine site, produced scientific knowledge, and valued the cost of reclamation amid attempts to offset the environmental impacts of mining. Through an analysis of historic and contemporary documentary evidence relating to Nanisivik's opening and closure, this chapter first briefly introduces the history of mine development at Nanisivik and describes some of the impacts of mining on the nearby community of Arctic Bay. In a critique of the mining imaginary, the next section explains how the mine company CanZinco attempted to cast the closed Nanisivik townsite as a valuable and useful site. The final section examines how the cost of reclamation was valued and contested by various parties during the reclamation of the tailings at Nanisivik. In particular, this final section argues that these debates involved generating objective, authoritative, and neutral knowledge to legitimize different claims about the environment and verify contesting valuations of the cost of reclamation. Far from an unproductive, valueless, or lifeless space after closure, this chapter outlines the ongoing but different ways Nanisivik continued to be productive after the mine's operations ceased.

NANISIVIK'S DEVELOPMENT AND COMMUNITY RESPONSES TO CLOSURE

Located 750 kilometres north of the Arctic Circle, the Nanisivik lead-zinc mine opened by Mineral Resources International (MRI) was the first mine north of the Arctic Circle and the northernmost mine in Canada at the time of its establishment in 1976. The Nanisivik site was comprised of a purpose-built town with a school, church, post office, recreational centre, dining hall, nearby airstrip, and dock constructed on Strathcona Sound, approximately twenty-five kilometres from the Inuit community of Arctic Bay. Nanisivik's infrastructure was partially financed by the federal government in the hope that this experimental project would test the feasibility of operating in the High Arctic and pave the way for expanded mining across Canada's northern resource frontier. As the vice president of CanZinco Ltd. (then-current owner of the Nanisivik property) explained in one public hearing:

> . . . one of the visions was that this would be a pilot project. It may not be successful, but if it was, what a wonderful way to

find out if we could do natural resource exploitation in the north. In 2007, there was $1-and-a-half billion that came through the north in mining, and Nanisivik was the first one north of the Arctic Circle and a pioneer breaking the way for all those others that have followed.[6]

As this quote suggests, government support of the Nanisivik venture was not driven by profitability alone, but also various social and political objectives. In particular, the government "saw benefit in the [Nanisivik] project as a 'pioneer project' that without setting precedents might enable large scale experimentation in Arctic mining techniques and transportation."[7] Like previous Inuit employment projects at the Rankin Inlet nickel mine (see Keeling and Boulter, this volume) and at DEW (Distant Early Warning) line stations, the government hoped that Nanisivik would introduce some Inuit residents in the Baffin region to wage labour in an industrial setting.

To ensure the success of Nanisivik, the federal government entered into the Strathcona Agreement with MRI in 1974, signed by the minister of Indian Affairs and Northern Development, the president of MRI, and a local witness by the name of I. Attagutsiak.[8] Under the agreement, the government invested $18.3 million into townsite, dock, and airstrip development in return for an 18 per cent stake in the company and representation on the company's board of directors.[9] For its part, MRI pledged compliance with the government's social, environmental, and economic objectives for the North. One key objective of the Strathcona Agreement was ensuring that Inuit workers comprised 60 per cent of the workforce at Nanisivik. The agreement also sought to minimize the environmental impacts of mining through environmental studies and reclamation activities.[10] Other conditions of the Strathcona Agreement included:

> Provisions of vocational training for northern residents, comprehensive environmental studies and planning, preference for the use of Canadian material and equipment and Canadian shipping, company exploration programs to increase ore reserves and possible further processing of mine concentrates in Canada.[11]

Deemed a progressive and unprecedented approach to northern resource development, the agreement clearly sought to enact the government's commitment to the well-being of northerners and "optimize experience and benefits obtainable from this pilot Arctic mining venture."[12]

This experiment proved successful from the point of view of the government and mine company: the Nanisivik mine operated profitably for twenty-six years until its closure in 2002 and typically employed a workforce of two hundred people. In line with the mining imaginary, CanZinco's Vice President of Environment and Sustainability Bob Carreau presented closure as an inevitable stage in the life course of the mine, albeit with a notably positive spin:

> Unlike many businesses where closure often means failure, closure of a mine is, in fact, a measure of success. It means that you have gone through all the stages of a mine, and you have reached closure and reclamation, at least a plan in closure and reclamation. If you didn't do that, you would be doing abandonment, and that's not the case with Nanisivik. We have reached this final stage, closure and reclamation, it is a measure of success . . . Now, as we enter the final stage of the project, we culminate the success with the closure of the mine and the townsite. Closing a mine is never a happy event. And in the case of Nanisivik where this means the community will cease to exist, it is that much harder. However, as stated at the outset of this introduction, the closure of the mine is inevitable, and planned reclamation, it is the final milestone of that achievement.[13]

In spite of these proclamations that Nanisivik had succeeded as a pilot project, the mine failed to achieve its target of a 60 per cent Inuit workforce. Instead, typically only 20 to 25 per cent of the workforce was Inuit, a figure that dropped to 9 per cent in the final years of the mine's operation.[14] The failure to employ higher levels of Inuit labour led consultants Hickling-Partners to conclude in one report that "the mine has not succeeded in the role for which it was intended—as an experimental prototype."[15]

For the community of Arctic Bay, the mine's closure left behind many uncertainties: no one knew whether other economic activities could be undertaken at Nanisivik, and concerns grew over the environmental impacts of mining (such as soil and water contamination and the disposal of tailings waste). In public meetings, community members expressed concern about the destiny of the Nanisivik townsite, the level of community involvement in reclamation activities, and the impacts of mining on local wildlife and the land upon which the Inuit depended for hunting. Kunuk Oyukuluk explained in one public hearing how wildlife had been impacted by mining at Nanisivik:

> In early spring, when it was still March or May, when there is still ice, they would break the ice. And because it is our wildlife area—and so my concern is that seals, we rely on the seal meat; and they have a breeding ground on the ice, that the ship went through the breeding ground of the seals. And in July when Arctic Bay residents were out Norwhale [narwhal] hunting, the ship also went through the hunting ground, the hunting area. And during the Norwhale hunting, Norwhales would be scattered away by the ship. So every year they did that through the ice . . . So I need more help so that our generation— next generation, that they will have to have food to eat. And because we were brought up from the country food, so—and they are best food and makes you stronger, and we will be weaker population on other kinds of food.[16]

In a similar narrative, Moses Akumalik described how this environmental change impacted traditional lifestyles:

> I'm not trying to look big but we were living off the land when we were young. Now children when they grow up will lean more towards the civilized life as opposed to the nomadic life. In 1978, the ships would come in to load concentrate and they break the ice. Hunters lost their machines that were on the ice. That's why I'm asking for compensation because there have been impacts . . . They should thank the community for supporting their mining activity for all those years. A public apology with a thank you in money would be good. More than 20 skidoos were lost and all of their hunting equipment.[17]

As this quote suggests, some community members raised concerns regarding the cultural and environmental impacts of mining and requested an apology from the mine company. Additionally Mucktar Akumalik described how, despite co-operating with the mine, the community had been detrimentally affected by it, and he called for the community of Arctic Bay to be compensated:

> I want some kind of an apology, I guess, from the company because they did—they did their own activity without considering what the Arctic Bay community wants. And, you know, they didn't even ask the community how they feel about their activity, whether to, you know—Arctic Bay residents were concerned that—they were anxious for an apology, I guess, and they all just leave the area without apologizing to us.[18]

While some community members requested monetary compensation, others called for compensation in the form of old furnishings and equipment from the Nanisivik townsite or employment in future reclamation activities. In whatever form, these requests embodied appeals for justice: justice for harming the land, justice for impacting hunting activities, and justice for failing to reach Inuit employment targets.

NEGOTIATING CLOSURE AND RECLAMATION AT NANISIVIK

Community appeals for justice, inclusion, and empowerment during the closure of the Nanisivik mine reflected, in part at least, a dramatic change in the political context between the founding and closure of the mine. With the creation of "the new territory of Nunavut and, with it, the expectation that Inuit would become the managers of their own destiny," Inuit awareness of what power they could exercise increased significantly in the period leading to the closure of the mine.[19] In addition, the newly formed Government of Nunavut was cognizant that many mining companies had, in the past, abandoned northern mining projects without dealing with the environmental impacts of these activities, and was conscious that the livelihoods of Aboriginal northerners had been severely affected by changes to the environments on which they depend. As such, the "Mine Site Reclamation Policy for Nunavut" attempted to empower northern communities and provide "the Inuit a 'clean slate' to develop the kind of resource management regime they want to take with them into the new millennium."[20]

Whereas the costs associated with environmental degradation had been largely externalized by mine companies and paid by the government in the past, the "Mine Site Reclamation Policy" applied the "polluter pays" principle, enforceable through security bond arrangements written into water licences, land leases, and other regulatory instruments. At Nanisivik, this meant that a water licence administered by the Nunavut Water Board (NWB) set the terms of reclamation, and the board assumed the primary responsibility for regulating and enforcing reclamation efforts. As part of the security bond arrangements, CanZinco, the NWB, and other intervening parties present at public hearings had to agree on the value of the bond, based on the projected costs of reclamation. In addition, as part of their commitment to forge a positive legacy for this Arctic experiment, CanZinco and the Government of Nunavut worked with the community of Arctic Bay to produce the "Closure and Reclamation Plan" for Nanisivik.

Newspaper stories documenting this process reveal something intriguing about Nanisivik's closure: while the mine had closed and its production had stopped, Nanisivik continued to be valued, but these valuations were contested by CanZinco, the government, and the community.[21] These valuations were estimates of the cost of reclamation, specifically, the amount that would be held in a security bond to ensure that CanZinco completed Nanisivik's reclamation in line with the "Mine Site Reclamation Policy for Nunavut." Huge valuations were suggested (and contested) by each party: initially, Indian and Northern Affairs Canada (INAC) suggested that reclamation would cost $27,536,028, while CanZinco's consultants estimated reclamation would cost $9,224,608, a figure almost three times lower than the INAC estimate.[22]

Two interesting features of this valuation process unsettle the notion that mining landscapes are devalued after their closure. First, as payee of the cost of reclamation, CanZinco attempted to inscribe the Nanisivik townsite with value. The importance of the Nanisivik townsite—as a potential cost to CanZinco if it had to be destroyed due to contamination—was evident through CanZinco's attempts to attribute a high economic value to the site in order to offset the costs of reclamation. CanZinco commissioned an engineering firm in Toronto that estimated that it would cost more than $100 million to rebuild Nanisivik.[23] Declaring that "we have long taken and continue to take the view that it would be a tragedy if this facility were destroyed as part of the reclamation exercise,"[24] CanZinco worked hard to find future uses for the infrastructure, and undertook negotiations with companies interested in Nanisivik's production assets. CanZinco believed that given the "sheer number of companies currently exploring for diamonds within the immediate vicinity of Nanisivik . . . it would be foolhardy to destroy the existing industrial complex . . . when such a complex may serve as an inducement to one or more of these companies to establish a base at Nanisivik."[25] CanZinco eventually sold the mill, concentrate storage facility, power generation installation, conveyors, and ship-loading equipment to Wolfden Resources (owners of a property in Nunavut), who, in return, performed environmental cleanup in the area of the mill and storage facilities.[26] CanZinco's efforts to recoup monetary value from the closed townsite and mine illustrates how some of Nanisivik's infrastructure continued to possess both use-value and exchange-value after the mine's closure.

A second key feature of the closure process is the way that scientific knowledge was produced and mobilized to legitimate particular valuations of the cost of reclamation. A proliferation of studies undertaken by government scientists—and more frequently scientists, engineers, and technical consultants working for private environmental consulting firms—sought to provide an authoritative basis for resolving the dispute over the cost of reclamation. For CanZinco, this scientific knowledge was important in determining the amount of money the company would have to pay for reclamation. Consequently, both the government and CanZinco hired their own scientific experts to ensure that the knowledge produced was accurate and rigorous. These studies examined the extent of soil contamination, tested the stability and impact of tailings, contributed to various environmental site assessments and the Human Health and Ecological Risk Assessment (HHERA), and measured the level of contamination of the townsite infrastructure.[27]

Studies on soil contamination were among some of the most important in determining the cost of reclamation, as well as some of the most contentious, because establishing the level of contamination was decisive in determining whether the townsite would be destroyed. In one study, the Government of Nunavut hired consultants EBA Engineering to conduct a soil-sampling program to determine the extent of contamination at Nanisivik—research that cost over $49,000.[28] Because of the high costs involved in destroying a townsite due to contamination, CanZinco also hired privately owned environmental consulting firm Lorax Environmental Services. Lorax observed the work of EBA, and represented CanZinco's interests through collecting duplicate samples following the same methodology as EBA.[29] The economic importance of these study results is evident in a letter written by CanZinco disputing reports that surface soils at the Nanisivik townsite were "toxic."[30] In this letter to the NWB, CanZinco asserted:

> The parties realized that there would have to be some amount of clean up performed at the Nanisivik townsite before the transfer of infrastructure assets from CanZinco to the GN [Government of Nunavut] could be completed. It is hoped, though, that the introduction of the word "toxicity" (by all accounts an inaccurate inference) has not derailed those discussions and caused irreparable harm and considerable expense to CanZinco Ltd. It is also hoped that the use of the word can be put in its proper context and that before becoming unduly alarmed the Ecological and Human Health Risk Assessment (which the group correctly points out was proposed by CanZinco Ltd.) is allowed to serve its designed purpose—to provide concrete information from which to act in a reasonable manner.[31]

In this letter, branding soil at Nanisivik as toxic represented a "considerable expense to CanZinco" because it raised reclamation costs while threatening to devalue the townsite, worth up to $100 million in CanZinco's eyes. Producing and legitimizing scientific evidence, then, was critical for CanZinco to finalize the cost of reclamation and the value of its property at the site.

Together, these two features illustrate how CanZinco mobilized a counter-discourse to the "mining imaginary." Although Nanisivik was no longer a site for the production of valuable ores (and in fact, the site was a financial liability for the mine company), CanZinco cast Nanisivik as a useful and valuable site while subduing claims regarding the severity of contamination of the town. Furthermore, the Nanisivik minescape had become a space of intensive scientific investigation as soil samples were collected, water quality monitoring stations were established, and various field projects were initiated in order to legitimize different valuations of the cost of reclamation. However, as the next section will argue, these valuations were contested by the community of Arctic Bay and other intervenors, and the authority of scientific knowledge itself came under scrutiny.

THE SCIENCE AND POLITICS OF NANISIVIK'S TAILINGS COVER

The depth of an engineered tailings cover was perhaps the most contentious issue during Nanisivik's reclamation, and an issue that demonstrates how the scientific knowledge-making central to determining the cost of reclamation at Nanisivik was contested. The tailings at Nanisivik were the material by-product from the extraction and transformation of ores into lead-zinc concentrates. Within these tailings, *Thiobacillus* bacteria catalyzed the transformation of reactive sulphide minerals to generate outflows of acidic water containing high concentrations of heavy metals, in a process known as acid mine drainage. Even long after mining, these tailings continued to produce acid mine drainage—described as "poison water"[32] by the community—that the community and the government viewed as harmful to the surrounding environment. For instance, Elder Leah Oqallak commented in two public hearings, "So snow bunting, little bird landed on the tailings and it died right away, and it got—I got scared that I saw the bird die, so that's why it is my big concern."[33]

As part of progressive reclamation efforts undertaken during Nanisivik's operation, a field-monitoring program from 1990 investigated how acid mine drainage could be mitigated. Research conducted on behalf of Nanisivik mines indicated that *Thiobacillus* bacteria catalyzed the production of metals at a slower rate at lower temperatures.[34] The field-monitoring program sought to test the optimum conditions under which freeze-up of the tailings would occur using "test cell" covers.[35] Shale covers were constructed of varying levels of compaction and saturation, with thermocouples and frost gauges used to monitor temperatures. It was hoped that constructing a cover over the tailings at Nanisivik would thermally insulate the exposed tailings and promote freeze-up.[36] Once incorporated into the permafrost regime, these freezing conditions would reduce oxygen diffusion to make contaminants inert, preventing the contamination of surface water.[37] The extreme Arctic climate thus offered a "natural" method by which acid mine drainage could be prevented; in the words of CanZinco, this "reclamation work [was] focused on utilising the natural conditions to provide for the secure, long-term closure of the mine."[38]

Data from this field monitoring program, in combination with other studies conducted during the closure of the mine, were critical to informing the design of the engineered cover that would limit acid mine drainage. Data collected by CanZinco indicated that test cell 1, constructed from shale without compaction or saturation, had an average thaw depth of 0.92 metres.[39] To ensure that the tailings would remain frozen even under worst-case climate warming scenarios, geothermal models predicted thaw of 1.0 metres in a one-year period in the event of an extreme weather scenario (1-in-100 year warm event) and thaw of 1.22 metres at the end of one hundred years under a global warming scenario.[40] Whereas worst-case climate scenarios predicted by the Intergovernmental Panel on Climate Change (IPCC) and the Panel on Energy Research and Development (PERD) estimated warming of 3.5°C to 4.5°C respectively, CanZinco's modelling assumed a change of 5.5°C in order to mitigate against thaw.[41] Based on the test cover results and geothermal models, CanZinco asserted in its 2002 "Mine Closure and Reclamation Plan" that a 1.25-metre cover depth was sufficient, comprising 1.0 metres of shale and 0.25 metres of armour surfacing.

Throughout the closure and reclamation process, however, much debate surrounded the depth—and thus cost—of the engineered tailings cover proposed by CanZinco. Some Arctic Bay residents asserted that the cover should have been 10 metres deep at the dock area and 5 metres deep at the industrial site, areas (correctly) perceived as the most contaminated.[42] Though the rationale behind these estimates is unclear from the archival record, public hearing transcripts reveal that the Hamlet of Arctic Bay regarded the tailings depth as an important issue and the community asserted its disappointment with the lack of information they had received regarding the tailings. In one hearing, the mayor of Arctic Bay, Joanasie Akumalik, explained:

> In the past we know that there was monitoring happening of the water and the tailings pond and even the air. We have also been aware of tailings monitoring devices that have not worked for long periods of time. We have not received the results from these activities. It is important that the local people in Arctic Bay become fully involved in this long term monitoring work and be trained to undertake this activity. It is important that the local people trust the results of these activities.[43]

This quote suggests some residents felt excluded from these scientific activities during Nanisivik's closure, in similar ways to how the community had felt marginalized during the mine's operation. To rectify this, some residents hoped that the community could observe the reclamation work undertaken at Nanisivik. An elder commented that:

> There should be someone observing when you are burying the tailing so that they can share their story and the information that they observe. Back in 1959 I was working for the Bay store. We used to hide things from the Manager before they came to the store so that the Manager would know it was a good store. I want someone there to observe the burying of tailings. If you tell me straight [it] will not contaminate the people and environment, I will believe I won't mind if you cover it. It is a concern without someone telling me that it won't have impact on my life. I want someone to observe. There will be work for Arctic Bay residents

to work on the clean-up but when you are covering the tailings I want someone too. I want to see the picture of the tailings on the side of it. I'm serious here. People are serious here. We should ask all kinds of questions here.[44]

Rigorous monitoring was important for many residents to trust that the impacts of mining on their health and livelihoods had been offset. As well, these recommendations positioned community members as independent observers who could fill employment positions during reclamation and confirm whether work was being conducted correctly.[45]

While the cover depth issue was important for the health and well-being of the residents of Arctic Bay, it was equally important for CanZinco in determining the total amount of the security bond—a figure disputed by CanZinco and the Nunavut Water Board. On behalf of INAC, Brodie Consulting initially estimated that a cover depth of 1.75 metres was required, based on the fact that one of the test cells had experienced thawing to a depth of 1.59 metres.[46] Brodie later suggested that a 1.5-metre cover depth was required, still costing $1.25 million more than CanZinco's estimate of 1.25 metres. These cover depth estimations were of utmost importance to CanZinco, as they represented significant sums of money needed to pay for the surface covering—at the very least, $1.25 million was at stake.

CanZinco asserted the legitimacy of its estimate by presenting its cover depth as a "scientifically sound" estimate. CanZinco stressed that a depth of 1.25 metres was sufficient to keep the tailings frozen by highlighting that the data input into the geothermal model was more conservative than the estimates used by world-renowned scientific panels such as the IPCC. Emphasizing the authority of its scientific facts, CanZinco declared:

> We have calculated with the warming effect, so that's calculated in there. Global warming, as you mentioned, is a concern, and so we had, as I mentioned, included modelling that takes the worst-case scenario that Environment Canada offers you now over the years, we include that in the mine. And like any engineering we do, that's the best you can do, it has to be based on some scientific data, and that is based on *sound scientific data*.[47]

As this quote suggests, the scientific method not only produced knowledge about the environment, but this method in itself was presented as an authoritative and reliable source for the production of knowledge.

Indeed, CanZinco heavily relied on arguments based in notions of scientific expertise to validate its estimate and protest the valuations made by Brodie Consulting and the community. Throughout the closure and reclamation period at Nanisivik, CanZinco had urged the intervening parties to use "good science to come up with the best answers."[48] In public hearings, CanZinco introduced scientific and technical consultants as "independent and outside professionals,"[49] neutral parties external to the politics of reclamation and without bias. This is not to say that one estimate was more accurate than another, but rather CanZinco sought to present its rationale as "scientifically sound" in order to legitimize its estimate of the cost of the cover depth. For instance, CanZinco wrote in one letter to the NWB that:

> The intervening parties who are saying 1.25m is insufficient are not supporting this with any concrete information. They are *simply* and quite *arbitrarily* saying that they *intuitively* assume that 1.25 metres is not enough, and more cover should be added. If the intervening parties are able to take their *rationale* for additional coverage, at the very least a *meaningful technical* debate could ensue, and CanZinco is confident that it would prevail. CanZinco is currently at a disadvantage, though, where it presents scientifically defensible information and the only rebuttal is "we want more."[50]

In this quote, non-scientific estimates are cast as "arbitrary" and "intuitive," whereas scientific expertise is "meaningful" and "rational." Again, this is not to argue that the science behind each estimate was correct (or incorrect), but rather that this discourse inscribed science with the power to adjudicate and validate competing claims over reclamation, in such a way that at times it delegitimized the non-scientific estimates suggested by the community. Indeed, some residents were disappointed that suggestions they made in public hearings were not acted upon.[51] In this way, CanZinco-sponsored research not only produced an economic valuation (of the cost of reclamation), but necessarily reproduced the authority of science: an explicit example of the way that the closed Nanisivik mine was a site for the production of contested valuations (of the cost of reclamation) and the scientific knowledge that legitimized these valuations.

CONCLUSION

After many meetings and much technical debate between the intervening parties, it was agreed that a 1.25-metre cover depth would be appropriate. The security bond was finally set at $17.6 million, and CanZinco's closure and reclamation plan was approved in 2004. It had become increasingly clear that the Nanisivik townsite and infrastructure would have to be demolished, as efforts to find alternate uses for the site were unsuccessful and contamination proved a costly problem. Many buildings had exceeded their lifespan, and those still in usable condition required as much as $50 million over four years for renovation.[52] After reclamation was completed in 2008, the security bond was reduced to $2 million to cover a five-year post-closure monitoring period. CanZinco estimated in a 2009 public hearing that the company had spent $17 million and that Wolfden had spent $12 million on reclamation at the site.[53]

Now, Nanisivik's mining infrastructure, housing, and support facilities are all but gone from the site. This was not an inevitable outcome of Nanisivik's life cycle, however. Some of Nanisivik's mining infrastructure continued to embody value: it was dismantled and reconfigured at other mine sites in Northern Canada. Furthermore, CanZinco mobilized a counter-discourse to the mining imaginary—the finality of mine closure—by casting the Nanisivik minescape as a valuable (and uncontaminated) space. In fact, after the closure of the mine, Nanisivik became a landscape of data production and economic valuation: scientific and technical consultants were hired from several external engineering firms, and technological infrastructures were erected in order to mine data from the environment. The closed Nanisivik minescape had become a hive of new activity that produced scientific knowledge and informed valuations of the cost of reclamation, which were disputed by CanZinco (as payee of the reclamation) and the Government of Nunavut (as regulator of the reclamation). The most fascinating aspect about this scientific production process is that this knowledge making embodied scientific authority and neutrality that was used to assert the cost of reclamation by these different parties. These efforts not only generated scientific knowledge about the environment at Nanisivik, but the intervening parties cast this knowledge as being neutral, external, and unbiased—the most reliable knowledge for determining the cost of reclamation. Efforts to legitimize scientific knowledge concurrently legitimized valuations of the cost of reclamation. Thus, despite appearing to be an economically worthless post-productive space—as popularly imagined of closed mines under the mining imaginary—Nanisivik was a site of the production of both scientific knowledge and valuations of the cost of reclamation.

As such, this chapter favours a reading of Nanisivik's closure as an important historical-geographical event in the life of the mine, set amidst Nunavut's transforming political and regulatory context, and thereby negotiated and navigated by various actors (such as community members, government officials, and mine company representatives) with different, and at times, conflicting interests in the mine's closure and reclamation. Far from an unproductive, valueless, or lifeless space, the townsite continued to be valued in different (and contested) ways after the mine's closure and during reclamation. The case of Nanisivik prompts a reconsideration of the political-economic function and character of "post-productive" landscapes to account for the ongoing, and often contested, historical-geographical reconfiguration of such landscapes after resource extractive activities have formally ended.

NOTES

1. "Nanisivik to Die Four Years Early," *Nunatsiaq Online*, 2002, accessed April 2011, http://www.nunatsiaqonline.ca/archives/nunavut011102/news/nunavut/11102_1.html.
2. Mr. Oqituq, "Nanisivik Mine Type A Water License Renewal and Amendment Application," Nunavut Water Board Official Public Hearing Transcripts, February 2009, 184, accessed April 2011, ftp://nunavutwaterboard.org/1%20PRUC%20PUBLIC%20REGISTRY/ 1%20INDUSTRIAL/1A/1AR%20-%20Remediation/1AR-NAN0914/2%20ADMIN/ 4%20HEARINGS/2%20HEARING/2008%20Hearing/.

3   David Robertson, *Hard as the Rock Itself: Place and Identity in the American Mining Town* (Boulder: University of Colorado Press, 2006). See also Gavin Bridge, "Contested Terrain: Mining and the Environment," *Annual Review of Environment and Resources* 29 (2004): 205–59.

4   Ben Marsh, "Continuity and Decline in the Anthracite Towns of Pennsylvania," *Annals of the Association of American Geographers* 77, no. 3 (1987): 337–52; William Wyckoff, "Postindustrial Butte," *Geographical Review* 85, no. 4 (1995): 478–96; Robertson, *Hard as the Rock Itself*.

5   Trevor Barnes, "Borderline Communities: Canadian Single Industry Towns, Staples, and Harold Innis," in *B/ordering Space*, eds. Henk van Houtum, Olivier Kramsch, and Wolfgang Zierhofer (Burlington, VT: Ashgate Publishing, 2005); Trevor Barnes, Roger Hayter, and Elizabeth Hay, "Stormy Weather: Cyclones, Harold Innis, and Port Alberni, BC," *Environment and Planning A* 33 (2001): 2127–47; John Bradbury, "Towards an Alternative Theory of Resource-Based Town Development in Canada," *Economic Geography* 55, no. 2 (1979): 147–66; John Bradbury, "Declining Single-Industry Communities in Québec-Labrador, 1979–1983," *Journal of Canadian Studies* 19, no. 3 (1984): 125–39; Gavin Bridge, "Resource Triumphalism: Postindustrial Narratives of Primary Commodity Production," *Environment and Planning A* 33 (2001): 2149–73; Gavin Bridge, "Material Worlds: Natural Resources, Resource Geography, and the Material Economy," *Geography Compass* 3, no. 3 (2009): 1217–44; David S. Trigger, "Mining, Landscape, and the Culture of Development Ideology in Australia," *Cultural Geographies* 4, no. 2 (1997): 161–80.

6   Bob Carreau, Nunavut Water Board Official Public Hearing Transcripts, 2009, 18.

7   Hickling-Partners Inc., "Evaluation of the Nanisivik Project" (Canada Department of Indian Affairs and Northern Development, 1981), 6.

8   "Strathcona Agreement," in Robert Gibson, "The Strathcona Sound Mining Project: A Case Study in Decision Making" (Science Council of Canada, No. 42, 1978).

9   Department of Indian and Northern Affairs, "North of 60: Mines and Mineral Activities 1976" (Mining Division, Northern Non-Renewable Resources Branch, 1976).

10 For example, BC Research, "Lethal and Sublethal Bioassays on Tailing from Nanisivik Mine Ore" (Prepared for Strathcona Mineral Services, Project 1552, Progress Report No. 3, 1975); C. T. Hatfield and G. L. Williams, "A Summary of Possible Environmental Effects of Disposing Mine Tailings into Strathcona Sound, Baffin Island" (Hatfield Consultants Limited Arctic Land Use Research Program, 1976); S. Reedyk, "Profiles of Baffin Communities Affected by Closure of the Nanisivik Mine and a Summary of the Impact of Closure on

the Communities" (Indian and Northern Affairs Canada, Mining Management and Infrastructure Division, 1987).

11   Department of Indian and Northern Affairs, "North of 60," 50.

12   Cabinet Committee on Government Operations, "Record of Committee Decision: Strathcona Sound Project," Mineral and Metal Commodities Branch Registry Files, RG21 Acc 1996–97/723 box 10 file 1.50.15.3 vol. 5., p. 10, Library and Archives Canada; Gibson, "The Strathcona Sound Mining Project."

13   Bob Carreau, "Nunavut Water Board Final Closure and Reclamation of the Nanisivik Mine," Nunavut Water Board Official Public Hearing Transcripts, June 2004, 13–16, accessed April 2011, ftp://nunavutwaterboard.org1%20PRUC%20PUBLIC%20REGISTRY /1%20INDUSTRIAL/1A/1AR%20-%20Remediation/1ARNAN0914/2%20ADMIN/ 4%20HEARINGS/2%20HEARING/2004/.

14   Brubacher & Associates, "The Nanisivik Legacy in Arctic Bay: A Socio-economic Impact Study," prepared for the Department of Sustainable Development, Government of Nunavut, 2002.

15   Hickling-Partners Inc., "Evaluation of the Nanisivik Project," 36.

16   Kunuk Oyukuluk, Nunavut Water Board Official Public Hearing Transcripts, 2004, 121.

17   Moses Akumalik, "Public Hearing Minutes," Nunavut Water Board Official Public Hearing Transcripts, July 2002, 44, accessed April 2011, ftp://nunavutwaterboard.org1%20PRUC%20PUBLIC%20REGISTRY/1%20INDUSTRIAL/1A/1AR%20-%20Remediation/1ARNAN0914/2%20ADMIN/4%20HEARINGS /2%20HEARING/2002%20Renewal /PH%20July%2022-24,%202002/.

18   Mucktar Akumalik, Nunavut Water Board Official Public Hearing Transcripts, 2004, 100.

19   Indian and Northern Affairs Canada, "Mine Site Reclamation Policy for Nunavut," Minister of Indian Affairs and Northern Development, Ottawa, ON, INAC, 2002, 2. The changing context of regulation between the Strathcona Agreement and the Nunavut Land Claims Agreement is also noted in Léa-Marie Bowes-Lyon, Jeremy P. Richards, and Tara M. McGee, "Socio-Economic Impacts of the Nanisivik and Polaris Mines, Nunavut, Canada" in *Mining, Society, and a Sustainable World*, ed. J.P. Richards (Heidelberg: Springer-Verlag, 2009).

20   Indian and Northern Affairs Canada, "Mine Site Reclamation Policy for Nunavut."

21   "Nanisivik Owner Wants More Time to Post $17.6 Million Bond," *Nunatsiaq Online*, 2002, accessed April 2011, http://www.nunatsiaqnews.com/archives/nunavut021101/news/nunavut/21101_04.html; "CanZinco heading for showdown with NWB?," *Nunatsiaq Online*, 2002, accessed April 2011, http://

www.nunatsiaqonline.ca/archives/nunavut021108/news/nunavut/21108_01.html; "Arctic Bay Residents Want Details on Nanisivik Health Risks," *Nunatsiaq Online*, 2003, accessed April 2011, http://www.nunatsiaqonline.ca/archives/30905/news/nunavut/30905_05.html.

22   Breakwater Resources, "Letter from Breakwater to NWB regarding security bond arrangements," 2002, accessed April 2011, ftp://nunavutwaterboard.org/1%20PRUC%20PUBLIC%20REGISTRY/1%20INDUSTRIAL/1A/1AR%20-%20Remediation/1AR-NAN0914/2%20ADMIN/4%20HEARINGS/2%20HEARING /2002%20Renewal/Post%20Hearing%20Submissions/BWR/English/.

23   Bill Heath, "Exhibit 1: opening remarks by Bill Heath, vice-president of CanZinco," Public Meeting at Arctic Bay, 2002, accessed April 2011, ftp://nunavutwaterboard.org/1%20PRUC%20PUBLIC%20REGISTRY /1%20INDUSTRIAL/1A/1AR%20-%20Remediation/1AR-NAN0914/2%20ADMIN/ 4%20HEARINGS/2%20HEARING /2002%20Renewal/Exhibits%20Undertakings/.

24   Ibid.

25   Ibid.

26   B. Young, "Memo: Wolfden Resources: massive sulphide discovery at High Lake," 2003, accessed April 2011, ftp://nunavutwaterboard.org/1%20PRUC%20PUBLIC%20REGISTRY/1%20INDUSTRIAL/1A/1AR%20-%20Remediation/1AR-NAN0914/3%20TECH/10%20A%20and%20R%20(J)/2003/.

27   CanZinco, "Final Closure and Reclamation Report," 2004, accessed January 2012, ftp://nunavutwaterboard.org/1%20PRUC%20PUBLIC%20REGISTRY /1%20INDUSTRIAL/1A/1AR%20-%20Remediation/1ARNAN0914/3%20TECH/ 10%20A%20and%20R%20(G)(J)/2004/G3%20Final%20Closure%20Plan/2004%20Final%20AR%20plan/G.4%20Cover%20Report/G.4%20Report/.

28   EBA Engineering Consultants Ltd., "Soil sampling program report," 2002, accessed April 2011, ftp://nunavutwaterboard.org/1%20PRUC%20PUBLIC%20REGISTRY/1%20INDUSTRIAL/1A/1AR%20-%20Remediation/1AR-NAN0914/2%20ADMIN/4%20HEARINGS/ 2%20HEARING/2002%20Renewal/Post%20Hearing%20Submissions/GN/English/.

29   Dillon, "Dillon Consulting Final Assessment," 2003, accessed April 2011, ftp://nunavutwaterboard.org/1%20PRUC%20PUBLIC%20REGISTRY/1%20INDUSTRIAL/1A/1AR%20-%20Remediation/1AR-NAN0914/2%20ADMIN/0%20GENERAL/2003/.

30   EBA, "Soil sampling program report."

31   Nanisivik Mine, "Letter from Nanisivik mine to Thomas Kudloo of NWB regarding final submissions," 2002, 5, accessed April 2011, ftp://nunavutwaterboard.org/1%20PRUC%20PUBLIC%20REGISTRY/1%20INDUSTRIAL/1A/1AR%20-%20Remediation/1AR-NAN0914/2%20ADMIN/4%20

HEARINGS/ 2%20HEARING/2002%20Renewal/Post%20Hearing%20 Submissions/BWR/English/.

32   Bob Carreau, Nunavut Water Board Official Public Hearing Transcripts, 2009.

33   Leah Oqallak, Nunavut Water Board Official Public Hearing Transcripts, 2004; Nunavut Water Board Official Public Hearing Transcripts, 2002.

34   M. Kalin, "Ecological Engineering for Gold and Base Metal Mining Operations in the Northwest Territories," *Environmental Studies*, no. 59 (Ottawa, 1987); Bo Elberling, "Environmental Controls of the Seasonal Variation in Oxygen Uptake in Sulfidic Tailings Deposited in a Permafrost-Affected Area," *Water Resources Research* 37, no. 1 (2001): 99–107; Bo Elberling, "Temperature and Oxygen Control on Pyrite Oxidation in Frozen Mine Tailings," *Cold Regions Science and Technology* 41 (2005): 121–33; Curt Kyhn and Bo Elberling, "Frozen Cover Actions Limiting AMD from Mine Waste Deposited on Land in Arctic Canada," *Cold Regions Science and Technology* 32 (2001): 133–42.

35   BCG Engineering, "Engineering Design of Surface Reclamation Covers," 2003, accessed January 2012, ftp://nunavutwaterboard.org/1%20PRUC%20 PUBLIC%20REGISTRY /1%20INDUSTRIAL/1A/1AR%20%20Remediation/1ARNAN0914/3%20TECH/10%20A%20and%20R%20(G)(J)/2004/G3%20 Final%20Closure%20Plan/2004%20Final%20AR%20plan/G.4%20Cover%20 Report/G.4%20Report/G.4%20Cover%20Report%20Final.pdf.

36   CanZinco, "Final Closure and Reclamation Report."

37   CanZinco, "Final Closure and Reclamation Report"; BCG Engineering, "Engineering Design of Surface Reclamation Covers."

38   CanZinco, "Final Closure and Reclamation Report," ix.

39   CanZinco, "Final Closure and Reclamation Report"; BCG Engineering, "Engineering Design of Surface Reclamation Covers."

40   BCG Engineering, "Engineering Design of Surface Reclamation Covers."

41   Nunavut Water Board, Nunavut Water Board Official Public Hearing Transcripts, 2002.

42   Nunavut Water Board, "Public hearing meeting notes," 2004, accessed April 2011, ftp://nunavutwaterboard.org/1%20PRUC%20PUBLIC%20REGISTRY /1%20INDUSTRIAL/1A/1AR%20-%20Remediation/1AR-NAN0914/2%20ADMIN /4%20HEARINGS/0%20GENERAL/2004/.

43   Hamlet of Arctic Bay Working Group, "Submission to the Nunavut Water Board," 2002, 2, accessed April 2011, ftp://nunavutwaterboard.org/1%20PRUC%20PUBLIC%20REGISTRY /1%20INDUSTRIAL/1A/1AR%20-%20Remediation/1AR-NAN0914/2%20ADMIN/ 4%20HEARINGS/2%20HEARING/2002%20Renewal/Exhibits%20

Undertakings/020726NWB1NAN9702%20Exhibit11-Submission%20of%20 Hamlet%20AB%20Wrkg%20Group English.pdf.

44   Leah Oqallak, Nunavut Water Board Official Public Hearing Transcripts, 2002, 43.

45   Nunavut Water Board, "Public hearing meeting notes."

46   Brodie Consulting, "Nanisivik Mine: Reclamation Cost Estimate," prepared for Indian Affairs and Northern Development, Water Resources Division, June 2002, accessed April 2011, ftp://nunavutwaterboard.org/1%20PRUC%20 PUBLIC%20REGISTRY /1%20INDUSTRIAL/1A/1AR%20-%20Remediation/1ARNAN0914/2%20ADMIN/ 3%20SUBMISSIONS/2002%20Renewal/ DIAND/.

47   Bob Carreau, Nunavut Water Board Official Public Hearing Transcripts, 2004, 118. Emphasis added.

48   Bob Carreau, "Letter from Nanisivik mine to Thomas Kudloo of NWB regarding final submissions," 2002, 26, accessed April 2011, ftp://nunavutwaterboard.org/1%20PRUC%20PUBLIC%20REGISTRY/1%20INDUSTRIAL /1A/1AR%20-%20Remediation/1AR-NAN0914/2%20ADMIN/4%20HEARINGS/ 2%20HEARING/2002%20Renewal/Post%20Hearing%20Submissions/ BWR/English/.

49   Heath, "Exhibit 1," 3.

50   Nanisivik Mine, "Letter from Nanisivik mine to Thomas Kudloo of NWB regarding public hearing outcomes," 2002, 3, accessed April 2011, ftp://nunavutwaterboard.org/1%20PRUC%20PUBLIC%20REGISTRY/1%20INDUSTRIAL /1A/1AR%20-%20Remediation/1AR-NAN0914/2%20ADMIN/4%20 HEARINGS/ 2%20HEARING/2002%20Renewal/Post%20Hearing%20Submissions/BWR/English/. Emphasis added.

51   Léa-Marie Bowes-Lyon, "Comparison of the Socio-economic Impacts of the Nanisivik and Polaris Mines: Sustainable Development Case Study" (MSc thesis, University of Alberta, 2006).

52   Nunavut Water Board, Nunavut Water Board Official Public Hearing Transcripts, 2004.

53   Nunavut Water Board, Nunavut Water Board Official Public Hearing Transcripts, 2009.

| CHAPTER 11

"There Is No Memory of It Here": Closure and Memory of the Polaris Mine in Resolute Bay, 1973–2012

Heather Green

Industrial closure is about much more than decreased market value, capital loss, commodity decline, and economic disruption. It is also about individuals and communities. Though deindustrialization is a broad process that occurs worldwide, those impacted by closure experience an intimate and local connection to this process. In single industry towns especially, closure frequently starts a chain of unemployment, out-migration, population decline, and abandoned infrastructure. It is also common for post-industrial communities to suffer negative environmental impacts. Previous scholarship has studied the socio-economic, cultural, and environmental legacies of mine closure and deindustrialization in both Rust Belt zones and single industry towns in Southern Canada and the United States.[1] Mining and mine closures have also been prevalent in the Canadian North since the 1950s, and historians have recently begun paying attention to the impacts of closure in the North, as this volume attests. Each case of closure is unique, though scholars have identified key trends in southern industrial closures, including their economic, cultural, and social impacts, which have also been repeated in cases of northern deindustrialization.[2]

One of the more recent topics in this literature is the connection between collective memory and closure.[3] Much scholarly work currently available about mining and collective memory is concerned with how communities react to closure and decline, and how in this reaction communities form a group/collective memory or a mining heritage. This literature provides case studies of mine closure and collective memory formation around the globe. What is striking about these cases, which examine different types of mining, different demographics of mine workers, and different geographic spaces, are the similarities they share in terms of both the economic and social importance of mining and the collective memories these local communities develop in retaining their mining heritage. What is further striking for my research concerning the Polaris mine in Resolute was how much the Polaris mine and the community of Resolute diverge from this post-closure narrative. Throughout this chapter I will attempt to provide an explanation for this divergence.

Studying coal-mining heritage in Britain, Rosemary Power explains that mining heritage is defined in a community in terms of "what has been lost, what needs to be retained, and what needs to be preserved to benefit future generations."[4] She says mining heritage includes local community organizations that gather written records, and abandoned equipment and artifacts that are set within the community as symbols of honour. Mining heritage involves both physical artifacts and "community spirit."[5] Finally, to be considered as having mining heritage, a community must identify as a mining or former mining community (even in cases where the mine has been closed for a period of time).[6] Such desire to commemorate mining heritage comes from the social and economic factors these single industries brought to local communities. In most cases, these towns revolved around mining, and the secondary economic development was based on providing services and products for mining. Further, Power argues that tight-knit social communities came from the structure of mining lifestyles, particularly in terms of gender roles and class consciousness.[7]

Scholars have also argued that the formation and retention of a collective mining memory has served political purposes. For example, in their work, Mellor and Stephenson outline the attempts of Durham mining communities to maintain their mining heritage through continuing the Durham Miners' Gala. The authors argue that the gala represented a political platform for the community to defend itself from the marginalization faced by post-industrial single industry towns.[8] In his work, Ben Marsh also provides an understanding of power struggles and power structures that commonly existed in such small towns industrialized from an outside force.[9] The inhabitants and workers in the small anthracite mining towns he discusses came from other continents in the early twentieth century. Though these places were created by companies, the workers felt these places were "theirs," and they claimed a sense of place on their own terms. This is important to understanding the development of community strength and loyalty to place.[10] Forming a collectively shared memory for the community helps in claiming legitimacy for future political issues, such as demands for economic development or government support for the deindustrialized area.

As this literature suggests, the memories that communities form about industry after closure are largely influenced by the degree to which a community participated in the industrial activity and the extent of its impacts, both positive and negative. In the Canadian North, these memories also include the experiences of Aboriginal communities, impacts on traditional land use, and the penetration of outside mining companies into the region. There has been less scholarly attention to mine closures and heritage in the North and, more specifically, in Aboriginal communities, though this area of scholarship is growing. Tara Cater and Arn Keeling study the ongoing influence the North Rankin Nickel Mine has in the community's built environment and cultural landscape since closure in 1962. They argue that the community of Rankin Inlet is "(re)staking its claims to its industrial past, as part of contemporary efforts to manage the costs and benefits of new mineral development in the region."[11] Once again, the case of the Polaris mine and Resolute Bay community widely differs from Rankin Inlet. The town of Rankin Inlet was created because of the mine. Community members not only worked at the mine, but it was the town's sole source of income. Finally, in Rankin, the former mine

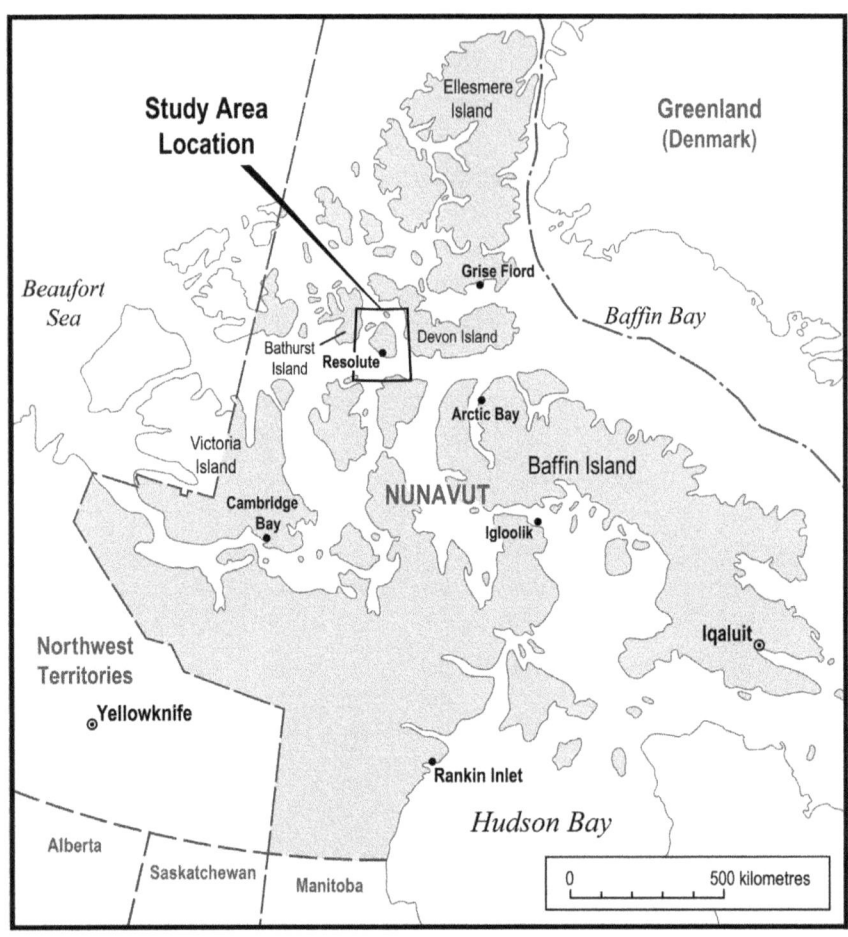

FIGURE 1: Nunavut, showing study area location (detail in figure 2). Map by Michael Fisher.

site has become an attraction for the community; it is a landscape to base mining heritage and memory around.

This chapter will explore the connections between memory and closure of the Polaris lead-zinc mine (in operation from 1982 to 2002) in the community of Resolute Bay (Qausuittuq),[12] located about one hundred kilometres from the mine (Fig. 1). Because of the deeply personal nature of the connection between closure and memory (both individual and

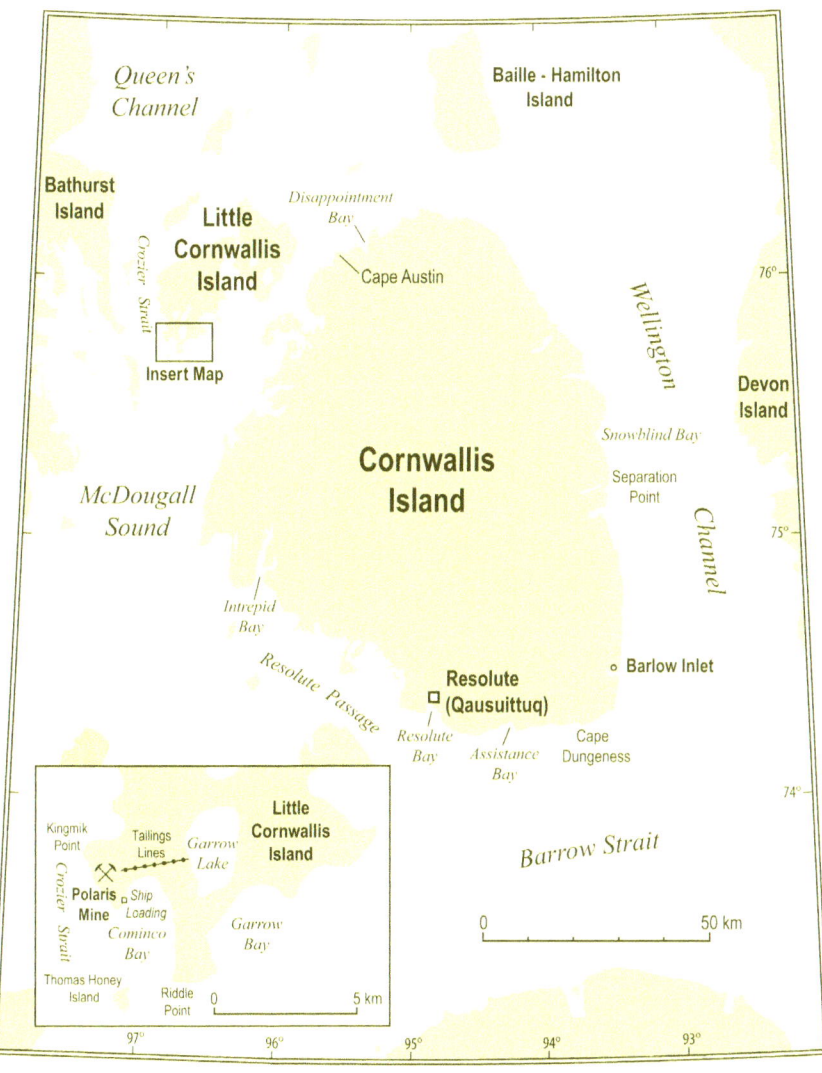

FIGURE 2: Cornwallis Island and Little Cornwallis Island indicating Resolute Bay and the Polaris mine site. Map by Michael Fisher.

collective), it is impossible to fully study mining memory without speaking with those who were affected by industry. Oral history interviews conducted with former Inuit Polaris workers and Resolute community members form the basis of this analysis. Interviewees spoke about their memories of the Polaris mine, from the time of its announcement to its closure. However, this research also found that, without asking directly about the mine, one would never guess there had been a mine nearby, even though Polaris closed only eleven years before this study. Due to the geographic isolation of the mine site from the town (Fig. 2), there are no physical remnants of industry in the town. There is no heritage site, no photographs in public buildings, and the youth are largely unaware that there had been a mine nearby. A *collective* memory of Polaris is absent in the hamlet.[13]

It may seem contradictory to state that Resolute lacks community memory of the Polaris operation and then proceed to discuss residents' memories of the mine. In *Oral History and Public Memories,* Hamilton and Shopes argue that the relationship between the individuals who do the remembering (which is the central concern in oral history) and the memory of a group has not yet been resolved nor analyzed in-depth. When I began this research project, I was initially interested in exploring the community memory of mining in Resolute Bay. However, as I began talking to more people and exploring the area, it became clear that the *community* did not have a specific mining memory, or a collective memory specific to a mining past. It is important to note that those interviewed were a select few from the total population of Resolute (9 interviewees from a population of 240), and those interviewed spoke of their *individual* memories and experiences with Cominco, the mining company that owned Polaris, and the Polaris mine, while indicating that a *collective* or *community* memory of the mine remains absent in Resolute.[14] Individual memories were quite strong, and each interviewee brought his or her own unique experiences, opinions, and memories to the narrative of the Polaris story. While many individuals shared similar personal memories, as I will expand upon below, these memories were not publicly codified in memorials, monuments, or events associated with mine work.

Previous scholarship reveals that community memory is often preserved within the deindustrialized landscape.[15] An overwhelming presence of industrial heritage (whether abandoned infrastructure or environmental legacies) tends to force people to remember and, sometimes, to engage with the industrial past. By contrast, the residents of Resolute do not face such physical reminders (in part because of their distance from the actual mine site), which has contributed to the lack of mining memory in the town. Even if individuals mostly enjoyed their experience at the mine, those interviewed were in agreement that Cominco let the community down in not living up to its pre-development promises of employment and community benefits.

It is not enough to note the presence or absence of a collective mining memory; understanding why this is the case and what factors influence the formation of collective memory is critical to this story. Collectively, the stories from Resolute Bay suggest that the lack of involvement of Resolute Inuit in the Polaris development, from consultation to operation to closure, strongly affected the way that residents remember deindustrialization and their mining past. Cominco began planning and developing Polaris in 1973, opening the mine in 1982. The mine site was located in an area traditionally used by Resolute Inuit, which raised concerns about environmental impacts from the community in the planning phase. However, environmental concerns largely dissipated in the operational stage as Resolute residents became discontented over the lack of Inuit employed at Polaris. The Inuit employment rate, both in general and from Resolute specifically, remained low throughout operation. Some Resolute residents remain bitter about this, and many contend that their marginalization and lack of involvement in the Polaris mine explain the lack of socio-economic benefits the town received from mine development. It is clear that the absence of collective community memory of Polaris is rooted in the exclusion of Resolute Inuit throughout the lifespan of the mine, even though the Polaris operation was developed and operating at a time when Inuit political activism was becoming more recognized by the federal government and the mining industry.

❋❋❋

Northern mining development has always been pursued by forces from outside the region. Since the 1950s, the Canadian state has promoted the mineral industry as part of its agenda of northern modernization.[16] David Trigger, using an Australian example similar to Canada in the 1950s to 1970s, highlights the prevalent historical belief that mining was moral progress, bringing "value" to the land and allowing people to "maintain a standard of living" through industrial opportunities.[17] As a result, mining companies often ignored the value of the land to Aboriginal peoples and rarely accounted for the consequences industry may bring to northern Aboriginal peoples specifically. By the 1970s, this attitude began to shift in Northern Canada as political activist groups, such as the Indian Eskimo Association (IEA), developed among southerners concerned about the plight of Inuit. Inuit-initiated political activism such as the Inuit Tapirisat of Canada (ITC) and the Committee for Original People's Entitlement (COPE) occurred parallel to new mine developments.[18] Geologist Robert McPherson has linked the changing political climate and growing Inuit political activism in the 1970s to resource issues, as Inuit asserted their rights to be active participants in northern economic development, to have their concerns and opinions considered and respected, and to become owners of their land. Inuit activism achieved some success, especially when political agitation led to the Nunavut land claims negotiations throughout the 1970s and the 1990s, culminating in the creation of Nunavut as its own territory in 1999.[19] However, during the 1970s and 1980s, at the same time that land claims were negotiated between the federal and territorial governments and Inuit organizations, companies such as Cominco continued to overlook Inuit concerns about mineral development in the development decision-making process.

The development of Polaris began when Bankeno Mines originally discovered mineralization on Little Cornwallis Island in 1960 and staked the first claim.[20] Cominco Ltd., one of the largest Canadian natural resource companies at the time, bought these claims in 1964 and, upon further exploration, discovered the Polaris lead-zinc ore body in 1971.[21] Unlike previous mine developments in the North, the company included community consultation in its planning process. In 1973, Cominco sent consultant J. E. Barrett to some Inuit communities to interview residents about their potential interest in working at Polaris.[22] In the planning stage, Cominco directed most of its attention (however marginal) to Resolute, the community nearest to the Polaris operation, holding two community meetings (both in May 1980) before opening the site. The company promised the Resolute Inuit employment opportunities and described other economic benefits that the town would gain from having a mine nearby.[23] Interviews with Resolute residents suggest that Resolute expected Cominco would draw an Inuit labour force from the community and that the town would gain services and economic growth from the mine. According to interviewees, Cominco said the mine would help the town accumulate capital, and "it would be easier [to] build up a little bit of the community."[24] They agreed that Resolute residents were optimistic about the arrival of industry.

Though Cominco took consultation further than any previous operation had, it is important to point out that this consultation took the form of information sessions, rather than co-planning or negotiation. There were no legal requirements to provide communities with decision-making power, and throughout planning, Cominco did not exceed its obligations under an informal "Socioeconomic Action Plan" signed between the company and the Government of the Northwest Territories (GNWT), which simply required the company stay in contact with communities.[25] In 1976, Barrett and some company officials returned to seven communities from which Cominco believed it would likely draw Inuit employees.[26] After his second visit to Resolute in 1976, Barrett reported, "It seemed to the Consultant that at the meeting [in Resolute] Inuit were feeling a little threatened by the thought of the mine development. It was coming closer and becoming more of a reality."[27] This concern persisted as the mine opening drew closer, and was evident at a further community meeting on May 23, 1980. One remark made at this meeting was that Resolute people felt like "Cominco is rushing the Inuit."[28] Resolute Inuit wanted more communication so that they could be better informed and make certain that their concerns would be addressed. This poor consultation process reflected the mining industry at this time, which exhibited ignorance of and apathy to Inuit needs and desires.[29]

Among the biggest concerns Resolute residents had in the planning stages were the environmental impacts of the mine. Some residents also expressed concerns about the impacts of the mine on the subsistence economy, migration patterns, and animal populations on Little Cornwallis Island. Resolute Inuit traditionally hunted in that area as they crossed between Resolute and Bathurst Island.[30] The Inuit consulted made clear that they wanted to continue hunting and trapping while working at the mine, and they wanted to continue hunting in the area around the mine.[31] Residents were concerned that the white-whale wintering colony along the shore near the mine site and the sea bird population might be scared away by the noise and shipping.[32] They also expressed concern about disposing solid waste and sewage in the sea.

At the public community meeting on May 23, 1980, Resolute Inuit expressed their concern about Cominco's tailings disposal plan. Originally, Cominco had planned to dispose of mine tailings in Crozier Strait;[33] however, a feasibility study commissioned in 1974 advised against marine disposal of tailings from the Polaris Mine.[34] Instead, the consultants recommended Garrow Lake, a permafrost-bound, hypersaline lake two miles away from the mine, as an alternative disposal site (Fig. 2).[35] Further reports confirmed that the bottom of Garrow Lake was concentrated salt water, that there was no plant or fish life in the lake, and that the hydrogen sulphide in the water would precipitate any soluble heavy metals deposited in the lake.[36] However, local residents still worried about the possibility that Garrow Lake could overflow and carry tailings, consisting of mine waste and lead and zinc ion concentrates, into the surrounding marine environment.[37] They also believed that, because the lake was saline, it must have an underground channel from the sea, and this concerned them. Cominco assured residents that the tailings would not leave Garrow Lake.[38]

Resolute residents were not the only group concerned about the Polaris development. A Northwest Territories Water Board public hearing held in Resolute on May 22, 1980, provided a venue for the Canadian Arctic Resources Committee (CARC) and Inuit Tapirisat of Canada to voice their opposition to the Polaris development. Founded in 1971 and operating out of offices in Yellowknife and Ottawa, CARC was a public interest group comprised mainly of southern academics dedicated to the environmental and social well-being of Northern Canada and its peoples. The organization emerged as part of the Mackenzie Valley Pipeline Inquiry and presented alternative opinions about industry and resource development projects; through criticizing the problems posed by resource development, CARC sought to bring public attention to issues that impacted the people and environment of the north.[39]

During the Polaris development process, CARC criticized Cominco's actions in the pre-development stage. CARC's concerns about the Polaris project were largely environmental in nature, but it also presented social and economic concerns while working closely with ITC.[40] CARC demanded that the development undergo a federal Environmental Assessment and Review Process (EARP).[41] Cominco never did undergo the EARP process, but instead conducted its own environmental assessments.[42] In 1975, Cominco commissioned BC Research to conduct an environmental study of the mine. BC Research acknowledged that possible and probable environmental effects included direct destruction of vegetation and animal habitat, habitat avoidance due to human activity, and chemical pollutants including sulphur dioxide and nitrogen oxide compounds.[43] Like Resolute residents, CARC was also concerned about the Garrow Lake tailings disposal plan and criticized Cominco for not considering the possible impact on marine environments.[44] In spite of these concerns, the NWT Water Board granted Cominco a water licence for the mine's water supply and tailings disposal effective November 1981.[45] CARC also criticized Cominco for avoiding any discussion of compensating the Inuit for environmental damage caused by the mine or loss of income due to reduced resource base.[46]

ITC supported CARC's environmental criticisms, but the Inuit organization was principally concerned with the economic and political aspects of the Polaris project. The transcript of the May 23 public hearing reveals that ITC was not consulted or involved in the discussion and planning process, despite its efforts to foster contact with Jake Epp, minister of Department of Indian Affairs and Northern Development (DIAND). ITC was critical of Cominco's dismissal of land claims negotiations and the action plan signed between government and Cominco. Largely initiated by the GNWT, the "Socioeconomic Action Plan" primarily focused on employment and training assistance programs for Inuit workers, as well as the dissemination of information, consultation with communities and governments, and maximization of business opportunities for northerners. Though ITC did not go into specific detail in a letter it submitted for reading at the NWT Water Board hearing, it declared the action plan an overall failure, and was particularly dissatisfied with an absence of Cominco Inuit training or hiring programs.[47]

The cooperation of these two groups in opposing the mining industry exemplifies the growing importance that both southern and northern political activists attributed to Inuit rights in the 1970s. Cominco's exclusion of ITC from Polaris discussions was typical of the industry's attitude at this time. Though CARC and ITC shared some anxieties with the people of Resolute Bay, specifically in terms of possible environmental impacts, their attitudes differed from local concerns about mine development. CARC and ITC acted in what they thought were the best interests of the northern environment and people in their criticisms, but their opinions overshadowed the concerns and opinions of those most directly affected by development—the people of Resolute Bay. For example, CARC claimed that the people of Resolute stood in outright opposition to the mine development.[48] While Resolute residents did have concerns about development, as outlined above, they mostly supported it.

During planning, Cominco told the community it planned to hire local people, but unlike the Nanisivik mine, which opened near the community of Arctic Bay in 1976, Cominco did not commit to a formal Inuit employment agreement with the government. Nanisivik was one of the first mines in Canada to have an agreement with the government specifying a quota for Aboriginal workers (the company pledged that 60 per cent of its workforce of 219 would be Inuit within the first three years).[49] According to Robert McPherson, by the time Polaris was developing, the government realized that Nanisivik's agreement was unrealistic and had given up on imposing employment quotas.[50] Instead, Cominco signed informal memorandums of understanding with DIAND in 1980 and 1981 that did not include specific employment targets for Native hiring. Dan McKinnon of DIAND's Northern Resource and Economic Planning branch stated that Polaris would not require any formal agreement with the government, largely because of the lack of state financial support and involvement in the Polaris project.[51] Cominco committed only that it would advertise jobs in the Northwest Territories first and that "whenever possible preference will be given to NWT residents."[52]

It is difficult to know whether Resolute residents were aware that there was no hiring agreement. Interviewees were knowledgeable about the Nanisivik mine and may have assumed that Polaris would be similar in terms of Inuit employment. Most of the correspondence consulted regarding employment was private between the state and the company. The public meeting transcripts do not necessarily suggest that Resolute had concerns or anxiety about employment possibilities; for the most part, according to interviewees, Resolute residents took Cominco at its word that it intended to hire locals. Overall, the hamlet looked forward to the employment opportunities they believed the operation would bring to the community. Once Polaris began operation in 1982, however, Resolute soon realized that significant employment had not materialized.

Once the mine began operation (Fig. 3), previous environmental concerns dissipated when Inuit workers saw little damage to the surrounding area, though one interviewee remembered that dust and chemical ash coming from the mill in the summer months covered the surrounding land.[53] Another interviewee reported that, as he was handling ore in the mill, he noticed ore concentrates going into the ocean while ships were being loaded.[54] Furthermore, interviewees reported that animal populations decreased during operation, though they did note that populations returned after closure.[55]

Employment rather than the environment remained the major point of contention for Resolute during the operational stage. Inuit comprised fewer than thirty (of 250) Polaris employees at peak employment periods, making up less than 10 per cent of the mine's total workforce.[56] Interviewees could recall only ten people from Resolute employed at Polaris over the twenty years it was in operation.[57] Former workers from Resolute stated that the majority of mine workers came from Southern Canada, including a large number from Springdale, Newfoundland, and many from Alberta and Manitoba. They also remembered non-local Aboriginal people working at the mine, including Dene from Dettah (near Yellowknife) and Inuit from other areas in the Northwest Territories.[58]

Interviewees commented on the difficulty of getting the better jobs, which tended to be filled by non-Natives because there was no training offered to Inuit for positions requiring skilled labour.[59] Those from Resolute did a variety of jobs. Most were general labourers (all started out in this position, with the exception of one person who was hired as a guide in the development stage), mill workers (reported as the worst job

Figure 3: The Polaris Mine. NWT Archives/Northwest Territories Dept. of Public Works and Services fonds/G-1995-001: 1525

because of the dust and ash from the mill), surface crew, heavy equipment operators, and polar bear monitors. Female employees worked as housekeepers in the accommodation facilities, although one moved up to become a heavy machine operator after one year.[60]

The low number of employees from the community helps to explain why Resolute residents felt ignored by Cominco. It also adds to our understanding about the lack of collective mining memory in Resolute; so few people from the town were employed at the mine that working there did not become a significant part of the community identity. Interviewees agreed that they would not consider Resolute a mining community either then or now. Furthermore, because of the nature of their positions, the Inuit employees did not form individual identities as miners or mine workers. Many described working at Polaris, though they enjoyed the work, as "just [another] job."[61]

The lack of benefits to the community from the operation also reinforced the town's disconnection from the mine. Before the mine opened, residents believed that the mine would bring spinoff industries to the town. However, such spinoffs failed to materialize, and some interviewees stated that residents were left feeling fooled and betrayed.[62] In their study of post-closure Polaris, Bowes-Lyon et al. suggested residents realized some minor short-term economic benefits from mining operations, but they identified very few long-term benefits.[63] Short-term benefits included increased income for those individuals employed at the mine, more frequent and less expensive jet services in and out of Resolute, and cheaper grocery prices due to the frequency of air traffic coming into the community. When asked if there were any major changes to the community as a result of the mine, all interviewees said the mine had "no real impact" or benefits for the community, other than for the individuals who worked there. Some residents had hoped that Polaris' fly-in/fly-out structure would stimulate extra spending in the community co-op store and the hotels while incoming workers waited for the company charter to fly them to the mine for their rotation. However, these workers stayed at a company hotel next to the airport and very rarely came into the town.[64]

****

Polaris closed in 2002 because of declining ore grades and profitability. Interviewees noted that they knew well in advance that the operation would cease. In fact, the *Nunatsiaq News* reported that Cominco initially intended to close in 2001, but managed to get another year out of the mine.[65] At closure, out of 225 employees, only twenty from the North and only one from Resolute Bay were still working at the mine.[66] Naturally, that one individual was disappointed to lose his job, but interviews clarified that, collectively, Resolute residents felt no sense of loss when they discovered the mine would cease operation. Economically, since there had been little spinoff business as a result of the mine, there was no significant service sector loss or economic disruption upon closure. The biggest impact on the community was the loss of jet services, as Resolute lost all service from Canadian North airlines. First Air is now the only airline with service to Resolute Bay.[67]

Those interviewed expressed mixed feelings about the closure of Polaris. Some reported they were glad to see the mine close because of "less pollution" on Little Cornwallis Island and an increase in animal populations. Some former workers reported that working at the mine was the best time of their lives. One stated, "I am still homesick for that place," and another reported that when the company was closing it up and demolishing the buildings he did not want to see it happen so he chartered a plane to leave early. When interviewees spoke of their personal experiences and memories of the mine, their stories usually related to social events or work. For the most part, although there were some negative memories, individuals generally emphasized the positive aspects of their mining experiences. They all spoke of how much fun they had working at Polaris, and recalled the activities available to them during their time off such as swimming, karate demonstrations, and passing time at the gym. One interviewee recalled, "there was always something to do . . . but lots of work too." Another remembered the baseball games when the Polaris team played the Nanisivik team, as well as teams from the airport and from the hamlet of Resolute. "It had a big impact on me, that mine," one interviewee reflected, and saying if he had the chance he would love to work in a mine again.

In contrast to these strong personal memories, the exclusion and marginalization of Resolute itself left the community with no strong ties to Polaris. Cominco largely ignored residents' concerns about environmental impacts and failed to conduct any further environmental assessments. Some community members were critical of the lack of communication and involvement in the planning stages:

> It was good but it would have been better if we talked to them more and worked with them more by communicating [with] each other. But we leave them alone; we were so Inuk that . . . Inuit way is leave things alone. Live down here, let the people live up there, on top of you. Don't harass and ask around. If they ask you then, "Ok, thank you."[68]

This particular interviewee also suggested that the biggest disappointment that she, personally, had with the company and with northern mining was the lack of employment opportunities. If Inuit had been more involved in the planning process, she believes, they would have received more benefits locally from the mine, and many more people from Resolute would have been hired at Polaris.[69]

Together, the lack of involvement in both planning and employment at the mine, the marginalization of Resolute Inuit concerns about development, and the limited benefit (and loss) created by operation (and closure) suggest why Resolute does not retain heritage or collective mining memory in the town. In addition, Resolute Bay's distance from the mine—being one hundred kilometres away from the mine site and physically separated by water—has kept its physical legacies hidden. The structure of its fly-in/fly-out rotation schedule meant that often the only people who ever spent significant amounts of time in and around the mine were those who worked there. One interviewee stated, "If you'd never gone to that mine, you'd never know who's working there."[70] The only time Resolute Inuit would have occasion to see the mine infrastructure was during freeze-up when they made their way across Little Cornwallis Island on the sea ice to hunt on Bathurst Island, and even then they usually only saw Polaris as a light in the distance directing them west.

Unlike many former mining communities in the North and in the South, Polaris left few industrial ruins on the landscape that might provide reminders of the region's industrial past. Residents do not walk past the old mine site every day. They do not see it in the distance from their front porch or their window. At the site itself, the landing strip is the only relic of the former mine that remains in the area. Decommissioning and reclamation of the site began immediately after closure and was completed in 2004. This process consisted of removing buildings and the dock, disposing of metal-contaminated soil, and decommissioning the tailings dam. Infrastructure such as the mill, mill equipment, and mining fleet were buried underground.[71]

Furthermore, there is a general absence of visible ecological changes left on the land. At most abandoned mines, one is likely to find tailings piles, pollutants, industrial waste, open pits, abandoned infrastructure, destroyed migration zones, and (potentially) adverse human health effects.[72] Analyzing Schefferville, Jean Sébastien Boutet argues, "For the Innu and Naskapi, the post-industrial environment acts as an incessant material reminder of three decades of intensive land exploitation."[73] At Polaris, aside from twenty million tonnes of mining tailings that have been dumped into the Garrow Lake, there is little evidence of environmental degradation around the mine site. In Resolute, in the absence of any persistent or urgent environmental impacts, there are few lingering anxieties about the old mine site, aside from some ongoing concern about Garrow Lake. Teck Cominco's environmental monitoring ended in 2012, and, when the author was in Resolute, one interviewee was hopeful that the company would update them on the condition of the tailings. Though there has been no reported or suspected damage, the concern is still present, especially when hunting in the area. He stated, "I'd like to know. I just want to be safe."[74]

In Resolute, the town did not think it critical to preserve a public, collective memory of the mine. There is no plaque or memorial to the mine. Every now and then, when hunting in the area, some notice that the buildings are no longer there. As one interviewee stated, "It's gone now, there is no memory of it here in the community."[75] The youth in the community are largely unaware that there had been a mine near Resolute (unless a parent or relative had worked there).[76] Those interviewed without exception reported that working at the mine was just a job. None of the former workers I interviewed identified themselves as mine workers; rather, they self-identified as hunters. I heard repeatedly (sometimes in laughter) that Resolute is certainly not a mining town today, and it was not during the operational phase either.[77] The absence of commemoration means that the youth (and outsiders) fail to learn about that aspect of the community's past.

The absence of commemoration speaks to the marginalization that Resolute Inuit, like communities before them, felt at a time when it was expected and normal for an outside force to make decisions that would affect a community, without including the community in the process. Just as commemoration can tell us much about what a community wants to remember, the absence of commemoration can teach us about what a community may choose to forget. The lack of commemoration suggests that the Polaris mine was not deemed an important site to the community for good reason. The formation of a collective mining memory served no purpose for the residents of Resolute Bay. Unlike southern mining towns (and other northern mining towns such as Rankin Inlet), Polaris lacked a strong labour force from within the local community; Resolute

Bay existed before the mine came, and the town did not rely on the mine for its major income. Finally, a collective memory served no political, economic, or social purpose for the town. Arguably, the Polaris mine operation *itself* was not a part of Resolute's history as a community. It is a part of individual persons' histories, for those who worked there or those who were actively engaged in the planning phase. Aside from private, individual reminiscence, Resolute residents largely do not engage with memories of a process in which they were largely slighted, excluded, and marginalized.

ACKNOWLEDGMENTS

I would sincerely like to acknowledge the Resolute Hamlet Office and those individuals in Resolute Bay who agreed to sit down and discuss their stories about mine development and mine work with me. I must thank the various funding organizations that have made my travels to Resolute and to various archives possible: Social Sciences and Humanities Research Council, ArcticNet, Northern Scientific Training Program, and Memorial University of Newfoundland's Department of History. Finally, thank you to John Sandlos and Arn Keeling, who helped shape this project from its earliest stages and who have been meticulous editors of this collection.

NOTES

1   Stephen Brazen and Tony Thirlwall, *Deindustrialization* (Oxford: Heinemann Educational Books Ltd., 1992); Steven High, *Industrial Sunset: The Making of North America's Rust Belt, 1969–1984* (Toronto: University of Toronto Press, 2003). Some works that deal with the closure of mines specifically include Cecily Neil, Markku Tykklainen, and John Bradbury, *Coping with Closure: An International Comparison of Mine Town Experiences* (New York: Routledge, 1992); Timothy LeCain, *Mass Destruction: The Men and Giant Mines That Wired America and Scarred the Planet* (New Brunswick, NJ: Rutgers University Press, 2009); David Robertson, *Hard as the Rock Itself: Place and Identity in the American Mining Town* (Boulder: University Press of Colorado, 2006);

Richard Francaviglia, *Hard Places: Reading the Landscape of America's Historic Mining Districts* (Iowa City: University of Iowa Press, 1991).

2   See James Connolly, ed., *After the Factory: Reinventing America's Industrial Small Cities* (Toronto: Lexington Books, 2010); High, *Industrial Sunset*; Jefferson Cowie and Joseph Heathcott, eds., *Beyond the Ruins: The Meaning of Deindustrialization* (Ithaca, NY: Cornell University Press, 2003).

3   This recent trend in the literature has a focus on the social histories of closure. Cowie and Heathcott's *Beyond the Ruins*, Robertson's *Hard as the Rock Itself*, and Steven High and David Lewis's *Corporate Wasteland: The Landscape and Memory of Deindustrialization* (Toronto: Between the Lines, 2007) move beyond the quantitative-type studies to widen the discussion of deindustrialization to include memory, culture, and politics of deindustrialization. Common themes in these works include commemoration, legacy, identity, and connection to place.

4   Rosemary Power, "'After the Black Gold': A View of Mining Heritage from Coalfield Areas in Britain," *Folklore* 119, no. 2 (Aug. 2008): 160.

5   Ibid.

6   Ibid., 170.

7   Ibid., 162.

8   M. Mellor and C. Stephenson, "The Durham Miners' Gala and the Spirit of Community," *Community Development Journal* 40, no. 3 (2005): 343–51.

9   Ben Marsh, "Continuity and Decline in the Anthracite Towns of Pennsylvania," *Association of American Geographers* 77, no. 3 (1987): 339.

10 Ibid., 344.

11 Tara Cater and Arn Keeling, "That's where our future came from": Mining, Landscape, and Memory in Rankin Inlet, Nunavut," *Etudes/Inuit/Studies* 37, no. 2 (2013): 59.

12 Qausuittuq means "place with no dawn" in Inuktitut.

13 I use "collective memory" and "community memory" interchangeably throughout this chapter referring to the memory of a group of people, typically of an event or era that holds meaning to the group, and often passed down over generations.

14 It is important to state that these opinions are those of a small percentage of the population. I interviewed seven (out of a possible ten) former Polaris workers from Resolute. Of the remaining three, one no longer lived in Resolute, one was away at the time I was there, and the third did not wish to be interviewed. The accounts I rely on in this narrative are the accounts of those people who agreed to be interviewed. I cast a wide net open to anyone who was in the community at the time of development, operation, and closure,

workers or not, to talk about their memories of the mine; it was surprising how many people did not believe they were a worthy candidate because they did not work at the mine and, hence, said they had no connection to the mine. Relying on ethnographic sources can be a slippery slope, and I initially worried that such a small percentage of interviewees would create dangerous generalizations. Frankly, I had to trust in and rely on my interviewees' stories about the mine and their assurances that there is no collective mining memory in Resolute. I feel confident in stating that there was an absence of a collective memory after my time spent in the community and both the formal and informal conversations I had with residents while there. Though I cannot cite informal conversations, the general collective impression agrees with the interviewees' stories.

15   See High and Lewis, *Corporate Wasteland*, and Robertson, *Hard as the Rock Itself*.

16   John Sandlos and Arn Keeling, "Claiming the New North: Development and Colonialism at the Pine Point Mine, Northwest Territories, Canada," *Environment and History* 18, no. 1 (2012): 5–34.

17   David Trigger, "Mining, Landscape, and the Culture of Developmental Ideology in Australia," *Cultural Geographies* 4, no. 2 (1997): 164–65.

18   Robert McPherson, *New Owners in Their Own Land: Minerals and Inuit Land Claims* (Calgary: University of Calgary Press, 2003), 58–65.

19   For more, see McPherson, *New Owners in Their Own Land*.

20   Mary Josephine Taylor, "The Development of Mineral Policy for the Eastern Arctic, 1953–1985" (MA thesis, Carleton University, 1985), 124.

21   Formerly known as Consolidated Mining and Smelting Company (CM&S), Vancouver-based Cominco Ltd. (now Teck Cominco) was one of the largest Canadian natural resource companies with mine sites internationally. Cominco owned and operated the Con Mine in Yellowknife (1938), Pine Point in the Northwest Territories (1964), and Black Angel in Arctic Greenland (1973) at the time they developed Polaris in Nunavut.

22   J. E. Barrett, "Employment of the Inuit at Polaris, Little Cornwallis Island: A Feasibility Study Requested by Arvik Mines Ltd.," 1973.

23   NWT Water Board Public Hearing transcript, May 22, 1980, RG12 1985-86-578 Box 10 9004-6-2, Library and Archives Canada (hereafter LAC).

24   Interviewee 7, oral history interview with author, Resolute Bay, June 2, 2012.

25   It is unclear as to the exact date that the Socioeconomic Action Plan was signed; however, it was approximately around the same time as the Memorandum of Understanding in 1981 (RG12 1985-86-578 Box 10 9004-6-2, LAC).

26   These communities were Spence Bay, Pelly Bay, Gjoa Haven, Cambridge Bay, Holman Island, Coppermine, and Resolute. J. E. Barrett and Associates, "The

Polaris Project and the Inuit: An Assignment Concerned with Involving the Inuit in the Polaris Mine Development Requested by Arvik Mines Ltd.," 1976.

27 Barrett, "The Polaris Project," 36.
28 Cominco Ltd., "Polaris Mine Project: Transcript of Public Meeting in Resolute Bay, NWT," May 23, 1980, RG12 1985-86-578 Box 10 9004-6-2, LAC.
29 McPherson, *New Owners in Their Own Land*, xvii.
30 Interviewee 3, oral history interview with author, Resolute Bay, June 1, 2012. Other concerns in Resolute included accommodation, schooling, and the loss of experienced workers from the community to the mine (RG12 1985-86-578 Box 10 9004-6-2, LAC); as well, residents were concerned with possible in-migration of southerners settling in Resolute to be closer to the mine (Barrett, "The Polaris Project," 36).
31 RG12 1985-86-578 Box 10 9004-6-2, LAC.
32 Derek V. Ellis and Jack L. Littlepage, *Feasibility Study for Marine Disposal of Tailings at the Polaris Mine-site* (Victoria: University of Victoria, 1974), 16.
33 Ibid., 1.
34 Ibid., 17.
35 Ibid., 21.
36 RG12 1985-86-578 Box 10 9004-6-2, LAC.
37 RG12 1985-86-578 Box 10 9004-6-2, LAC.
38 Barrett, "The Polaris Project," 15–16.
39 For more about CARC, see Everett Peterson and Janet Wright, eds., *Northern Transitions: Northern Resource and Land Use Policy Study* (Edmonton: CARC Publishing Programme, 1978).
40 Correspondence, papers, June 1976 – September 1979, Wilfrid Laurier University Archives (WLUA) CARC fonds file 3.13.2.2.5; and Northwest Territories Water Board public hearing: application for a water licence by Cominco Ltd., Polaris Mine Project, 1980, WLUA CARC fonds file 3.13.2.2.26.
41 Though EARP has been applied by the Canadian government since 1974, it was not required until 1984 and then only mandatory for proposals requiring federal involvement. The records are not clear as to what extent the federal government was a part of the Polaris operation. Therefore, Cominco may not have been legally obliged to undergo EARP. An article in the *Northern Miner* in 1979 reported that the ITC and CARC wanted to halt development of the mine until up-to-date environmental studies were completed. The editorial claimed that Cominco had already conducted thirteen studies (environmental, economic, and social) (*Northern Miner*, November 1979).
42 In 1973, Cominco and Sheritt-Gordon carried out a research and development program to investigate hydrometallurgical processes to ensure that the

mine's processing would meet environmental standards. Polaris Mine Project, R1526-30-4-E vol. 269 file 243-18, LAC.

43   BC Research, "Environmental Study of the Polaris Mine, Little Cornwallis Island" (1975), WLUA CARC fonds file 3.13.2.2.20. BC Research reported that "any major chemical pollutant input into the terrestrial ecosystem will be from air-born emissions" and that such emissions were deemed to only be important in the summer months when the mill was active (BC Research, "Environmental Study of the Polaris Mine," 5).

44   Papers, newspaper clippings, January–May 1980, WLAU CARC fonds file 3.13.2.2.3

45   NWT Water Board, WLUA File 0262, Nov. 1, 1981.

46   Papers, newspaper clippings, January–May 1980, WLUA CARC fonds file 3.13.2.2.3. Though CARC criticized Cominco, no archival records or oral interviews suggest that Inuit of Resolute asked for compensation. They believed that they would receive employment opportunities from the mine, which is what they looked forward to about development.

47   Unfortunately I was unable to locate a copy of the Socioeconomic Action Plan, and the information provided above is the only available information coming from company and government reports. Furthermore, ITC was not specific about particular criticisms it had with the action plan, only general concerns as noted in this text. RG12 1985-86-578 Box 10 9004-6-2, LAC.

48   Papers, newspaper clippings, January–May 1980, CARC fonds file 3.13.2.2.3.

49   However, there were only forty in the first three years, and in 1979 there were only ten Inuit employees at the Nanisivik mine. McPherson, *New Owners In Their Own Land*, 92, 109. See also Chapter 10, this volume.

50   McPherson, *New Owners In Their Own Land*, 126.

51   McKinnon said Nanisivik had a formal agreement because the federal government granted the development $21 million in funding, indicating that Polaris did not receive such financial support. Correspondence, papers, October 1979 – April 1980, WLUA CARC fonds file 3.13.2.2.6. See also Katherine Graham, *The Development of the Polaris Mine* (Kingston: Queen's University Centre for Resource Studies , 1982), v.

52   Polaris Vision, MS-2500 Box 409 file 6, Royal British Columbia Museum Archives.

53   Interviewee 5, oral history interview with author, Resolute Bay, June 2, 2012.

54   Interviewee 8, oral history interview with author, Resolute Bay, June 4, 2012; interviewee 5, oral history interview with author, Resolute Bay, June 2, 2012.

55   One interviewee reported that before the mine, beluga whales were plentiful along the shores of Cominco Bay, but when mining activity started there were

much fewer. He says, "The Inuks relied on those populations, so that was one negative effect [of development]. After the mine was gone, the population of belugas increased." Interviewee 8, oral history interview with author, Resolute Bay, June 4, 2012; interviewee 5, oral history interview with author, Resolute Bay, June 2, 2012.

56   Léa-Marie Bowes-Lyon, Jeremy P. Richards, and Tara M. McGee, "Socio-Economic Impacts of the Nanisivik and Polaris Mines, Nunavut, Canada," in *Mining, Society, and a Sustainable World*, ed. Jeremy Richards (Berlin/Heidelberg: Springer-Verlag, 2009), 371–96.

57   I have found no company documents with statistics or data of Polaris employees, where they came from, and how long they were at the mine. However, those I spoke with in Resolute in June 2012 identified only ten people within the community who had worked at the mine. Of course, there may be others, but those I spoke with all repeated the same ten people (or fewer).

58   Interviewees 3, 5, 6, 8, 9, oral history interviews with author, Resolute Bay, June 1–4, 2012.

59   Interviewee 8, oral history interview with author, Resolute Bay, June 2, 2012.

60   Of the ten Inuit employees at Resolute, two were female. Interviewee 3, oral history interview with author, Resolute Bay, June 1, 2012.

61   Interviewee 3, oral history interview with author, Resolute Bay, June 1, 2012.

62   Interviewee 7, oral history interview with author, Resolute Bay, June 2, 2012.

63   Bowes-Lyon, Richards, and McGee, "Socio-Economic Impacts," 371.

64   Interviewee 3, oral history interview with author, Resolute Bay, June 1, 2012.

65   *Nunatsiaq News*, July 2000 and May 11, 2001.

66   Interviewee 4, oral history interview with author, Resolute Bay, June 2, 2012.

67   Bowes-Lyon, Richards, and McGee, "Socio-Economic Impacts," 383.

68   Interviewee 7, oral history interview with author, Resolute Bay, June 2, 2012.

69   Interviewee 3, oral history interview with author, Resolute Bay, June 1, 2012. The Mary River development in the Baffin Region (160 kilometres from Pond Inlet) will be an open-pit mine extracting iron ore. The *Globe and Mail* hailed the project as "the most ambitious mining venture ever undertaken in the Arctic." People in Resolute expressed support for the way the development has been handled in terms of an environmental impact statement and negotiations of an impact and benefit agreement. Though people in Pond Inlet expressed concerns and were divided in their opinions, Resolute residents believed that the consultation process and the ability to present local concerns had improved from the development of Polaris (Paul Waldie, "A Railway to Arctic Riches: Economic Boom, Environmental Threat?," *Globe and Mail*, May 14, 2011).

70   Interviewee 3, oral history interview with author, Resolute Bay, June 1, 2012.

71   The *Northern Miner,* November 11, 2002.

72   For more on this northern literature, see Arn Keeling, "'Born in an Atomic Test Tube': Landscapes of Cyclonic Development at Uranium City, Saskatchewan," *The Canadian Geographer / Le G´eographe canadien* 54, no. 2 (2010): 228–52; Jean-Sébastien Boutet, "An Innu-Naskapi Ethnohistorical Geography of Industrial Iron Mining Development at Schefferville, Québec" (MA thesis, Memorial University of Newfoundland, 2011); Sandlos and Keeling, "Claiming the New North"; Arn Keeling and John Sandlos, "Environmental Justice Goes Underground? Historical Notes from Canada's Mining Frontier," *Environmental Justice* 2 (2009): 117–25. To read more about the environmental impacts of mining outside of the northern Canadian context, see LeCain, *Mass Destruction* and Brett Walker, *Toxic Archipelago: A History of Industrial Disease in Japan* (Seattle: University of Washington Press, 2010).

73   Boutet, "An Innu-Naskapi Ethnohistorical Geography," 217.

74   Interviewee 9, oral history interview with author, Resolute Bay, June 2, 2012.

75   Interviewee 7, oral history interview with author, Resolute Bay, June 2, 2012.

76   However, it is not always the case that the youth know if their relatives worked there. One youth I interviewed said she did not know of anyone who worked at the mine, though I later discovered from her grandmother that three of her uncles had worked there. She made the following statement about what kind of history she learns from her community and what is deemed important parts of the past for the younger generation to know: "I learned some skills from my grandmother and heard her history about her family and how they would survive. Like, what they would use for their huts and I learned how to make an igloo and I learned how to make mitts, what stitches to do and how hard they would work every day and how they connected to each other. I learned about those from my grandmother and she's still teaching me more." She later said that these types of things are more important to her personal history than the mine and more common for kids to learn about in Resolute (Interviewee 1, oral history interview with author, Resolute Bay, May 31, 2012).

77   Interviewee 3, oral history interview with author, Resolute Bay, June 1, 2012.

| CHAPTER 12

Liability, Legacy, and Perpetual Care: Government Ownership and Management of the Giant Mine, 1999–2015

Kevin O'Reilly

INTRODUCTION

Yellowknife is a place of contrasts and transition.[1] The oldest rocks on earth were found about three hundred kilometres north of Yellowknife on the Canadian Shield, and the rock around Yellowknife itself is about 2.6 billion years old.[2] Geographically, it is on the edge of the tenth largest freshwater lake on the planet, Great Slave Lake. Near the northern edge of the Canadian Subarctic, the tree line is only about two hundred kilometres north of Yellowknife. The Yellowknives Dene and their ancestors have lived in the area for at least 7,000 years; their name comes from their use of native copper in making tools. The fur trade for the area was mostly handled out of Fort Rae, about one hundred kilometres northwest of Yellowknife, until a small outpost opened at Dettah near the present city of Yellowknife in 1920. Prospectors were drawn to the area as early as the Klondike gold rush in the 1890s, when traces of gold were found near Yellowknife. Full-blown prospecting and a producing mine waited until the 1930s. There was a lull in gold mining during the Second World War when labourers were drawn away from the industry, but a full-scale boom followed shortly afterwards.[3]

The population of Yellowknife grew from a few hundred in the 1930s to almost a thousand in 1940 and 3,200 by 1961. In 1967 Yellowknife became the first capital of the Northwest Territories, and many federal workers moved north with their families to work for the new territorial government. By 1981 the population had grown to about 9,500, and today almost 20,000 people live in the city. Until the 1990s, Yellowknife was still known primarily as a gold-mining centre, but with the closure of the gold mines, the economy has transitioned to diamond mining and services.[4]

An uneasy relationship long existed between the non-Aboriginal and Aboriginal populations. From the beginning of the mining era, few jobs were available to members of the Yellowknives Dene First Nation, and they received no revenues from the gold mines in their traditional territory. In the 1950s, federal Indian Affairs officials convinced the Yellowknives to move from scattered camps to centralized settlements around the trading post at Dettah and on part of Latham Island in the Old Town. The Yellowknives, as part of the Akaitcho Territory Government, are still negotiating their broader rights to lands and resources in the region with the federal and territorial governments. The lack of local control over land and resources also created feelings of marginalization on the part of many non-Aboriginal northerners, as the federal government has maintained authority over natural resources from the early twentieth century to the present, with a devolution agreement completed in 2014.[5]

The long-standing issue of arsenic pollution from the Giant Yellowknife Gold Mine provides one of the worst examples where external pressure to develop northern resources, and a corresponding failure to meaningfully involve local people in decision making, produced

FIGURE 1: Aerial view of the Giant Mine complex, Yellowknife, NWT, 2009. Photo by Kevin O'Reilly.

disastrous short- and long-term consequences for the environment and human health. The mine operated from 1948 to 1999, when its owner, Royal Oak Mines, went into receivership (see Table 1 for some key dates regarding the Giant Mine). Underground production continued until 2004 (with the ore being processed at the nearby Con Mine), when it finally shut down and became a huge public liability. The gold ore at Giant was found in arsenopyrite rock, which was heated or "roasted" to drive off sulphur and arsenic. The arsenic was released as arsenic trioxide, which takes the form of a gas when heated, but dust at normal room temperature. As part of the ore processing, 237,000 tonnes of arsenic trioxide were blown underground into old mine workings and some purpose-built chambers. The thirteen chambers are approximately equivalent to the volume of seven ten-storey buildings. Arsenic trioxide is a proven non-threshold carcinogen and dissolves easily in water. Currently, the chambers are leaking arsenic into the mine water, which is then treated before it eventually reaches Yellowknife Bay. There are ninety-five hectares of contaminated mine tailings on the surface, the equivalent of about three hundred football fields, ten to fifteen metres deep. Eight open pits and a roaster complex on the surface contain another 4,900 cubic metres of arsenic trioxide and other waste.[6]

Gold mining in Yellowknife, particularly the histories of the three mines closest to the city (Giant, Con, and Negus Mines), is often portrayed in a positive light as part of the Northwest Territories' "founding" story (see Table 1). But beginning in the 1950s, water and airborne arsenic pollution from all these mines (with Giant contributing the vast majority) had very severe health and environmental consequences, particularly for the nearby Yellowknives Dene First Nation communities.[7] The Yellowknives Dene First Nation citizens have described the area of the mine site as a previously very productive valley full of blueberries and fish, an important gathering area that the mining operation completely destroyed. The Yellowknives Dene also suffered disproportionate risks from arsenic due to extensive contamination of snow and water used for drinking. At least one First Nation child died from drinking meltwater made from snow in 1951, and several other community members became sick. The mine owner paid the family $750 in compensation for the death of their child.[8]

Ironically, it was the pollution control equipment installed in October 1951 in response to air pollution issues that created the underground storage problem, as ton after ton of captured toxic arsenic trioxide dust had to be deposited somewhere on the mine site. Up until the 1980s, the theory was that permafrost would contain the arsenic stored underground permanently. The mine owners and the government were of the view that underground storage of the arsenic trioxide produced by the roasting operation would be safe. An engineer working at the mine, W. M. Gilchrist, warned that this method might not work, writing that "the advantage of underground storage in the area considered over surface storage is not marked enough to warrant it, when the possible effects of operations in the area and the lack of control over the material once it is place in an underground chamber are considered."[9] The available records suggest, however, that few, if any, other government and mine officials said "stop" or "maybe we should think this through." Careful monitoring

TABLE 2: Chronology of Key Events for Giant Mine

Date	Event
1935	Original mining claims staked
1948	Giant Mine goes into production, tailings initially dumped directly into Great Slave Lake
1949	Gold roasting operations begin at Giant Mine
February 1951	Giant Mine tailings begin to be dumped directly into smaller lakes near the site
October 1951	Air pollution control equipment installed, captured arsenic trioxide dust blown underground into mined out areas
1957	Treatment of tailings and water discharges begins at Giant Mine
December 1977	Canadian Public Health Association Task Force reports on Giant Mine
1978	First water licence issued for Giant Mine
1999	Giant Mine owner Royal Oak Mines goes into receivership
July 2004	All mining at Giant Mine stops
March 15, 2005	Cooperation Agreement signed between federal and territorial governments for Giant Mine
June 15, 2005	Subsurface rights under Giant Mine are withdrawn by federal Order in Council SI/2005-55
December 31, 2005	Water Licence for Giant Mine expires
October 7, 2007	Giant Mine Remediation Plan water licence application submitted
February 21, 2008	Mackenzie Valley Land and Water Board completes a preliminary screening of the Giant Mine water licence and sends it forward without an environmental assessment
March 31, 2008	Giant Mine water licence application referred for an environmental assessment by the City of Yellowknife
July 22-23, 2008	Mackenzie Valley Environmental Impact Review Board holds scoping hearing
October 2010	DIAND and GNWT submit Developer's Assessment Report as part of the environmental assessment of the Giant Mine Remediation Project
September 10-14, 2012	Public Hearings held as part of the Environmental Assessment of the Giant Mine Remediation Project
June 20, 2013	Review Board issues its Report of Environmental Assessment and Reasons for Decision on the Giant Mine Remediation Project

of the underground storage process was recommended by the Canadian Public Health Association special task force that examined the arsenic issue in Yellowknife in the 1970s, advice that obviously was not followed.[10]

Canadian taxpayers have spent over $160 million to look after the Giant Mine site since 1999.[11] The government's remediation plan will cost $903 million and will require perpetual care at an estimated cost of $2 million a year—forever.[12] A 2002 study by government officials estimated that seven million ounces of gold were produced from Giant Mine with a value of about $2.7 billion.[13] The owners made $867 million in profit, while the government only collected $454 million ($360 million from workers as income tax, $78 million as corporate taxes, and only $16 million in royalties, while providing direct subsidies of $47 million). Canadians should seriously question whether this was a wise investment and who really benefited from a resource development that has become one of the worst toxic liabilities in the country.[14] Canadians should also question how the federal government has handled the remediation of the mine, particularly the proposal to freeze the underground arsenic and create a perpetual care scenario with unknown risks and potential liabilities that extend for an unimaginable amount of time into the future. As one means of contributing to this debate, this chapter will provide a history and critical review of the period of government ownership and management of the Giant Mine from 1999 to 2015, including the development of the remediation plan, the environmental assessment of the proposed frozen block method of arsenic containment and the negotiation of a legally-binding environmental agreement to set up an independent oversight body.

PRELUDE TO RECEIVERSHIP, 1991–1998

Pollution controversies at Giant predate the closure of the mine. They are worth briefly recounting because they reveal a lax federal regulatory regime for air and water pollution in the Northwest Territories in the decades leading up to closure of Giant Mine and Royal Oak's tendency to resist stubbornly the strict regulation of contaminants. Moreover, in the few years before Giant Mine went into receivership, the noose started to tighten in terms of increased public pressure for some form of improved pollution control, particularly for sulphur dioxide and arsenic emissions. These pollution controversies provided forums for the expression of increasing public and official concern over the underground arsenic issue.

In April 1991, two Yellowknife citizens (Chris O'Brien and the author), pressed for an investigation into the impact of air emissions from Giant on the environment and human health. The request proceeded under the newly minted NWT Environmental Rights Act, passed as a private member's bill by the NWT Legislative Assembly in 1990.[15] The initial response from the territorial government was to ask for more evidence of arsenic emissions, so the requesters sent relevant documents, including a report generated by the government itself. More than two years after the request for an investigation, the territorial government released a report confirming that trees were being damaged by sulphur dioxide.[16] The federal Department of Health and Welfare conducted a human health assessment and concluded that sulphur dioxide levels posed no health hazards to Yellowknifers other than mild and reversible impacts among those with respiratory issues such as asthma. For arsenic, however, the report emphasized that no safe threshold level could be imposed: "Canadian and international regulatory agencies have classified arsenic as a known human carcinogen. For this reason alone, exposure to arsenic should be reduced to the lowest possible level."[17]

At that time, estimates suggested that twenty-six kilograms of arsenic trioxide per day were still going up the stack at Giant Mine. There was no clearly enforceable air quality legislation or regulations covering the Northwest Territories, due in part to the fact that air quality legislation is being left to provincial authorities in Southern Canada and so some confusion over jurisdiction occurred.[18] The territorial government was also reluctant to take on new responsibilities and related monitoring and enforcement costs. Both governments were also extremely reluctant to regulate Giant Mine as the primary point source for airborne arsenic in the Yellowknife area. By the early 1990s, pollution concerns had been pushed to the sidelines as the horrendous labour dispute at the Giant Mine—during which nine replacement workers were killed—thrust Giant Mine into the national media spotlight. Even here, remarkably, the federal government was reluctant to intervene in Royal Oak's internal affairs, ensuring that there was little action on both the environmental and labour fronts through the early 1990s.[19]

In 1994 the Canadian Environmental Protection Act (CEPA) identified inorganic arsenic compounds, including the arsenic trioxide as found at the Giant Mine, as a priority toxic substance that required an action plan to reduce exposure. In May 1995 the Standing Committee on the Environment and Sustainable Development travelled to Yellowknife, but nothing had happened.[20] In June, the Committee recommended that Environment Canada identify what action it would take on arsenic by December 1995.[21] The Committee was particularly concerned about perceptions of arsenic contamination among the Yellowknives Dene, suggesting that public consultations had revealed "the elders' resulting loss of confidence in the government's ability to protect their environment and health. Nowhere was this loss of trust more apparent than on the issues of arsenic pollution."[22]

In response to growing pressure from the public and municipal authorities, the territorial government adopted an unenforceable ambient air quality guideline for maximum levels of sulphur dioxide on June 24, 1994, that was identical to the federal guideline. In August 1995, Yellowknife City Council adopted a motion calling on the federal and territorial governments to "take immediate steps to introduce enforceable, binding regulations dealing with sulphur dioxide and arsenic."[23] That same month, the federal and territorial governments set up a task force that included officials from federal and territorial departments responsible for health, environment, and mining.[24] Following a series of reports and consultations, the task force held a final workshop in July 1997 that recommended negotiating an agreement with the owners of Giant Mine to impose arsenic emission reductions over a fixed period of time with penalties for non-compliance.[25] Amid the task force deliberations on arsenic, the territorial government released draft Gold Roasting Discharge Control Regulations (in May 1996) that would have required the owners of Giant Mine to monitor and reduce sulphur dioxide by 90 per cent within ten years (i.e., toward the end of the then predicted mine life). The owner, Royal Oak Mines, openly threatened to shut down the mine if the government introduced air-quality regulations.[26] Nothing further happened with the proposed sulphur dioxide or arsenic regulations, and the ultimate solution was to wait for the mine to shut down.

Local concerns about arsenic and water pollution in Yellowknife Bay and Back Bay also persisted into the 1990s despite the fact the intake for potable water was moved upstream of Giant Mine in 1969.[27] Giant Mine's use of water, including any pollution of local waterways, was first regulated in 1975 through a water licence issued by the newly created NWT Water Board. Renewals of the water licence took place (roughly every five to six years), and transcripts of the hearings reveal persistent and ongoing concerns about the impact of arsenic issuing from tailings ponds into Baker Creek, and then draining out into Back Bay. In February 1975 the federal government fined the company $2,000 under the federal Fisheries Act for a tailings spill the previous April in Back Bay, where the heavy loading of cyanide and arsenic proved toxic to fish.[28] By 1993 ammonia emissions (a by-product of heavy explosive use in mining) were also regulated for the first time under the water licence due to this substance's acute toxicity to aquatic life. The company could not meet the ammonia limits due to increased sewage from the camp as replacement workers were being housed at the mine during the strike, but Royal Oak successfully petitioned for an emergency amendment to raise the limits from 2 parts per million (ppm) to 15 ppm, and then a longer term increase to 19.5 ppm in May 1994.[29]

Amid all the various pollution controversies and the evolving regulatory regime surrounding Giant Mine in the early 1990s, federal and territorial regulators began to express at least some concern about stability and long-term liabilities associated with Giant Mine's underground arsenic. The renewal of the water licence issued in May 1993 contained a few new provisions about the underground storage of arsenic and ultimate closure and reclamation of the site. An investigation and evaluation of the underground arsenic storage vaults, including rock mechanics, geohydrology, geochemistry, permafrost, and risk assessment, as part of abandonment and reclamation were required, with a terms of reference to be submitted for approval to the NWT Water Board by June 1993. The final report on this work was due in 1997 but not completed. A letter from the company indicated that Royal Oak was now pursuing removal, refining, and export of pure arsenic to the United States for use in

TABLE 3: DIAND's proposed options for remediating underground arsenic at Giant Mine in 1997 study

METHOD	DESCRIPTION	ESTIMATED COST
Secure Landfilling	removal and storage in an engineered landfill	$205-225 million
Offsite Treatment	removal and shipment by truck to another facility	$425-445 million
Off-site Incineration	removal and shipment by truck to another facility	$347-368 million
Bioremediation	use of biological agents for treatment, not proven	$1.542 billion
Cement Stabilization	removal and mixing with cement for deposition into an engineered landfill	$25-31 million
Phytoremediation	use of plants for treatment, not proven	$10-180 million
Conversion to Ferric Arsenate	use of a pressure oxidation unit to convert the arsenic into a less toxic form	$1.684 billion
Conversion to Arsenic Sulphide	conversion to a less toxic form of arsenic	$725 million
Marketing	reprocessing or arsenic for other uses such as pesticides or wood preservative	$10-30 million in profits from sales

manufacturing wood preservatives. The company suggested that leaving the arsenic underground would produce a perpetual care scenario, something that regulators and the general public would find unacceptable. The company proposed a new date of May 1, 2000, for a detailed proposal for removing and reprocessing the underground arsenic.[30]

At the same time, the federal Department of Indian Affairs and Northern Development (DIAND) began its own study of the underground arsenic storage, fearful that it would likely inherit the site because Royal Oak would not or could not carry out a proper reclamation program. On October 6, 1997, DIAND released a report that examined options for the underground arsenic with costs ranging from $10 million to $1.7 billion (see Table 2).[31]

Company, federal, territorial, and City of Yellowknife officials held a technical workshop on the issue in October 1997, but no representatives were invited from the Yellowknives Dene First Nation or non-governmental organizations. Workshop participants made no firm decisions but did complete a preliminary evaluation of options that rated continued storage underground as the least expensive option.³²

To partially fulfill previous requirements of its water licence, Royal Oak Mines submitted a rough proposal in December 1997 to the NWT Water Board for above-ground storage of any newly generated arsenic trioxide from continued gold-roasting operations, and re-mining and re-processing the underground arsenic trioxide for commercial resale.³³ The Water Board issued another water licence renewal for five years beginning in June 1998. This new licence required an arsenic trioxide management plan be submitted to the board by October 1, 1999, with quarterly progress reports beginning on October 1, 1998. Financial security (funds the mining company provided to offset the costs of remediation) remained the same under this licence ($400,000), but this low amount would be gradually increased to $7 million at the end of the term of the licence. The company filed its first quarterly report on the arsenic trioxide management plan in February 1999, reporting that they had made little headway on its plan to remove the underground arsenic. The company also indicated that it had submitted a proposal to the federal Department of Indian Affairs and Northern Development for a grant to fund the work, hoping to complete the plan by March 31, 2000.³⁴

Despite the very serious environmental issues posed by the underground arsenic, governments at all levels seemed more intent on propping up the failing finances of Royal Oak and squeezing just a few more years out of a mine that was clearly in decline rather than addressing the issue of who would pay the potentially massive costs of remediation after closure. In the fall of 1998, the price of gold had fallen to $280 an ounce from about $440 an ounce, and Royal Oak was seeking subsidies to remain in operation. The territorial government developed a seven-year program to subsidize further exploration and development for gold mines by making a per capita contribution for each employee. An annual grant was to be provided up to $1.5 million, based on $5,000 per employee, repayable if the price of gold moved above $365 an ounce for more than nine months.³⁵ Senior levels of government placed political pressure on the City of Yellowknife to participate in the subsidy program, and it agreed to contribute $150,000 per year on a matching basis to the GNWT funds using the same criteria for the two gold mines operating within city limits (Giant and Con Mines).[36] The city's subsidies actually flowed through the territorial government rather than directly to the mining companies, as it was illegal to make a direct financial contribution to private interests under the territorial legislation governing municipal governments.[37]

In the prelude to the closure of the Giant Mine in 1999, it is very clear that governments at all levels, even those responsible for regulating the mine, were not prepared to press the owners for any significant or substantive changes to reduce air or water pollution, or even to require better reclamation planning. The attitude of the Department of Indian Affairs and Northern Development, the main regulator, was summarized by one commentator as, "We try to work with a company to find solutions because frequent court actions don't give you the solution you want."[38] The company used the deteriorating gold market as a pretext for inaction on planning or implementing remediation measure for underground arsenic. As Larry Connell, Royal Oak's manager of environmental services, explained to the NWT Water Board in 1999, "in reality little progress was made during the fourth quarter of 1998 on advancing proposals for the extraction and recovery of the baghouse material from the existing underground storage vaults . . . The current low gold and copper prices have created a severe cash flow and liquidity problem at Royal Oak."[39] In the years immediately prior to the closure, the public interest in addressing the difficult remediation issues at Giant Mine was always subservient to the economic interests at play.

RECEIVERSHIP AND GOVERNMENT MANAGEMENT, 1999–2007

In addition to falling gold prices, the root cause of Giant Mine going into receivership on April 16, 1999, was a rapidly expanding mining company that could not adequately finance all of its ongoing operations. The company had devoted huge amounts of finance capital to opening the large Kemess Mine in northern British Columbia, investing $470 million but ending in 1997 with a $135 million loss, even after closing a couple of other mines in its portfolio.[40] Minutes after Royal Oak was put into receivership, the company's entire board of directors (including the controversial Peggy Witte, also known as Margaret Kent) resigned. Price Waterhouse Coopers was appointed as interim receiver to try to arrange sale of the assets to pay some of the creditors.[41] The City of Yellowknife was owed $737,000 in property taxes at the end of 1998 from Giant Mine, and began to withhold further subsidies to the operation as previously negotiated with the territorial government.[42] By May 1999, Royal Oak owed over $14 million to local creditors, including the City of Yellowknife ($1.078 million for property and school taxes), Government of the NWT ($1.088 million) and the NWT Power Corporation ($1.585 million).[43]

Two major developments came out of the bankruptcy and closure of Giant Mine. First, DIAND arranged for the sale of the Giant property for only $10 to Miramar Mining Corporation, which operated the Con Mine on the south side of the city. Second, all levels of government continued to provide subsidies and concessions to keep the mine operating. As part of a deal that was reached December 14, 1999, Miramar agreed to operate the mine with trucking of the ore across town to its processing facilities at the Con Mine.[44] A reduced work force of about fifty employees would be needed to operate Giant. Miramar also agreed to place a small levy on gold production into a Reclamation Security Trust, with the ability to write off various operating and management costs. It is unclear whether Miramar actually made any contributions to the trust.[45] Miramar set up a new shell company to operate the mine, and regulators granted the newly branded Miramar Giant complete and total indemnification from all environmental liabilities related to the previous operations at the site.

Though absolved of responsibility for long-term environmental issues at the Giant site, Miramar Giant did agree to keep the property in compliance with environmental regulations. The company's highest expense was the approximately $2 million a year required to pump water from the mine, a procedure needed to continue mining and to keep the underground arsenic from being dissolved. Responsibility for the longer-term issues associated with underground arsenic issue had now passed from private to public hands. The federal and territorial governments would pay for development of the arsenic trioxide management plan and subsequent reclamation activities at the site. On December 14, 2001, Miramar Giant exercised its right to terminate the agreement with government, and so the costs of maintaining the entire property became a permanent public liability. The federal government even agreed to assume immediately the company's expenses for pumping water and all environmental compliance costs, approximately $300,000 a month by this point, despite the fact the agreement required six months' notice prior to the government taking on maintenance costs of the site.[46]

FURTHER SUBSIDIES AND CONCESSIONS FOR GIANT MINE

Remarkably, even as the gold mines near Yellowknife became increasingly marginal operations, DIAND and GNWT agreed to provide matching funds up to $500,000 per year for exploration at the Giant and Con Mine properties. Miramar and the federal and territorial governments agreed among themselves that there should also be concessions from the City of Yellowknife: the company could take a two-year tax holiday and forgo tax arrears now totalling about $1.7 million.[47] Yellowknife Mayor David Lovell suggested that "they essentially came to an agreement that suited their needs and then threw the ball in our court for final approval," and further, "the City and the Education Districts are being asked to make the biggest relative sacrifices and we are, by far, the smallest players." The annual loss to the city and school boards amounted to about $700,000. A deal was eventually reached where:

- the city agreed to "buy" a portion of the surface lease covering the Giant Mine site where there was a townsite and a potential boat launch area, for a portion of the value of the property taxes ($233,000 in 2000 and $177,000 in 2001);
- a cap on property taxes for Giant Mine was imposed;
- the city received a release from GNWT and DIAND for any existing environmental contamination in the area;
- the federal and territorial governments agreed to remediate the area as part of the overall program for the property.[48]

Initially the city attempted to secure an indemnification for the lands to be leased by the city, similar to what had already been secured by Miramar for its use of Giant Mine, but both the federal and territorial governments refused to offer the same deal to Yellowknife. The city also sought to have the surface remediation of the area completed to a mutually agreeable standard but dropped this after receiving verbal assurances. The deal for the leased area and the release agreement was finally formalized in October 2000.[49] Less than a year later, the federal and territorial governments reneged on the understanding that city's leased area would be remediated to a standard agreeable to the city. When the city attempted to develop a boat launch in the area, the federal and territorial governments informed the city that they would only remediate areas within the leased lands to an "industrial" standard and that any other desired use would be Yellowknife's responsibility.[50]

DEVELOPMENT OF AN ARSENIC TRIOXIDE MANAGEMENT PLAN AND THE REMEDIATION PLAN

Once it became clear that the owners of the Giant Mine had little or no intention of developing a plan for the underground arsenic, the federal government began to conduct its own research, perhaps in anticipation of the site becoming a public liability. This work began in 1997, as discussed above, and there was a workshop on the subject that did not include parties outside of government and the company. DIAND hired a technical adviser in 1999 to provide advice on how to manage the underground arsenic. There was a series of successive workshops run as consultation sessions where the federal government and its consultants presented findings and options but with very little public input in-between and little or no involvement in the development of evaluation criteria and selection of preferred alternatives. Materials were often not provided ahead of time, no participant funding was provided to help parties obtain independent technical advice, and there was very little flexibility shown by the government in fully assessing new or preferred alternatives as expressed by workshop participants. In no way could this process be compared to principles of free, prior, and informed consent or consultation and accommodation in terms of the federal government's fiduciary obligations to Aboriginal peoples.[51]

The City of Yellowknife outlined its principles for the remediation of underground arsenic trioxide in an October 2001 letter to the minister of Indian Affairs and Northern Development as follows:

- DIAND's proposed in-situ option [freezing] be considered only as an interim measure;
- DIAND should commit to fund ongoing research and development to identify a solution and regularly review new advances in technology for a permanent solution;
- A community trust should be established and funded by the federal government to manage cleanup and containment and conduct research and development into a long-term, permanent solution;
- DIAND should consider all options available with preference given to methods that offer a long-term permanent solution with the lowest possible risks;
- The public should be consulted for input and that it be in the best interests of the citizens of Yellowknife;
- Timelines for completion of interim and permanent solutions should be established and adhered to;
- Local employment should be maximized; and
- All costs should remain the responsibility of the federal government.[52]

DIAND seems to have had some difficulty finding the funds to carry out initial assessment and planning work on the underground arsenic.[53] In early January 2003, DIAND advertised that funding up to $2,000 was available for Yellowknife-based organizations to make presentations at a workshop held that month and at other community consultations.[54] The funding could not be used to support additional technical review, critique of the DIAND's final report on arsenic trioxide, or for travel. DIAND made funding available only for internal consultations and for preparing a written and oral presentation at the workshop (maximum of ten minutes). Given the narrow scope of eligibility, no organizations secured any of this limited funding. The only organization to formally present at the May 2003 workshop was a group of Yellowknife MLAs, who generally supported the freezing option after a private meeting with DIAND and GNWT staff.[55] Following the final workshop held in June 2003, the federal government decided to proceed with the freezing option for the underground arsenic.

THE COOPERATION AGREEMENT

Mining ceased for good at Giant in July 2004 and the mine was scheduled to be returned to DIAND from Miramar in January 2005 (extended to July 2005 to allow the federal government to arrange a contract for care and maintenance). DIAND's consultants prepared a draft remediation plan in January 2005. Given the uncertainty over the respective costs and responsibilities of federal and territorial governments, an agreement of some sort had to be worked out before much in the way of substantive remediation could begin. The two governments signed a Cooperation Agreement in March 2005. In exchange for limiting its liability to $23 million, the territorial government agreed that the site would only be remediated to an "industrial" standard for arsenic and hydrocarbon spills and that the underground arsenic would be frozen in place forever. The two governments also agreed to become co-proponents in any environmental assessment and regulatory proceedings, and agreed to establish an "oversight committee" to coordinate their activities and to finalize a remediation plan. The agreement stipulated that the parties had to discuss with the City of Yellowknife the care and maintenance and remediation of the area under lease, municipal services to be provided to the mine site, and any other developments in the remediation.[56]

The day before the agreement was signed, territorial and federal officials met with city council members on what was about to be signed and indicated that the Yellowknives Dene First Nation Chiefs had been briefed three days earlier.[57] City council expressed its disappointment with the lack of consultation and nominal role afforded the city in the agreement. The city was also concerned because it had entered into a lease of the townsite and marina area for "municipal" and "recreational" purposes with the understanding that this area would be remediated to permit such uses, but the other orders of government now would only remediate to an "industrial" standard.[58]

ENVIRONMENTAL ASSESSMENT OF THE GIANT MINE REMEDIATION PLAN

At the end of the limited consultative process in 2003, DIAND's technical adviser suggested that their recommendation for freezing the underground arsenic be subject to an environmental assessment.[59] The federal government ignored this advice, attempting to leapfrog ahead to the regulatory process without final designs in hand, despite significant public concern about an approach that was perceived to be "freeze it and forget it." The government and the regulators refused to automatically refer to environmental assessment a remediation that proposed to maintain in situ a massive amount of toxic material immediately adjacent to a community of 20,000 people. The federal government's hand was forced only when the City of Yellowknife made a mandatory referral under the Mackenzie Valley Resource Management Act.[60] The city government took this action based on a request from the author, representations from a Yellowknives Dene First Nation Chief, and petitions from the local MLA of the NWT. This request for an environmental assessment represented an extraordinary convergence of interests and resistance from organizations and individuals that had not worked together effectively in the past. Ironically, these interests acted together in part because of their marginalization from the decision-making process surrounding the remediation of Giant Mine. Although the public's exclusion from meaningful consultation about Giant Mine might have continued indefinitely, concerned groups found a powerful ally in the city government, one that held the power to trigger a mandatory referral in face of indifference by the federal and territorial governments.

The federal and territorial governments resisted the environmental assessment, showing disregard for the process from the beginning and managing to drag it out for longer than five years. For example, Aboriginal Affairs and Northern Development Canada (AANDC, formerly DIAND) requested extensions or produced delays thirteen times during the process for a total of more than 1.6 years.[61] During the scoping hearing, the federal government promised to consider proposals for participant funding on a case-by-case basis. But when the Yellowknives Dene First Nation, the City of Yellowknife, and a private citizen (the author) submitted a proposal for $40,000 to study independent oversight in October 2009, AANDC rejected it eleven months later because the work was to be done by the government.[62] When the government's submission was filed with the review board more than a year later, there was nothing in it regarding independent oversight. The Oversight Working Group was established outside of the process to attempt to reach some consensus on a legally binding environmental agreement containing provisions for an independent oversight body, much like the arrangements already in place for three northern diamond mines. After twelve meetings, six drafts of a discussion paper, and eight drafts of an environmental agreement, the most the governments could do was to say to the review board that they generally agreed with the concept of oversight but wanted to discuss it further.[63] Documents obtained under federal access-to-information legislation indicate that AANDC secretly developed a "Site Stabilization Plan" to move forward with remediation outside of the environmental assessment in November 2011.[64] Finally, AANDC officials provided misleading cost figures for the overall project in August 2012 and at the public hearing, stating that implementation of the remediation plan would total $449 million.[65] This claim was made despite the fact that the federal Treasury Board approved a revised cost estimate for the project in March 2012 for $903 million based on a submission from AANDC. The fact that AANDC's projection of a near doubling in remediation costs for Giant Mine was only revealed through an access-to-information request by the author suggests, perhaps more than any other incident, the department's indifference to the public consultation requirements of the environmental assessment.[66]

FIGURE 2: Pipes constructed to freeze one of the underground arsenic chambers as part of a Freeze Optimization Study begun in 2009. Photo by Kevin O'Reilly.

Even at the public hearings stage of the environmental assessment, the federal and territorial governments also did not engage fully with the process. On the first day of public hearings held by the Mackenzie Valley Environmental Impact Review Board in September 2012, the chair had to urge the government representatives several times to answer the questions being put to them.[67] Federal government officials stated, on more than one occasion, that the focus of their work was on stabilizing the site; thus the big picture issues such as perpetual care and oversight had not been addressed very well. The government made vague commitments on oversight and perpetual care at the very last possible moment, and filed presentations just before the public hearings began.[68]

In contrast, any doubt that public concern with the government's remediation plan existed was removed by over twenty presentations made by private citizens during the hearings. Over fifty people waited during a power outage one evening before the lights came on again to make many of these submissions directly to the review board. Public participants raised many key issues through the many days of hearings, including:

- Lack of perpetual care planning and independent oversight for the project;
- The current plan for the Giant Mine is really a "stabilization plan" and not a "remediation plan" that truly reflects the views and needs of the community;
- As the proposed plan requires perpetual care and is not a real solution, strategic investment for ongoing research and development into a more sustainable and permanent outcome is required;
- There is still an opportunity to work together, but a written, legally binding agreement is needed to firm up the vague commitments made by the governments;
- Proper management of the project will cost a lot more money, and community leaders will need to work together with the project team to convince decision makers and funders that this is the case;
- There is more than enough evidence for the review board to conclude that there is significant public concern with the remediation plan and that there is the potential for significant adverse environmental impacts. These two findings are necessary to establish the legal basis for the board to recommend binding measures before the project can proceed.

In its final submission to the review board in October 2012, the government continued to conclude that the project "poses no significant risks of adverse environmental effects" and "is not the source of long standing concerns."[69] The government concluded that the legacy of the site is the source of public concern and not the incomplete stabilization proposals in its remediation plan. The government actually stated in this letter that "during the hearings, there was wide support for the ground freezing approach"—this despite not one person offering support for the plan during the public hearing. The letter denies the concerns expressed and appears to be motivated foremost by the desire to "get on with it." The government responded to public concern in its closing comments to the review board by stating that it was committed to "continuing and even increasing our engagement efforts so that concerned individuals, groups and the public can feel confident, as we do, that the proposed Remediation Plan presents the best available approach."[70] But more of the same consultation without consideration of the approach is simply not going to work. The government needs to devote a lot more resources to the development of a real remediation plan that is based on a foundation of trust and working together. The government also should heed calls for a public apology and compensation to the Yellowknives Dene First Nation for the suffering they experienced to due arsenic contamination and the permanent damage that Giant Mine has caused to their traditional territory. Furthermore, at this point in the process frequent calls for a real commitment to develop a comprehensive perpetual care plan and a legally binding environmental agreement (one that contains provisions for access to information, independent oversight, investment in strategic ongoing research, and development of a more permanent solution) all fell on deaf ears.

Nevertheless, the federal government began a formal process to exempt or split out from the ongoing environmental assessment the demolition of the most highly contaminated buildings on the site, the roaster complex, intending to start this work in June 2013.[71] This is the most contaminated mine feature above ground and contains about 4,000 tonnes of toxic arsenic trioxide and other contaminants. Initially there were no plans for air monitoring of arsenic stirred up by the demolition, and no clear limits for ambient airborne arsenic to protect the health of workers or those using the Ingraham Trail, a popular public highway running right through the Giant Mine site.[72] The work on the roaster demolition was completed in 2014. Despite the fact that the work was not contracted using the federal government's emergency authority, the federal and territorial governments wanted the regulators to consider it an "emergency" situation so it could be exempted from the scrutiny and protections that will come from the environmental assessment.

The review board released its Report of Environmental Assessment and Reasons for Decision on the Giant Mine Remediation Project on June 20, 2013. The board found that there is significant public concern with the project and the potential for significant adverse environmental impacts. This was a crucial finding that then gave the board the authority to recommend binding measures to reduce or eliminate these concerns and impacts. The board recommended an unprecedented twenty-six binding measures that would have to be adopted by anyone carrying out the project and incorporated as terms and conditions on any licences or permits issued by the federal and territorial governments. There are a further sixteen suggestions in a variety of areas. It would be fair to say that the board listened very carefully to what the communities had to say and made every effort to clearly reflect the concerns and issues they heard. Virtually all of the recommendations from the Yellowknives Dene First Nation and civil society were adopted in various ways.[73]

One of the most significant recommendations from the board was that the project should only proceed as a reversible "interim solution" for a maximum of one hundred years, rather than the perpetual care approach of the government. This shifts the onus to come up with a permanent solution onto this and a few subsequent generations. This is reinforced with a mandatory twenty-year comprehensive and independent review of the entire project, with the involvement of the communities. The government was also required to create a multi-stakeholder research agency (as part of a legally binding environmental agreement) to facilitate active research on a permanent solution. There was also a requirement for a legally binding environmental agreement that would pick up where negotiations left off in 2012 and would create a community-based independent oversight body with provisions for dispute resolution.[74]

Further measures recommended an independent quantitative risk assessment and an improved human health assessment and human health-monitoring program focused on arsenic. The review board also recommended that the government investigate long-term funding options, including an independent, self-sustaining trust, to support the long-term management of the site. The government is required to consider the results of the human health risk assessment and consult with the

Yellowknives and City of Yellowknife in determining suitable end uses of the mine site.

Once the report was received by the federal (Aboriginal Affairs and Northern Development, Fisheries and Oceans, and Environment) and territorial (Environment and Natural Resources), government ministers they had to decide on one of the following courses of action:

1. Accept the recommendation that the project can proceed with the recommended measures, which means that the measures then become binding on regulators and those that carry out the work; or

2. Refer the recommendation that the project can proceed with the recommended measures back to the review board for further consideration; or

3. Reject the recommendation that the project can proceed with the recommended measures, and send the project for a higher level of assessment (an impact review, by the same board, where alternatives and other factors must be considered); or

4. Invoke a murky process called "consult-to-modify" where the ministers write the review board and proved a rationale to change (and/or weaken) the measures to make them "acceptable." The role, if any, for the public is not clear.

A period of intense lobbying took place in the summer and fall of 2013 encouraging the relevant federal and territorial ministers to accept the review board report.[75] Meanwhile, the Giant Mine Remediation Project staff released their own critique of the review board that basically advised rejection of key recommendations because of perceived delays they may cause, additional costs and threats to the accountability and authority over the project.[76] Despite these objections, the responsible ministers entered into a "consult-to-modify" process with the review board on December 23, 2013 that actually resulted in strengthened legally binding measures when the ministers finally accepted the measures on August

11, 2014.[77] Among the measures was a requirement for a legally binding environmental agreement to set up an independent oversight body for the Giant Mine Remediation Project that would also have responsibility for research and development of a permanent solution to the underground arsenic. After seven negotiating sessions from September 2014 to April 2015, an agreement was reached and signed in the summer of 2015.[78] In a bold and unusual move, Alternatives North was recognized as a full party with roles and responsibilities the same as the Yellowknives Dene First Nation and the City of Yellowknife. The Giant Mine Oversight Body is expected to be operational in the fall of 2015.

CONCLUSIONS

What have we learned from the Giant Mine story? Unfortunately, it seems very little. The governments who manage environmental issues in the Northwest Territories apparently still feel they are more in the business of promoting northern development than protecting the public interest. There is nothing in federal or territorial law that would prevent another Giant Mine: no regulations on air quality in the NWT, no mandatory financial security for mines, and no mandatory requirement for closure plans for mines.[79] Only through the actions of individual citizens (including the author) and a few regular MLAs was the territorial government embarrassed into amending its land management legislation to require mandatory financial security for industrial and commercial surface leases over the small pockets of land it controls, beginning in February 2011 through revisions to the Commissioner's Lands Act.[80]

If Giant Mine is symbolic of what other communities facing remediation of orphan sites might face, here are a few findings and suggestions:

- There needs to be an acknowledgement of failure and an apology for wrongdoing (not necessarily blame), to provide a basis for reconciliation and moving forward;
- Governments are driven by cost avoidance and will usually gravitate toward least expensive options;

- Governments will often attempt to manufacture consensus and avoid public scrutiny;
- Short-term technical fixes will often be proposed and prevail without organized resistance;
- Governments have infinite confidence in their internal capacity, commitment, and longevity and will portray perpetual care as manageable and sustainable;
- Communities need to organize and seek independent technical assistance;
- Marginalization in decision making often provides opportunities for new alliances and convergence of interests.

With the current federal government dismantling decades of work to establish environmental assessment to prevent impacts to human health and the environment, and the dramatic weakening of the most important federal environmental protection legislation, namely the Fisheries Act, we are likely to see more Giant Mines. The Government of the Northwest Territories acted as a silent accomplice throughout much of the Giant Mine tragedy. This does not bode well for the much-vaunted devolution of authority over resource development to the territorial government.

An ounce of prevention is worth a pound of cure. As a nation, a society, and a species, we do not seem capable of limiting ourselves. It will take individual citizens and civil society to hold governments accountable. We still have a chance to get it right with what is left at Giant Mine, to build a true partnership based on trust, learning from our mistakes, and communicating to future generations what we have done. On the bright side, the report from the Mackenzie Valley Environmental Impact Review Board sets the standard for how the remediation should be carried out and is a rare example of environmental assessment working and responding to Aboriginal interests, citizens, and civil society. It is still unclear why the responsible ministers went against the advice of their own staff on the Giant Mine Remediation Project and accepted the review board's measures. Perhaps it was the remarkable convergence of public opinion and the common interests of the affected First Nation, municipal and non-governmental organizations that finally overcame resistance. Significant progress has been made since the environmental assessment, largely reflecting a change in project management staff and attitude. Time will tell whether the Giant Mine Oversight Body is able to fulfill its mandate as an effective environmental watchdog and manager of ongoing research into a long-term solution. For northern residents, one lesson is to never give up and never downplay the ability of a small group of committed citizens to create change.

NOTES

1. Portions of this chapter appeared as Kevin O'Reilly, "Giant Mine, Giant Legacy," *Northern Public Affairs* 1, no. 2 (2012): 50–53.
2. For an excellent natural history of the Yellowknife area, see Jamie Bastedo, *Shield Country: The Life and Times of the Oldest Piece of the Planet*, Komatik Series 4 (Calgary: Arctic Institute of North America, 1994).
3. For an overview, see Morris Zaslow, *The Northward Expansion of Canada, 1914–1967* (Toronto: McClelland and Stewart, 1988).
4. For a general overview, see Jerald Sabin and Frances Abele, *State and Society in a Northern Capital: Yellowknife's Social Economy in Hard Times* (Ottawa: Carleton Centre for Community Innovation, 2010).
5. A "Northwest Territories Lands and Resources Devolution Agreement" was reached on June 25, 2013 and came into effect on April 1, 2014, but left out the Akaitcho Territory and Dehcho First Nations governments, see http://devolution.gov.nt.ca/wp-content/uploads/2013/09/Final-Devolution-Agreement.pdf accessed August 10, 2015.
6. Further details on the proposals to remediate Giant Mine are found in Indian and Northern Affairs Canada and Government of the Northwest Territories, "Giant Mine Remediation Project Developers' Assessment Report," October 2010, accessed February 11, 2013, http://www.reviewboard.ca/upload/project_document/EA0809-001_Giant_DAR_1328896950.PDF.
7. For a very helpful historical overview of Giant Mine, see John Sandlos and Arn Keeling, "Giant Mine: Historical Summary, Report submitted to the Mackenzie Valley Environmental Impact Review Board," August 12, 2012, accessed February 11, 2013, http://www.reviewboard.ca/upload/project_document/EA0809-001_Giant_Mine__History_Summary.PDF.

8 A. K. Muir, General Manager, Giant Yellowknife Gold Mines, Ltd. to G.E.B. Sinclair, Director, Northern Administration and Lands Branch, Department of Resources and Development, no date. RG 85, vol. 40, file 139-7, pt. 1, Library and Archives Canada (hereafter LAC).

9 W. M. Gilchrist [Mining Engineer and Mine Superintendent] to A. K. Muir, General Manager, Giant Mine, Giant Yellowknife Gold Mines, November 22, 1950. Author's collection as provided by Ben Nordahn, Aboriginal Affairs and Northern Development Canada on February 24, 2011.

10 Canadian Public Health Association, "Final Report," Yellowknife, NWT: Canadian Public Health Association Task Force on Arsenic, 1977. See especially pages 63–64 where the task force makes a number of assumptions and recommendations regarding continued disposal of arsenic trioxide underground.

11 Joanna Ankersmit, Aboriginal Affairs and Northern Development Canada, Opening Remarks at the Giant Mine Remediation Project Environmental Assessment Public Hearing, Hearing Transcript for September 10, 2012, 25, accessed February 11, 2013, http://www.reviewboard.ca/upload/project_document/EA0809-001_Giant_Mine_hearing_transcripts_-_September_10__2012.PDF.

12 AANDC reported the cost for the remediation plan as $479 million in the 2010 Developer's Assessment Report. In August 2012, AANDC reported the cost to implement the remediation plan at $449 million. Alternatives North obtained documents under Access to Information that show the cost to the federal government for developing and implementing all remediation measures at Giant Mine will be $903 million. See Appendices 1 and 2 in http://www.reviewboard.ca/upload/project_document/EA0809-001_Letter_from_Alternatives_North_on_2013_IR_on_Water_Treatment.PDF, as accessed March 29, 2013.

13 These and the following figures are taken from the paper by Warwick Bullen (A/Senior Mining Advisor, Department of Resources, Wildlife and Economic Development, Government of the NWT) and Malcolm Robb (Manager, Mineral Development Division, Department of Indian and Northern Affairs), "Socio-Economic Impact of Gold Mining in the Yellowknife Mining District," 2002, accessed February 11, 2013, http://www.miningnorth.com/docs/Socio-Economic%20Impacts%20of%20Gold%20Mining%20in%20Yellowknife%202002.pdf.

14 Office of the Auditor General, "2002 Report of the Commissioner of the Environment and Sustainable Development," Chapter 3, *Abandoned Mines in the North*, accessed March 10, 2013, http://www.oag-bvg.gc.ca/internet/English/parl_cesd_200210_03_e_12409.html, and Office of the Auditor General, "Spring 2012 Report of the Commissioner of the Environment and Sustainable Development," Chapter 3, *Federal Contaminated Sites and Their Impact*,

84-85, accessed March 10, 2013, http://www.oag-bvg.gc.ca/internet/docs/parl_cesd_201205_03_e.pdf.

15   C. O'Brien and K. O'Reilly to Hon. Titus Alloloo, Minister of Renewable Resources, Government of the Northwest Territories, April 22, 1991. Author's collection.

16   Government of the Northwest Territories, "An Investigation of Atmospheric Emissions from the Royal Oak Giant Yellowknife Mine," Department of Renewable Resources, June 1993.

17   J. R. Hickman, Director General, Environmental Health Directorate, Health and Welfare Canada, to Dr. Ian Gilchrist, Medical Director, Northwest Territories Health, Government of the NWT, July 6, 1993. Author's collection.

18   Standing Committee on Environment and Sustainable Development, "Evidence of Meeting #123," May 11, 1995, accessed January 27, 2013, http://www.parl.gc.ca/content/hoc/archives/committee/351/sust/evidence/123_95-05-11/sust-123-cover-e.html.

19   The author was engaged as an activist on Giant Mine issues during this period, and these observations represent his own recollections of this period.

20   Standing Committee on Environment and Sustainable Development, "Evidence of Meetings #122 and 123," May 11, 1995, accessed January 27, 2013, http://www.parl.gc.ca/content/hoc/archives/committee/351/sust/evidence/122_95-05-11/sust-122-cover-e.html and http://www.parl.gc.ca/content/hoc/archives/committee/351/sust/evidence/123_95-05-11/sust-123-cover-e.html.

21   "It's About Our Health! Towards Pollution Prevention, CEPA Revisited," Report of the House of Commons Standing Committee on Environment and Sustainable Development, June 1995.

22   Ibid., 196.

23   Minutes of Yellowknife City Council, August 28, 1995, Motion #0417-95. Author's collection.

24   Resource Futures International, "Socio-Economic Analysis of Three Management Options to Reduce Atmospheric Emissions of Arsenic from Gold Roasting," September 9, 1996. Author's collection.

25   Environment Canada, "Workshop on Controlling Arsenic Releases into the Environment in the Northwest Territories, Final Workshop Report," October 1997. Author's collection.

26   Sadek E. El-Afy, Vice President of Operations, Royal Oak Mines to Mayor Dave Lovell, City of Yellowknife, October 4, 1995. Author's collection.

27   Canadian Public Health Association, "Final Report." See especially pages 49–51 where the municipal water supply for Yellowknife is discussed.

28 Ron Verzuh, "Giant Mine Fined $2,000 for Arsenic Pollution in Back Bay Last April," *News of the North*, February 26, 1975, 1–2.

29 The author gave a deposition to the NWT Water Board on the ammonia issue during these hearings.

30 Ed Szol, Executive Vice President and Chief Operating Officer, Royal Oak Mines, to Gordon Wray, Chair, NWT Water Board, August 18, 1997. NWT Water Board Public Registry.

31 Dillon Consulting Limited, "Arsenic Trioxide Management Feasibility Study," prepared for Indian and Northern Affairs Canada, October 6, 1997. Author's collection.

32 Dillon Consulting Limited, "Giant Mine Arsenic Trioxide Management Technical Meeting Proceedings," prepared for Department of Indian Affairs and Northern Development, October 1997, 28–30.

33 Royal Oak Mines and EBA Engineering Limited, "Arsenic Trioxide Surface Storage and Handling Project Scoping Document," December 1997. Author's collection.

34 Larry Connell, Manager of Environmental Services, Royal Oak Mines Inc. to Gordon Wray, Chairman, NWT Water Board and Bob Overvold, Regional Director General, Indian and Northern Affairs Canada, February 5, 1999. NWT Water Board Public Registry N1L2-1563.

35 "Royal Oak Receives Subsidy, GNWT Approves $1.5 Million, City Gives Additional Dollars," *Yellowknifer*, October 2, 1998, A9.

36 "Royal Oak Losses, Company in the Red in Third Quarter," *Yellowknifer*, January 13, 1999.

37 Hon. Manitok Thompson, Minister of Municipal and Community Affairs, Government of the Northwest Territories to Dave Lovell, Mayor, City of Yellowknife, November 23, 1998. Author's collection.

38 Ian MacKinnon, "Royal Oak May Leave Taxpayers on Hook: Cleanup at NWT Mines Likely to Top $256 Million," *National Post*, March 9, 1999, C1.

39 Larry Connell, Manager of Environmental Services, Royal Oak Mines Inc., to Gordon Wray, Chairman, NWT Water Board, and Bob Overvold, Regional Director General, Indian and Northern Affairs Canada, February 5, 1999. NWT Water Board Public Registry N1L2-1563.

40 Allan Robinson, "Losses Mount at Royal Oak. Mine Closings, Writedowns and Currency, Commodity Dealings Lead to $135-Million Shortfall," *Globe and Mail*, April 1, 1998, B1.

41 Allan Robinson, "Royal Oak Board Quits, Firm in Receivership, Witte and Directors Resign as Debt Cripples Company; Miner Unable to Fine Rescue Plan," *Globe and Mail*, April 17, 1999.

42   City of Yellowknife press release, "City Receives Status Report on Royal Oak Mines," April 7, 1999.

43   Dane Gibson, "Cashless Mine Owes Big, Royal Oak Yellowknife Creditors Holding Their Breath on Getting What's Owed," *Yellowknifer*, May 7, 1999, A9.

44   Some details on the deal are found in a news release (99–15) put out by Miramar Mining Corporation on December 14, 1999, "Miramar Acquires Giant Mine Assets, Boosts Production and Lowers Costs."

45   Terriplan Consultants, "Giant Mine Underground Arsenic Trioxide Management Alternatives. Draft Workshop Summary Report, January 14–15, 2003," prepared for the Department of Indian Affairs and Northern Development, March 2003. See page 12 where the author attempted to secure information about the payments by Miramar into the Reclamation Security Trust. Robert Lauer, an official with the Department, later stated that no funds were added to the trust by Miramar as a result of its operation of the Giant Mine.

46   Miramar Mining Corporation, "Miramar and DIAND Finalize Amendment to Existing Giant Mine Agreement," November 13, 2001. Author's collection.

47   City of Yellowknife, "News Release," November 18, 1999.

48   Memorandum to Council, City of Yellowknife, November 26, 1999. Author's collection.

49   Release Agreement between Government of the NWT and City of Yellowknife, October 1, 2000, and Giant Mine Leasehold Purchase and Sale Agreement between Miramar Giant Mine Ltd. and City of Yellowknife, October 1, 2000. Author's collection.

50   Emery Paquin, Director, Environmental Protection Services, Government of the NWT, to Gary Craig, Director of Public Works and Engineering, City of Yellowknife, July 10, 2001, and D. E. Nutter, Senior Advisor, Giant Mine Project, Indian and Northern Affairs Canada, to Gary Craig, Director of Public Works and Engineering, City of Yellowknife, July 13, 2001. Author's collection.

51   The author participated in the process, and these statements represent personal recollections.

52   Gordon Van Tighem, Mayor, City of Yellowknife, to Hon. Robert Nault, Minister of Indian Affairs and Northern Development, October 22, 2001. Author's collection.

53   Richard Gleeson, "Cleanup Dollars in Short Supply, $3 Million to Cover this Year's Work at Both Giant and Colomac Mines," *News/North*, June 26, 2000, A5.

54   Bill Mitchell, "Giant Mine Remediation Project," to various recipients (e-mail) and attachment "Presenter Funding," January 4, 2003. Author's collection.

55 Terriplan Consultants, "Giant Mine Underground Arsenic Trioxide Management Alternatives, Moving Forward: Selecting a Management Alternative, Draft Summary Workshop Report," May 26–27, 2003, prepared for the Department of Indian Affairs and Northern Development.

56 "Cooperation Agreement Respecting the Giant Mine Remediation Project," Between Her Majesty the Queen in Right of Canada as Represented by the Minister of Indian Affairs and Northern Development and the Government of the Northwest Territories as Represented by the Minister of Resources, Wildlife and Economic Development, March 15, 2005, accessed February 10, 2013, http://db.maca.gov.nt.ca/resources/Cooperation_Agreement_Giant-mine_remediation.pdf.

57 Priorities, Policies, and Budget Committee Agenda, City of Yellowknife, March 14, 2005, and author's notes of the meeting.

58 Gordon Van Tighem, Mayor, City of Yellowknife, to Hon. Andy Scott, Minister of Indian Affairs and Northern Development, May 11, 2005, and Gordon Van Tighem, Mayor, City of Yellowknife, to Hon. Michael Miltenberger, Minister of Environment and Natural Resources, Government of the NWT, May 11, 2005. Author's collection.

59 See Developer's Assessment Report, page 6–10, where the technical adviser made the following recommendation: "The in situ alternative recommended by the Technical Advisor, namely Alternative B3 – Ground Freezing as a Frozen Block, should be adopted as the preferred approach for managing the arsenic trioxide dust stored underground at Giant Mine. Elements of the alternative should be modified to take into account suggestions made by the general public, the Yellowknives Dene, and the GNWT. The modified alternative should be described within a Project Description that presents a complete plan for final closure and reclamation of the Giant Mine site, including surface works. *The Project Description should then be submitted for formal environmental review, licensing and subsequent implementation.*" Emphasis added.

60 Gordon Van Tighem, Mayor, City of Yellowknife, to Gabrielle Mackenzie-Scott, Chair, Mackenzie Valley Environmental Impact Review Board, March 31, 2008, accessed February 11, 2013, http://www.reviewboard.ca/upload/project_document/EA0809-001_Letter_of_Referral_from_the_City_of_Yellowknife_1328900441.pdf.

61 Alternatives North, Record of Giant Mine Environmental Assessment Delays (Requests and Notices), August 27, 2012, accessed February 11, 2013, http://www.reviewboard.ca/upload/project_document/EA0809-001_Giant_Mine_EA_Schedule_Delays.PDF. An additional one-week delay took place in March 2013, as shown by http://www.reviewboard.ca/upload/project_document/

EA0809-001_Board_letter_granting_extension.PDF (accessed March 29, 2013).

62 See correspondence filed with the Mackenzie Valley Environmental Impact Review Board, accessed February 11, 2013, http://www.reviewboard.ca/upload/project_document/EA0809-001_E-mail_from_Kevin_O_Reilly_Regarding_Funding_Requests_for_Independent_Oversight_1328898680.PDF; and Hon. Chuck Strahl, Minister of Indian Affairs and Northern Development, to Bob Bromley, Member of the Legislative Assembly for Weledeh, Government of the NWT, January 28, 2010, accessed February 11, 2013, http://www.reviewboard.ca/upload/project_document/EA0809-001_Reply_from_INAC_to_Bob_Bromley__MLA_1328896942.PDF.

63 See the letter from Adrian Paradis, A/Manager, Giant Mine Remediation Project, Aboriginal Affairs and Northern Development Canada, and Ray Case, Assistant Deputy Minister, Environment and Natural Resources, Government of the NWT, to Richard Edjericon, Chair, Mackenzie Valley Environmental Impact Review Board, August 31, 2012, accessed February 11, 2013, http://www.reviewboard.ca/upload/project_document/EA0809-001_Letter_from_develoepr_re__Oversight.PDF.

64 See the Alternatives North, "Technical Report Giant Mine Remediation Project," submitted to the Mackenzie Valley Environmental Impact Review Board, July 2012, accessed February 11, 2013, http://www.reviewboard.ca/upload/project_document/EA0809-001_AN_Giant_Mine_EA_Technical_Report__Final_.PDF. See especially page 9 and appendix B. Aboriginal Affairs and Northern Development Canada finally filed the "Site Stabilization Plan" on August 10, 2012, with heavy redactions throughout, as seen at http://www.reviewboard.ca/upload/project_document/EA0809-001_Site_Stabilization_Plan_for_the_Giant_Mine_Remediation_Project.PDF (accessed February 11, 2013).

65 Adrian Paradis, A/Manager, Giant Mine Remediation Project, Aboriginal Affairs and Northern Development Canada, and Ray Case, Assistant Deputy Minister, Environment and Natural Resources, Government of the NWT, to Richard Edjericon, Chair, Mackenzie Valley Environmental Impact Review Board, August 10, 2012, accessed March 10, 2013, http://www.reviewboard.ca/upload/project_document/EA0809-001_Cover_Letter_for_Giant_Mine_Responses_to_Parties__Recommendations.PDF; and Giant Mine Remediation Project Environmental Assessment Public Hearing Transcripts, September 14, 2012, 165–167, accessed March 10, 2013, http://www.reviewboard.ca/upload/project_document/EA0809-001_Giant_Mine_public_hearing_transcript_-_September_14__2012.PDF.

66 See Appendices 1 and 2, accessed March 29, 2013, http://www.reviewboard.ca/upload/project_document/

EA0809-001_Letter_from_Alternatives_North_on_2013_IR_on_Water_Treatment.PDF.

67   See the transcripts of the public hearing held on September 10, 2012, especially pages 50–71, accessed February 11, 2012, http://www.reviewboard.ca/upload/project_document/EA0809-001_Giant_Mine_hearing_transcripts_-_September_10__2012.PDF.

68   See public hearing transcript, September 13, pages 107, 116, 151, accessed February 11, 2012, http://www.reviewboard.ca/upload/project_document/EA0809-001_Giant_Mine_public_hearing_transcript_-_Sept_13__2012.PDF; and public hearing transcript, September 14, pages 138–39, accessed February 11, 2012, http://www.reviewboard.ca/upload/project_document/EA0809-001_Giant_Mine_public_hearing_transcript_-_September_14__2012.PDF.

69   Joanna Ankersmit, Director, Contaminated Sites Program, Aboriginal Affairs and Northern Development Canada, and Ray Case, Assistant Deputy Minister, Environment and Natural Resources, Government of the NWT, October 12, 2012, accessed February 11, 2013, http://www.reviewboard.ca/upload/project_document/EA0809-001_Closing_comments-_Developer.PDF.

70   Giant Mine Remediation Closing Comments, Giant Mine Remediation Project Environmental Assessment, accessed August 5, 2013, http://reviewboard.ca/upload/project_document/EA0809-001_Closing_comments-_Developer.PDF.

71   See the application for a water licence for "Giant Mine Roaster Complex Deconstruction and Underground Stabilization Work" as filed with the Mackenzie Valley Land and Water Board on December 19, 2012, as found at http://mvlwb.ca/Boards/mv/Registry/Forms/FolderView.aspx?RootFolder=%2FBoards%2Fmv%2FRegistry%2F2012%2FM-V2012L8-0010&FolderCTID=0x012000E7E47FF72F0B4D43A5BB2A7B-7616D923&View={757F78CE-B22D-4021-9A5A-EA094ABF5B0B} (accessed February 11, 2013).

72   See http://www.mvlwb.ca/Boards/mv/Registry/2012/MV2012L8-0010/MV2012L8-0010%20-%20AANDC%20-%20CARD%20-%20Giant%20Mine%20site%20-%20Comment%20on%20Roaster%20Complex%20-%20Alternates%20North%20-%20Feb15-13.pdf (accessed March 29, 2013).

73   Mackenzie Valley Environmental Impact Review Board, "Report of Environmental Assessment and Reasons for Decision on Giant Mine Remediation Project EA0809-0001," June 20, 2013, accessed July 1, 2013, http://www.reviewboard.ca/upload/project_document/EA0809-001_Giant_Report_of_Environmental_Assessment_June_20_2013.PDF.

74 Mackenzie Valley Environmental Impact Review Board, "Report of Environmental Assessment and Reasons for Decision on Giant Mine Remediation Project EA0809-0001."

75 See the letters sent by the Yellowknives Dene First Nation on September 13, 2013 http://www.reviewboard.ca/upload/project_document/EA0809-001_YKDFN_letter_re_Giant_Report_of_EA.PDF; City of Yellowknife letter of August 6, 2013 http://www.reviewboard.ca/upload/project_document/EA0809-001_City_of_YK_letter_to_Responsible_Ministers.PDF; letter from three Yellowknife Members of the Legislative Assembly on August 7, 2013 http://www.reviewboard.ca/upload/project_document/EA0809-001_Letter_from_3_NWT_MLA_to_responsible_Ministers.PDF; and letter from the Western Arctic Member of Parliament dated October 7, 2013   http://www.reviewboard.ca/upload/project_document/EA0809-001_Bevington_Letter_of_Support_on_Giant_Mine_EA_Report.PDF accessed August 10, 2015.

76 Joanna Ankersmit, Director, Contaminated Sites Program, Aboriginal Affairs and Northern Development Canada, and Ray Case, Assistant Deputy Minister, Environment and Natural Resources, Government of the NWT, November 1, 2013, http://www.reviewboard.ca/upload/project_document/EA0809-001_Giant_Team_comments_on_REA.PDF accessed August 10, 2015,

77 Letter from Hon. Bernard Valcourt, Minister of Aboriginal Affairs and Northern Development to Richard Edjericon, Chairperson, Mackenzie Valley Environmental Impact Review Board dated December 23, 2013 HYPERLINK "http://www.reviewboard.ca/upload/project_document/EA0809-001_AANDC_Minister_letter_-_consultation_to_modify_measures.PDFaccessed August 10" http://www.reviewboard.ca/upload/project_document/EA0809-001_AANDC_Minister_letter_-_consultation_to_modify_measures.PDF accessed August 10, 2015; and letter from Hon. Bernard Valcourt, Minister of Aboriginal Affairs and Northern Development to JoAnne Deneron, Chairperson, Mackenzie Valley Environmental Impact Review Board dated August 14, 2014 HYPERLINK "http://www.reviewboard.ca/upload/project_document/EA0809-001_Letter_from_AANDC_Minister_to_MVRB_Chairperson.PDFaccessed August 10" http://www.reviewboard.ca/upload/project_document/EA0809-001_Letter_from_AANDC_Minister_to_MVRB_Chairperson.PDF accessed August 10, 2015.

78 For a full copy of the Giant Mine Environmental Agreement, see http://www.alternativesnorth.ca/Portals/0/Documents/Mining%20Oil%20and%20Gas/Giant%20Mine/Giant%20Mine%20Environmental%20Agreement%20%28Signed%29%20June%202015.pdf accessed August 10, 2015 and the news release issued by Alternatives North on June 16, 2015, http://www.alternativesnorth.ca/Portals/0/AN%20News%20Release%20on%20Giant%20

Mine%20Environmental%20Agreement%20%28final%29%20.pdf accessed August 10, 2015.

79   For a detailed critique of the Northwest Territories mining reclamation regime, see Michael Wenig and Kevin O'Reilly, *The Mining Reclamation Regime in the Northwest Territories: A Comparison with Selected Canadian and U.S. Jurisdictions*, Canadian Institute of Resources Law and Canadian Arctic Resources Committee, January 2005, accessed February 11, 2013, http://carc.org/pdfs/mining49_nwtminingreclam_final_21jan05.pdf. Another recent report on federal government management of financial security and inspections found that the existing system is completely inadequate, see Office of the Auditor General, "Fall 2012 Report of the Commissioner of the Environment and Sustainable Development, Chapter 2 Financial Assurances for Environmental Risk," 10-12, accessed February 11, 2013, http://www.oag-bvg.gc.ca/internet/docs/parl_cesd_201212_02_e.pdf.

80   See s. 3.1 of the Commissioner's Lands Act brought into force on February 15, 2011, accessed March 29, 2013, http://www.justice.gov.nt.ca/PDF/ACTS/Commissioners%20Land.pdf.

Conclusion

Arn Keeling and John Sandlos

Mining controversies are not new. The noted late-medieval scholar Georgius Agricola opened his classic text, *De Re Metallica* (1556), with an extensive defence of mining against critics who cited its environmental impact, hazardous work conditions, and economic instability. But he also suggested that opinions on mining had long been divergent: "There has always been the greatest disagreement amongst men concerning metals and mining, some praising, others utterly condemning them."[1] In a period when large-scale mining was just beginning to exploit the riches beneath indigenous lands in the New World, there remained no consensus on the relative value of minerals when weighed against the social and environmental costs of development.

The issues Agricola discussed nearly half a millennium ago continue to resonate throughout the world's mining regions. In Canada, there has always been a particular intensity and longevity to debates about mineral development, perhaps matched only by environmental battles over clear-cut logging in British Columbia in the 1980s and 1990s. Conflicts over mining past and present are often uniquely polarized, with miners derided as profiteers and despoilers of the earth, while mining critics are branded as "anti-development" types who would "lock up" resources in the name of environmental protection. Like all caricatures, there are uncomfortable elements of truth in these accusations, and presumably one can find extreme examples of each in the historical record (not to mention the daily news).

The case histories of mineral development presented in this volume prompt a more reflective examination of these positions, in light of the complex historical experience of communities adjacent to, or founded by, mining activities in the North. The Abandoned Mines project began as an attempt to understand the role of mineral development in the industrialization of Northern Canada, the impacts of mining in hitherto largely undisturbed northern environments, and the social and cultural changes experienced by local communities, particularly Aboriginal people, in the wake of these developments. What the many contributors to this volume discovered, however, was more than a simple tale of despoliation and decline rooted in the "bad old days" of lax regulation and industrial disregard for people and nature. Certainly the economic instability, community collapse, and environmental impacts associated with northern mining suggests the industry delivered far less in terms of social and economic development than the boosters of northern development in government and business promised time and time again. For some communities, however, mining presented an important economic stopgap against price shocks and resource declines in other sectors of the northern Aboriginal economy, especially hunting and trapping. Perhaps the most remarkable of our collective findings is the ability of many northern Aboriginal communities and individuals to accommodate and assimilate mine work, culture, and communities, while simultaneously engaging in a profound critique of mineral-intensive development strategies. Stories of northern mining encompass widely varying historical experiences of economic dislocation and opportunity, community development and collapse, memories that continue to shape widely divergent local attitudes to northern mineral development in the present day.

But even as the authors in this book reflect on the complexity of individual and local experiences, they by no means reject a critical examination of the industry and its social, environmental, and economic impacts. While never a historical source of violent conflict in Northern Canada (as it has been elsewhere in the world), mineral development has often reinforced the inequities associated with settler colonialism in the region. The injustices experienced by Aboriginal northerners associated with mineral development documented in these chapters range from territorial dispossession and social and economic marginalization to toxic contamination and degradation of important local environments. These injustices are best understood in connection with the broader suite of federal government neocolonial policies aimed at "modernizing" Aboriginal people and territories in the region in the twentieth century. There is also a strong link between these past experiences and the deep suspicion felt today by many toward the industry (and its promoters in federal and territorial governments). Signs of the physical changes wrought by mining, from open pits to toxic sites to industrial ruins on the landscape, remain potent material reminders of the mining experience, legacies that continue to haunt both the memories and the biophysical environments of local communities. These experiences, memories, and material conditions continue to inform present debates about the costs and benefits of mineral development today, as mining is touted as the economic salvation of supposedly dependent northern communities. Contemporary developments in the vicinity of abandoned mines may threaten to reawaken or reproduce the injustices associated with past developments, regardless of the goodwill of companies or the improvement of regulatory oversight. As several of the chapters also show, attending to the history of mining can raise uncomfortable questions about the persistence of colonial legacies and institutions in the region, even in this era of land claims, impact and benefit agreements, and corporate social responsibility.

Ultimately, we think that a better understanding of the scope and impacts of historical mineral development in the North, as well as its many ongoing legacies, can contribute to wider debates about resources and sustainability in Canada, and beyond. Historically, the goal of mining has only been tangentially (and at times only rhetorically) about the modernization and development of northern communities. Rather, the main aim of mining companies and their government promoters has been to fulfill the insatiable demand for minerals (and profits) in Southern Canada and for export markets. Thus, the story of mining in Northern Canada is one that connects all Canadians, whether consumers of mineral products or holders of mining stocks in investment portfolios, to the remote, often poorly understood northern places, people, and environments affected by mining. As seen in debates around bitumen development in northern Alberta, southern Canadians (and urbanites in particular) have often been willing to ignore the environmental and public health burdens associated with our material- and energy-intensive lifestyles. Perhaps a better understanding of the historical and contemporary impacts of mining can help lead to a better reckoning of the how the costs and benefits of resource extraction have been distributed among local communities, highly mobile capital, and distant consumers. Certainly the ephemeral local benefits that stem from northern mining— the historical boom and bust cycle that is emblematic of the industry—force Canadians to face the difficult question of what a sustainable economy might look like in the North and in the country as whole.

These histories also highlight the problem of who bears the long-term financial consequences of abandoned mines—in Canada, typically taxpayers have been left on the hook for billions in environmental liabilities associated with cleaning up past industrial developments. These long-term environmental legacies and financial liabilities—the "zombies" that stalk northern mine sites and communities—illustrate the fundamentally unsustainable nature of extractive industries such as mining. Based as they are on finite resources, extractive sites like mines must ultimately decline as deposits are exhausted or, more likely, as the costs of extraction begin to exceed the market value of the product. There is no renewable "resource cycle" associated with minerals, though technological change and market conditions can lead to the revaluation of formerly closed sites or unworked deposits. But as many of the chapters show, the cessation of extractive activity does not necessarily end the costs and controversies associated with mineral development. Toxic contaminants, ecological changes, and the collapse of local resource-dependent economies may resonate for decades or longer after the end of mining, scarring the landscape and displacing communities. Even in cases where some remediation has been undertaken, many of the financial costs associated with these impacts are borne not by the industry, but by the public, individually and collectively. Under new forms of environmental regulation, companies are often required to post financial security during the life of a mine to cover closure and clean-up costs, though time will tell if today's estimates match the actual cost of cleanup. What cannot be recovered from industry, however, is the massive cost of historic abandoned mines. Despite frequent calls from industry and recent calls from the Canadian government to cut regulatory red tape to speed the pace of northern resource development, the environmental liabilities associated with historic abandoned mines provide a potent reminder of the need for strict environmental assessment, public oversight, and regulation of new northern mineral projects in all phases of their operation.

In broader historical terms, there can be little doubt that mining was one of the most important drivers of social, cultural, and economic change in Northern Canada. This book has attempted to reckon with the precise meaning of these changes, and how they intersected with the broader suite of social, cultural, and economic changes sweeping the region in the twentieth century. In doing so we discovered histories of resource projects that were too often focused narrowly on mineral extraction to the exclusion of social development and environmental issues in adjacent Aboriginal (and in a few cases non-Aboriginal) communities. In some cases these communities responded creatively to the sudden social, economic, and environmental changes associated with northern mineral development. At other times the adjustments to new patterns of work, settlement, and economic exchange proved painful, as did environmental changes that brought toxic exposure, sickness, and sometimes death. As we enter a new era of promises and unbridled optimism about northern development based on minerals and offshore oil, all Canadians would do well to reflect on the opportunities and challenges these types of megaprojects have presented to northern communities in the past, and what kind of issues they might raise for them in the future.

NOTE

1 Georgius Agricola, *De Re Metallica*, trans. Herbert Clark Hoover and Lou Henry Hoover (New York: Dover Publications Inc., 1950 [1556]), 4.

Notes on Contributors

PATRICIA BOULTER is an MA graduate from Memorial University of Newfoundland. She studied environmental history, focusing primarily on the socio-economic and cultural impacts of mining in Canada's Arctic regions. When she is not out hiking, you can find her in a classroom, where she shares her passion for the environment and history with her students.

JEAN-SÉBASTIEN BOUTET holds a master's degree in geography from Memorial University. He now works as a mining policy analyst for the Nunatsiavut Government (Newfoundland and Labrador).

EMILIE CAMERON is an assistant professor in the Department of Geography and Environmental Studies at Carleton University. She is the author of *Far Off Metal River: Inuit Lands, Settler Stories, and the Making of the Contemporary Arctic* (UBC Press, 2015). Her current research focuses on geographies of resource extraction, empire, and labour in the contemporary North.

SARAH M. GORDON earned her PhD in folklore at Indiana University and her MA in comparative literature at University College London. Her major research interests include cultural adaptability in colonized communities, with an emphasis on performance theory as it pertains to oral traditions, folk narratives, and personal experience narratives. She has in the past collaborated with the Déline Knowledge Project in Déline, Northwest Territories, Canada, and Traditional Arts Indiana in Bloomington, Indiana, USA.

HEATHER GREEN is a doctoral student in the Department of History and Classics at the University of Alberta. She studies social, environmental, and indigenous history of Northern Canada, specifically mining and indigenous communities. Her dissertation examines the Klondike gold rush as a practice in economic colonialism that brought long-term environmental change to the landscape of the central Yukon and created lasting consequences for the Tr'ondëk Hwëch'in in the period 1895 to 1945.

JANE HAMMOND completed her work on Labrador while pursuing her master's degree in the Department of History at Memorial University of Newfoundland. She is currently completing her PhD in the Department of Geography at Western University in Ontario.

JOELLA HOGAN is Manager of Heritage, Culture, and Language with the First Nation of Na-Cho Nyäk Dun in Mayo, Yukon. Her mandate is the preservation, protection, and promotion of Northern Tutchone language, culture, and heritage. She has degrees in environmental planning (UNBC) and Native and rural development (UAF). Her research interests are in the relationships between the natural environment and cultural identity. She has worked on projects in the Circumpolar North and is passionate about preservation of her Northern Tutchone language and culture.

ARN KEELING is a historical-cultural geographer and associate professor at Memorial University in Newfoundland. His research and publications focus on the historical and contemporary encounters of northern indigenous communities with large-scale resource developments, domestic and industrial pollution, and environmental politics, as well as on the history of the conservation/environmental movement.

TYLER LEVITAN is a graduate of the Institute of Political Economy at Carleton University, where his research focused on the political economy of northern resource extraction and indigenous–state relations. He is currently working as the coordinator of a Canadian-based human rights organization.

HEREWARD LONGLEY is a PhD student at the University of Alberta, studying indigenous and environmental histories of hydrocarbon extraction in northeastern Alberta. Hereward's research examines how changing economic and political conditions shaped the development of the oil sands industry, and how bitumen extraction has changed human relationships with nature by recreating northeastern Alberta as a resource extraction zone. Hereward holds an MA in history from Memorial University of Newfoundland where he worked with the Abandoned Mines in Northern Canada project. Hereward also works as a research analyst with Willow Springs Strategic Solutions Inc. conducting traditional land use studies for indigenous communities affected by resource extraction and infrastructure projects in northern Alberta.

SCOTT MIDGLEY is a graduate from the Department of Geography at Memorial University of Newfoundland. His thesis examined the political economy of High Arctic mining and mine closure in Canada and Norway.

KEVIN O'REILLY has resided in Yellowknife, NWT, since 1985, working for Aboriginal, federal, and territorial government agencies on land use planning, environmental assessment, and resource management. He was the research director for the Canadian Arctic Resources Committee from 1995 to 2005. He now serves as the executive director for an independent environmental oversight body on Canada's first diamond mine. He is a founding board member for MiningWatch Canada, established in 1998, and he helped to form Alternatives North, a Yellowknife-based social justice group, in 1993. He served on the Yellowknife City Council from 1997 to 2006, during the time that the Giant Mine went into receivership. Kevin led the intervention by Alternatives North on the Mackenzie Gas Project and the environmental assessment of the Giant Mine. He holds a BES in environmental studies and an MA in urban and regional planning from the University of Waterloo. He received the Queen's Diamond Jubilee medal in July 2012 for his work on environmental issues, especially on Giant Mine.

ANDREA PROCTER is a post-doctoral fellow at the Labrador Institute, Memorial University of Newfoundland. Her work focuses on the relationships between resource development, indigenous autonomy, and settler colonialism, and she is a co-author of *Settlement, Subsistence, and Change among the Labrador Inuit: The Nunatsiavummiut Experience* (University of Manitoba Press, 2012).

JOHN SANDLOS is an associate professor of history at Memorial University of Newfoundland. His research addresses the historical politics of resource development and management in Northern Canada. He has written on the subjects of mining history, wildlife conservation, and human exclusions from national parks in Canada. With Arn Keeling, he has directed the Abandoned Mines in Northern Canada project, as well as a community-based project on the commemoration of arsenic contamination at Yellowknife's Giant Mine.

ALEXANDRA WINTON is a master's degree candidate in the Department of Geography at Memorial University of Newfoundland, where she has studied the impact of mine development and closure on northern communities. Her thesis explores the recurring closure and redevelopment of the Yukon's Keno Hill silver mine and the reactions of the communities of Mayo, Keno, and the First Nation of Na-Cho Nyäk Dun. Being from a mining community herself, she is fascinated with how communities reinvent themselves during times of economic and social change, how mineral development shapes the image of a community, and how people share these stories.

Bibliography

Abel, Kerry. *Drum Songs: Glimpses of Dene History*. Montreal and Kingston: McGill-Queen's University Press, 1993.

———. History and Provincial Norths: An Ontario Example." In *Northern Visions: New Perspectives on the North in Canadian History*, edited by Kerry Abel and Ken S. Coates, 127–40. Peterborough, ON: Broadview Press, 2001.

Abel, Kerry, and Ken S. Coates. "The North and the Nation." In *Northern Visions: New Perspectives on the North in Canadian History*, edited by Kerry Abel and Ken S. Coates, 7–21. Peterborough, ON: Broadview Press, 2001.

Abele, Frances. "Canadian Contradictions: Forty Years of Northern Political Development." *Arctic* 40 (1987): 310–20.

———. "Urgent Need, Serious Opportunity: Towards a New Social Model for Canada's Aboriginal People." *Canadian Policy Research Papers*. Ottawa, 2004. Accessed February 4, 2014. http://cprn.org/documents/28340_en.pdf.

Aboriginal Affairs and Northern Development Canada. Contaminants and Remediation Directorate. "Contaminated Sites Remediation: What's Happening in the Sahtu?" March 2009. http://publications.gc.ca/collections/collection_2010/ainc-inac/R1-27-2009-eng.pdf.

———. "History of Giant Mine." Accessed January 31, 2014. http://www.aadnc-aandc.gc.ca/eng/1100100027388/1100100027390.

———. "Joanna Ankersmit, Opening Remarks at the Giant Mine Remediation Project Environmental Assessment Public Hearing." Hearing

Transcript for September 10, 2012, 25. Accessed February 11, 2013. http://www.reviewboard.ca/upload/project_document/EA0809-001_Giant_Mine_hearing_transcripts_-_September_10__2012.PDF.

———. Major Mineral Projects and Deposits North of 60 Degrees. http://www.aadnc-aandc.gc.ca/DAM/DAM-INTER-HQ/STAGING/texte-text/mm_mmpd-ld_1333034932925_eng.html

———. Northern Contaminants Program. Last updated August 15, 2012. http://www.aadnc-aandc.gc.ca/eng/1100100035611.

Absolon, Kathy, and Cam Willett. "Putting Ourselves Forward: Location in Aboriginal Research." In *Research as Resistance: Critical, Indigenous, and Anti-oppressive Approaches*, edited by Leslie Allison Brown and Susan Strega, 97–126. Toronto: Canadian Scholars' Press/Women's Press, 2005.

Agricola, Georgius. *De Re Metallica*. Translated by Herbert Clark Hoover and Lou Henry Hoover. New York: Dover Publications Inc., 1950 [1556].

Aho, Aaro E. *Hills of Silver: The Yukon's Mighty Keno Hill Mine*. Madeira Park, BC: Harbour Publishing, 2006.

Alberta Environment. "Reclaiming Alberta's Oil Sands." Government of Alberta. Accessed April 5, 2013. http://www.environment.alberta.ca/02012.html.

Alberta Health and Wellness. "Alberta Oil Sands Community Exposure and Health Effects Assessment Program (HEAP) Summary Report." Edmonton: Health Surveillance, Alberta Health and Wellness, Government of Alberta, 2000.

Albo, Greg. "Neoliberalism, the State, and the Left: A Canadian Perspective." *Monthly Review* 54, no. 1 (2002): 46–55.

Alexco Resource Corporation. Alexco Implements Cost Savings Measures. News Release. Vancouver: Alexco Resource Corporation, 2013.

Alfred, Taiaiake. *Peace, Power, Righteousness: An Indigenous Manifesto*. Don Mills, ON: Oxford University Press, 1999.

Ali, Saleem H. *Mining, the Environment, and Indigenous Development Conflicts*. Tucson: University of Arizona Press, 2003.

Alternatives North. "Technical Report Giant Mine Remediation Project." Submitted to the Mackenzie Valley Environmental Impact Review Board, July 2012. Accessed February 11, 2013. http://www.reviewboard.ca/upload/project_document/EA0809-001_AN_Giant_Mine_EA_Technical_Report__Final_.PDF.

Andersen, William, III. "Address by William Andersen III at the Opening of Land Claims Negotiations, 22 January 1989." *Northern Perspectives* 18, no. 2 (1990): 4–5.

Andrews, Thomas G. *Killing for Coal: America's Deadliest Labor War.* Cambridge, MA: Harvard University Press, 2008.

ArcelorMittal. "Core Strengths, Sustainable Returns: Annual Report 2011." Luxembourg (Grand Duchy of Luxembourg), April 2012.

Asch, Michael. "From Calder to Van der Peet: Aboriginal Rights and Canadian Law, 1973–1996." In *Indigenous Peoples' Rights in Australia, Canada, and New Zealand*, edited by P. Havemann, 428–46. Auckland, NZ: Oxford University Press, 1999.

Athabasca Tribal Council v. Amoco Petroleum Co. Supreme Court of Canada. December 4–5, 1980, and June 22, 1981.

Atwood, Margaret. *Strange Things: The Malevolent North in Canadian Literature.* London: Virago Press, 2004 [1995].

———. *Survival.* Toronto: House of Anansi Press Ltd., 1972.

Aurora Energy Ltd. "Aurora Community Newsletter." Vol. 4 (July 2008). Accessed February 5, 2014. http://aurora-energy.ca/sites/default/files/newsletter/AXU_Community_Newsletter_V4_July_2008_Eng.pdf

Austin, John L. *How to Do Things with Words.* Cambridge, MA: Harvard University Press, 1975.

Baffinland Iron Mines. "Iron Ore Industry Trends and Analysis." August 31, 2009.

Baldwin, Correy. "Future Growth Built on Iron Ore Legacy." *CIM Magazine* 6, no. 2 (2011): 38–42.

Banerjee, Subhabrata Bobby. "Whose Land Is It Anyway? National Interest, Indigenous Stakeholders, and Colonial Discourses." *Organization & Environment* 13 (March 2000): 3–38.

Barnes, Trevor. "Borderline Communities: Canadian Single Industry Towns, Staples, and Harold Innis." In *B/ordering Space*, edited by Henk van Houtum, Olivier Thomas Kramsch, and Wolfgang Zierhofer. Burlington, VT: Ashgate Publishing, 2005.

Barnes, Trevor, Roger Hayter, and Elizabeth Hay. "Stormy Weather: Cyclones, Harold Innis, and Port Alberni, BC." *Environment and Planning* A 33 (2001): 2127–47.

Barnett, Clive. "Critical Review: The Consolidations of 'Neoliberalism.'" *Geoforum* 36, no. 1 (2005): 7–12.

Barrett, J. E. "Employment of the Inuit at Polaris, Little Cornwallis Island: A Feasibility Study Requested by Arvik Mines Ltd." 1973.

Barrett, J. E., and Associates. "The Polaris Project and the Inuit: An Assignment Concerned with Involving the Inuit in the Polaris Mine Development Requested by Arvik Mines Ltd." 1976.

Barrie, L. A. "The Fate of Particulate Emissions from an Isolated Power Plant in the Oil Sands Area of Western Canada." *Annals of the New York Academy of Sciences* 338 (1980): 434–52.

Bastedo, Jamie. *Shield Country: The Life and Times of the Oldest Piece of the Planet.* Komatik Series 4. Arctic Institute of North America, 1994.

BC Research. "Lethal and Sublethal Bioassays on Tailing from Nanisivik Mine Ore." Prepared for Strathcona Mineral Services, Project 1552, Progress Report No. 3, 1975.

BCG Engineering. "Engineering Design of Surface Reclamation Covers." 2003. Accessed January 2012. ftp://nunavutwaterboard.org/1%20PRUC/1%20INDUSTRIAL/1A/1AR%20-%20Remediation/1ARNAN0914/3%20TECH/10%20A%20and%20R%20(G)(J)/2004/G3%20Final%20Closure%20Plan/2004%20Final%20AR%20plan/G.4%20Cover%20Report/G.4%20Report/G.4%20Cover%20Report%20Final.pdf.

Bebbington, Anthony, Leonith Hinojosa, Denise Humphreys Bebbington, Maria Luisa Burneo, and Ximena Warnaars. "Contention and Ambiguity: Mining and the Possibilities of Development." *Development and Change* 39, no. 6 (2008): 965–92.

Beckwith, Karen. "Lancashire Women against Pit Closures: Women's Standing in a Men's Movement." *Signs: Journal of Women in Culture and Society* 21, no. 4 (1996): 1034–68.

Bell, J. Mackintosh. "Report on the Topography and Geology of Great Bear Lake and of a Chain of Lakes and Streams Thence to Great Slave Lake." In Annual Report. Ottawa: Geological Survey of Canada, 1901.

Bell, Lindsay A. "Economic Insecurity as Opportunity: Job Training and the Canadian Diamond Industry." In *Humanizing Security in the Arctic*, edited by Michelle Daveluy, Francis Lévesque, and Jenanne Ferguson, 293–304. Edmonton: Canadian Circumpolar Institute, 2011.

Ben-Dor, S. *Makkovik: Eskimos and Settlers in a Labrador Community.* St. John's, NL: ISER, 1966.

Berger, Thomas. *Northern Frontier, Northern Homeland: The Report of the Mackenzie Valley Pipeline Inquiry.* Vol. 1. Ottawa: Supply and Services Canada, 1977.

Berland, Jody. "Space at the Margins: Critical Theory and Colonial Space after Innis." In *Harold Innis in the New Century*, edited by Charles R. Acland and William J. Buxton. Montreal and Kingston: McGill-Queen's University Press, 1999.

Bernauer, Warren. "Uranium Mining, Primitive Accumulation, and Resistance in Baker Lake, Nunavut: Recent Changes in Community Perspectives." MA thesis, University of Manitoba, 2011.

Berube, Yves et al. "An Engineering Assessment of Waste Water Handling Procedures at the Cominco Pine Point Mine." Unpublished report. Ottawa: Department of Indian Affairs and Northern Development, 1972.

Bethell, Graeme. "Preliminary Inventory of the Environmental Issues and Concerns Affecting the People of Fort MacKay, Alberta." Brentwood Bay, BC: Bethell Management Ltd., May 1985.

Bielawski, Ellen. *Rogue Diamonds: Northern Riches on Dene Land.* Vancouver: Douglas and McIntyre, 2003.

Bielawski, Ellen (in collaboration with the community of Lutsel k'e). "The Desecration of Nanula Kué: Impact of the Talston Hydroelectric Development on Dene Soline." Unpublished report for the Royal Commission on Aboriginal Peoples, December 1993.

Blaser, Mario, Harvey Feit, and Glenn McRae. "Indigenous Peoples and Development Processes: New Terrains of Struggle." In *In the Way of Development: Indigenous Peoples, Life Projects, and Globalization*, edited by Mario Blaser, Harvey Feit, and Glenn McRae, 1–25. London: Zed Books, 2004.

Bleiler, Lynette R., Christopher Robert Burn, and Mark O'Donoghue. *Heart of the Yukon: A Natural and Cultural History of the Mayo Area.* Mayo, YT: Village of Mayo, 2006.

Bleiler, Lynette R., and Linda E. T. MacDonald. *Gold and Galena: A History of the Mayo District, with Addendum.* Mayo, YT: Mayo Historical Society, 1999.

Blondin, George. *When the World Was New: Stories of the Sahtú Dene.* Yellowknife, NWT: Outcrop, 1990.

Blondin-Perrin, Alice. *My Heart Shook Like a Drum: What I Learned at the Indian Mission Schools, Northwest Territories.* Ottawa: Borealis Press Ltd., 2009.

Blow, Peter. "Village of Widows." 52 min. Toronto: Lindum Films Inc., 1999.

Boland, Bobbie. "At a Snail's Pace: The Presence of Women in Trades, Technology, and Operations in Newfoundland and Labrador." St. John's, NL: Women in Resource Development Committee, April 2005.

Bothwell, Robert. *Eldorado: Canada's National Uranium Company.* Toronto: University of Toronto Press, 1984.

Boulter, Patricia J. "The Survival of an Arctic Boom Town: Socio-economic and Cultural Diversity in Rankin Inlet, 1956–63." MA major paper, Memorial University, 2011.

Boutet, Jean-Sébastien. "Développement ferrifère et mondes autochtones au Québec subarctique, 1954–1983." *Recherches amérindiennes au Québec* 40, no. 3 (2010): 35–52.

———. "An Innu-Naskapi Ethnohistorical Geography of Industrial Iron Mining Development at Schefferville, Québec." MA thesis, Memorial University of Newfoundland, 2011.

———. "Opening Ungava to Industry: A Decentring Approach to Indigenous History in Subarctic Québec, 1937–54." *Cultural Geographies* 21, no. 1 (2014): 79–97. doi: 10.1177/1474474012469761.

Bouthillier, P. H. "A Review of the GCOS Dyke Discharge Water." In *Great Canadian Oil Sands Dyke Discharge Water*. Edmonton: Alberta Department of the Environment, August 1977.

Bowen, Lynne. *Boss Whistle: The Coal Miners of Vancouver Island Remember*. Rev. ed. Nanaimo, BC: Rocky Point Books, 2002.

Bowes-Lyon, Léa-Marie. "Comparison of the Socio-economic Impacts of the Nanisivik and Polaris Mines: Sustainable Development Case Study." MSc thesis, University of Alberta, 2006.

Bowes-Lyon, Léa-Marie, Jeremy P. Richards, and Tara M. McGee. "Socio-Economic Impacts of the Nanisivik and Polaris Mines, Nunavut, Canada." In *Mining, Society, and a Sustainable World*, edited by Jeremy P. Richards, 371–96. Berlin/Heidelberg: Springer-Verlag, 2009.

Boyd, David R. *Unnatural Law: Rethinking Canadian Environmental Law and Policy*. Vancouver: UBC Press, 2003.

Brack, D. M., and D. McIntosh. "Keewatin Mainland Area Economic Survey and Regional Appraisal." Report for Industrial Division, Department of Northern Affairs and National Resources, March 1963.

Bradbury, John H. "Declining Single-Industry Communities in Quebec-Labrador, 1979–1983." *Canadian Studies* 19, no. 3 (1984): 125–39.

———. "The Impact of Industrial Cycles in the Mining Sector: The Case of the Quebec-Labrador Region in Canada." *International Journal of Urban and Regional Research* 8 (1984): 311–31.

———. "Some Geographical Implications of the Restructuring of the Iron Ore Industry, 1950–1980." *Tijdschrift voor Economische en Sociale Geografie* 73, no. 5 (1982): 295–306.

———. "Towards an Alternative Theory of Resource-Based Town Development in Canada." *Economic Geography* 55, no. 2 (1979): 147–66.

Bradbury, John H., and Isabelle St.-Martin. "Winding Down in a Quebec Mining Town: A Case Study of Schefferville." *Canadian Geographer* 27, no. 2 (1983): 128–44.

Brannstrom, Christian. "What Kind of History for What Kind of Political Ecology?" *Historical Geography* 32 (2004): 71–88.

Brantenberg, Terje. "Ethnic Commitments and Local Government in Nain, 1969–76." In *The White Arctic: Anthropological Essays on Tutelage and Ethnicity*, edited by Robert Paine, 376–410. St. John's, NL: ISER, 1977.

———. "Ethnic Values and Ethnic Recruitment in Nain." In *The White Arctic: Anthropological Essays on Tutelage and Ethnicity*, edited by Robert Paine, 326–43. St. John's, NL: ISER, 1977.

Brazen, Stephen, and Tony Thirlwall. *Deindustrialzation*. Oxford: Heinemann Educational Books, 1992.

Breakwater Resources. "Letter from Breakwater to NWB regarding security bond arrangements." 2002. Accessed April 2011. ftp://nunavutwaterboard.org/1%20PRUC/1%20INDUSTRIAL/1A/1AR%20-%20Remediation/1AR-NAN0914/Admin#/hearings/hearings/2002 hearing/post hearing submissions/BWR.

Brechin, Gray. *Imperial San Francisco: Urban Power, Earthly Ruin*. Berkeley: University of California Press, 1999.

Breen, David H. *Alberta's Petroleum Industry and the Conservation Board*. Edmonton: University of Alberta Press, 1993.

Brice-Bennett, Carol. *Reconciling with Memories: A Record of the Reunion at Hebron Forty Years after Relocation*. Nain, NL: Labrador Inuit Association, 2000.

———. "The Redistribution of the Northern Labrador Inuit Population: A Strategy for Integration and Formula for Conflict." *Zeitschrift fur Kanada-Studien* Nr. 2, Band 26 (1994): 95–106.

———, ed. *Our Footprints Are Everywhere: Inuit Land Use and Occupancy in Labrador*. Nain, NL: Labrador Inuit Association, 1977.

Bridge, Gavin. "Contested Terrain: Mining and the Environment." *Annual Review of Environment and Resources* 29 (2004): 205–59.

———. "Material Worlds: Natural Resources, Resource Geography and the Material Economy." *Geography Compass* 3, no. 3 (2009): 1217–44.

———. "Resource Triumphalism: Postindustrial Narratives of Primary Commodity Production." *Environment and Planning A* 33 (2001): 2149–73.

Bridge, Gavin, and Tomas Frederiksen. "'Order Out of Chaos': Resources, Hazards, and the Production of a Tin-Mining Economy in Northern Nigeria in the Early Twentieth Century." *Environment and History* 18, no. 3 (2012): 367–94.

Brodie Consulting. "Nanisivik Mine: Reclamation Cost Estimate." Prepared for Indian Affairs and Northern Development, Water Resources Division, June 2002. Accessed April 2011. ftp://nunavutwaterboard.org/1%20PRUC/1%20INDUSTRIAL/1A/1AR%20-%20Remediation/1ARNAN0914/2%20ADMIN/3%20SUBMISSIONS/2002%20Renewal/DIAND/.

Brosnan, Kathleen A. *Uniting Mountain and Plain: Cities, Law, and Environment along the Front Range*. Albuquerque: University of New Mexico Press, 2002.

Brownlie, Robin, and Mary-Ellen Kelm. "Desperately Seeking Absolution: Native Agency as Colonialist Alibi?" *Canadian Historical Review* 24, no. 4 (December 1994): 543–56.

Brubacher & Associates. "The Nanisivik Legacy in Arctic Bay: A Socio-economic Impact Study." Prepared for the Department of Sustainable Development, Government of Nunavut, 2002.

Brugge, Doug, and R. Goble. "The History of Uranium Mining and the Navajo People." *American Journal of Public Health* 92 (2002): 1410–19.

Brugge, Doug, Timothy Benally, and Esther Yazzie-Lewis. *The Navajo People and Uranium Mining*. Albuquerque: University of New Mexico Press, 2006.

Bullen, Warwick, and Malcolm Robb. "Socio-Economic Impact of Gold Mining in the Yellowknife Mining District." 2002. Accessed February 11, 2013. http://www.miningnorth.com/docs/Socio-Economic%20Impacts%20of%20Gold%20Mining%20in%20Yellowknife%202002.pdf.

Bureau, Martin, and Luc Renaud (dir.). "Une tente sur mars." 58 min. 06 s. Productions Thalie et Les Films, March 3, 2009.

Caine, Ken, and Naomi Krogman. "Powerful or Just Plain Power-Full? A Power Analysis of Impact and Benefit Agreements in Canada's North." *Organization & Environment* 23, no. 1 (2010): 76–98.

Cameco spot price. Accessed February 5, 2014. http://www.cameco.com/investors/markets/uranium_price/spot_price_5yr_history/.

Cameron, Emilie. "Copper Stories: Imaginative Geographies and Material Orderings of the Central Canadian Arctic." In *Rethinking the Great White North: Race, Nature, and Whiteness in Canada*, edited by Andrew Baldwin, Laura Cameron and Audrey Kobayashi, 169–90. Vancouver: UBC Press.

Cameron, Emilie, and Warren Bernauer. "Accumulation, Dispossession and Self-Determination: Wage Labour and the Expansion of Industrial Resource Extraction in Nunavut." Presentation at the Association of American Geographers Conference, New York, NY, February 26, 2012.

Cameron, Laura. "Listening for Pleasure." *Native Studies Review* 11, no. 1 (1996): 109–29.

———. *Openings: A Meditation on History, Method, and Sumas Lake*. Montreal: McGill-Queen's University Press, 1997.

Canada. "Iron Ore in Canada, 1886–1986." Mineral Policy Sector, Internal Report, MRI 88/2, ca. 1988.

———. "Speaking Notes for the Honourable Denis Lebel, Minister of Transport, Infrastructure and Communities." Annual conference of the Association of Canadian Port Authorities, Sept-Îles, QC, August 8, 2011. Accessed August 10, 2013. http://www.tc.gc.ca/eng/mediaroom/speeches-sept-iles-6413.htm.

Canada and Government of Alberta. "Joint Canada-Alberta Implementation Plan for Oil Sands Monitoring." Accessed February 7, 2014. http://environment.gov.ab.ca/info/library/8704.pdf.

Canada-Déline Uranium Table. "Canada Déline Uranium Table Final Report." Déline: Déline First Nation, 2005.

———. "Final Report Concerning Health and Environmental Issues Related to the Port Radium Mine." Ottawa: Indian and Northern Affairs Canada, 2005.

Canadian Institute of Resource Law. Independent Review of the BHP Diamond Mine Process. Calgary: CIRL, 1997.

Canadian Public Health Association. "Final Report." Yellowknife, NWT: Canadian Public Health Association Task Force on Arsenic, 1977.

CanZinco. "Final Closure and Reclamation Report." 2004. Accessed January 2012. ftp://nunavutwaterboard.org/1%20PRUC/1%20INDUSTRIAL/1A/1AR%20-%20Remediation/1ARNAN0914/3%20TECH/10%20A%20and%20R%20(G)(J)/2004/G3%20Final%20Closure%20Plan/2004%20Final%20AR%20plan/G.4%20Cover%20Report/G.4%20Report/.

Carlson, Hans. *Home Is the Hunter: The James Bay Cree and Their Land*. Vancouver: UBC Press, 2008.

Carr, Griselda. *Pit Women: Coal Communities in Northern England in the Early Twentieth Century*. London: Merlin Press, 2001.

Carreau, Bob. "Nunavut Water Board Final Closure and Reclamation of the Nanisivik Mine." Nunavut Water Board Official Public Hearing Transcripts, June 2004, 13–16. Accessed April 2011. ftp://nunavutwaterboard.org/1%20PRUC/1%20INDUSTRIAL/1A/1AR%20-%20Remediation/1AR-NAN0914/2%20ADMIN/4%20HEARINGS/2%20HEARING/2004/.

Castree, Noel. "From Neoliberalism to Neoliberalisation: Consolations, Confusions, and Necessary Illusions." *Environment and Planning* D 38 (2006): 1–6.

———. "Neoliberalism and the Biophysical Environment 1: What 'Neoliberalism' Is, and What Difference Nature Makes to It." *Geography Compass* 2, no. 12 (2010): 1725–33.

Cater, Tara, and Arn Keeling. "'That's where our future came from': Mining, Landscape, and Memory in Rankin Inlet, Nunavut." *Études/Inuit/Studies* 37, no. 2 (2013): 59–82.

Chastko, Paul. *Developing Alberta's Oil Sands: From Karl Clark to Kyoto*. Calgary: University of Calgary Press, 2004.

CIRL. Independent Review of the BHP Diamond Mine Process. Calgary, 1997.

City of Yellowknife. "City Receives Status Report on Royal Oak Mines." Press Release. April 7, 1999.

———. "News Release." November 18, 1999.

Clancy, Peter. "Working on the Railway: A Case Study in Capital-State Relations." *Canadian Public Administration* 30 (1987): 450–71.

Coates, Ken. *Best Left as Indians: Native-White Relations in the Yukon Territory, 1840–1973*. Montreal and Kingston: McGill-Queen's University Press, 1991.

Coates, Ken, and William Morrison. *Forgotten North: A History of Canada's Provincial Norths*. Toronto: James Lorimer and Company, 1992.

Cominco Report. "An Outline of the 'Arvik' Development for Discussions with the Government of Canada." March 1973.

Conference Board of Canada. "The Future of Mining in Northern Canada." Ottawa, January 2013.

———. "Mapping the Economic Potential of Canada's North." Ottawa, December 2010.

Connolly, James, ed. *After the Factory: Reinventing America's Industrial Small Cities*. Toronto: Lexington Books, 2010.

"Consensus Draft, Northwest Territories Lands and Resources Devolution Agreement." March 11, 2013. Accessed March 29, 2013. http://devolution.gov.nt.ca/wp-content/uploads/2012/04/NWT-DevolutionAgreement-March_11-2013.pdf.

"Cooperation Agreement Respecting the Giant Mine Remediation Project." Between Her Majesty the Queen in Right of Canada as Represented by the Minister of Indian Affairs and Northern Development and the Government of the Northwest Territories as Represented by the Minister of Resources, Wildlife, and Economic Development. March

15, 2005. Accessed February 10, 2013. http://db.maca.gov.nt.ca/resources/Cooperation_Agreement_Giantmine_remediation.pdf.

Cornell, Stephen. "Indigenous Peoples, Poverty, and Self-Determination in Australia, New Zealand, Canada, and the United States." In *Indigenous Peoples and Poverty: An International Perspective*, edited by Robyn Eversole, John-Andrew McNeish, and Alberto Cimadamore, 199–225. London: Zed Books, 2005.

Cowie, Jefferson, and Joseph Heathcott, eds. *Beyond the Ruins: The Meaning of Deindustrialization*. Ithaca, NY: Cornell University Press, 2003.

Cronon, William. "Kennecott Journey: The Paths out of Town." In *Under an Open Sky: Rethinking America's Western Past*, edited by William Cronon, George Miles, and Jay Gitlin. New York: W. W. Norton, 1992.

Cruikshank, Julie. *Do Glaciers Listen? Local Knowledge, Colonial Encounters, and Social Imagination* Vancouver: UBC Press, 2005.

———. *Life Lived Like a Story: Life Stories of Three Yukon Native Elders*. Lincoln: University of Nebraska Press, 1991.

———."Nature and Culture in the Field: Two Centuries of Stories from Lituya Bay, Alaska." In *Research in Science and Technology Studies: Knowledge and Technology Transfer*, edited by Marianne de Laet. Knowledge and Society 13. Oxford: Elsevier Science, 2002.

———. *The Social Life of Stories: Narrative and Knowledge in the Yukon Territory*. Lincoln: University of Nebraska Press, 1998.

Dailey, Robert C., and Lois Dailey. "The Eskimo of Rankin Inlet: A Preliminary Report." Ottawa: Northern Co-ordination and Research Centre, June 1961.

Damas, David. *Arctic Migrants, Arctic Villagers: The Transformation of Inuit Settlement in the Central Arctic*. Montreal: McGill-Queen's University Press, 2002.

de Leeuw, Sarah, Margo Greenwood, and Emilie Cameron. "Deviant Constructions: How Governments Preserve Colonial Narratives of Violence and Mental Health to Intervene into the Lives of Indigenous Children and Families in Canada." *International Journal of Mental Health and Addiction* 8, no. 2 (2009): 282–95.

Délįnę Uranium Team. *If Only We Had Known: The History of Port Radium as Told by the Sahtúot'įnę*. Délįnę, NWT: Délįnę Uranium Team, 2005.

DeLyser, Dydia. "Authenticity on the Ground: Engaging the Past in a California Ghost Town." *Annals of the Association of American Geographers* 89, no. 4 (1999): 602–32.

Dempsey, Jessica, Kevin Gould, and Juanita Sundberg. "Changing Land Tenure, Defining Subjects: Neoliberalism and Property Regimes on Native

Reserves." In *Rethinking the Great White North: Race, Nature, and Whiteness in Canada*, edited by Andrew Baldwin, Laura Cameron, and Audrey Kobayashi. Vancouver: UBC Press, 2011.

Dene Nation. "Dene Declaration." In *Dene Nation: The Colony Within*, edited by Mel Watkins. Toronto: University of Toronto Press, 1975.

Department of Indian and Northern Affairs. "North of 60: Mines and Mineral Activities 1976." Mining Division, Northern Non-Renewable Resources Branch, 1976.

Deprez, Paul. *The Pine Point Mine and the Development of the Area South of Great Slave Lake*. Winnipeg: Center for Settlement Studies, 1973.

Desbiens, Caroline. "Défricher l'espace de la nation: lieu, culture et développement économique à la baie James." *Géographie et cultures* 49 (2004): 87–104.

———. "Producing North and South: A Political Geography of Hydro Development in Quebec." *Canadian Geographer* 48, no. 2 (2004): 101–18.

Dickerson, Mark O. *Whose North?: Political Change, Political Development, and Self-Government in the Northwest Territories*. Vancouver: UBC Press, 1992.

Dillon Consulting Limited. "Arsenic Trioxide Management Feasibility Study." Prepared for Indian and Northern Affairs Canada, October 6, 1997.

———. "Dillon Consulting Final Assessment." 2003. Accessed April 2011. ftp://nunavutwaterboard.org/1%20PRUC/1%20INDUSTRIAL/1A/1AR%20-%20Remediation/1AR-NAN0914/Admin#/General/2003/.

———. "Giant Mine Arsenic Trioxide Management Technical Meeting Proceedings." Prepared for Department of Indian Affairs and Northern Development, October 1997.

Drache, Daniel, ed. *Staples, Markets, and Cultural Change: Selected Essays*. Innis Centenary Series. Montreal: McGill-Queen's University Press, 1995.

Driscoll, Jacqueline Jacques. "Development of a Labrador Mining Community: Industry in the Bush." PhD diss., University of Connecticut, 1984

Dublin, Thomas. *When the Mines Closed: Stories of Struggle in Hard Times*. Ithaca, NY: Cornell University Press, 1998.

Dunn, Lawrence. "Negotiating Cultural Identity: Conflict Transformation in Labrador." PhD diss., Syracuse University, 2002.

EBA Engineering Consultants Ltd. "Soil Sampling Program Report." 2002. Accessed April 2011. ftp://nunavutwaterboard.org/1%20PRUC/1%20INDUSTRIAL/1A/1AR%20-%20Remediation/1AR-NAN0914/

Admin#/hearings/hearings/2002hearing/posthearingsubmissions/GN.

Ebner, David, and Brenda Bouw. "Tata Joins Race for Canada's Iron Ore." *Globe and Mail*, March 8, 2011.

Elberling, Bo. "Environmental Controls of the Seasonal Variation in Oxygen Uptake in Sulfidic Tailings Deposited in a Permafrost-Affected Area." *Water Resources Research* 37, no. 1 (2001): 99–107.

———. "Temperature and Oxygen Control on Pyrite Oxidation in Frozen Mine Tailings." *Cold Regions Science and Technology* 41 (2005): 121–33.

Ellis, Derek V., and Jack L. Littlepage. *Feasibility Study for Marine Disposal of Tailings at the Polaris Mine-Site*. Victoria, BC: University of Victoria, 1974.

Environment Canada. "Workshop on Controlling Arsenic Releases into the Environment in the Northwest Territories, Final Workshop Report." October 1997.

Environmental Assessment Board. "Report of the Environmental Assessment Board, Brinex Kitts-Michelin Uranium Project." St. John's, NL: EAB, 1980.

Ericsson, Magnus, and Viktoriya Larsson. "E&MJ's Annual Survey of Global Mining Investment." *Engineering and Mining Journal* (January 2012): 2–7.

Evans, M. S., L. Lockhart, and J. Klaverkamp. "Metal Studies of Water, Sediments, and Fish from the Resolution Bay Area of Great Slave Lake: Studies Related to the Decommissioned Pine Point Mine." Burlington and Saskatoon: Environment Canada, National Water Research Institute, NWRI Contribution No. 98–87, 1998.

Evans, Peter. "Abandoned and Ousted by the State: The Relocations from Nutak and Hebron, 1956–1959." In *Settlement, Subsistence, and Change among the Labrador Inuit: The Nunatsiavummiut Experience*, edited by D. Natcher, L. Felt, and A. Procter. Winnipeg: University of Manitoba Press, 2012.

Evenden, Matthew. "The Northern Vision of Harold Innis." *Journal of Canadian Studies* 34 (1999): 162–86.

Farish, Matthew, and P. Whitney Lackenbauer. "High Modernism in the Arctic: Planning Frobisher Bay and Inuvik." *Journal of Historical Geography* 35 (2009): 517–44.

Feit, Harvey. "Neoliberal Governance and James Bay Cree Governance: Negotiated Agreements, Oppositional Struggles, and Co-governance." In *Indigenous Peoples and Autonomy: Insights for a Global Age*, edited by M. Blaser, R. de Costa, D. McGregor, and W. Coleman, 49–79. Vancouver: UBC Press, 2010.

Fidler, Courtney, and Michael Hitch. "Impact and Benefit Agreements: A Contentious Issue for Environmental and Aboriginal Justice." *Environments Journal* 35, no. 2 (2007): 45–69.

———. "Used and Abused: Negotiated Agreements." Paper presented at the Rethinking Extractive Industry: Regulation, Dispossession and Emerging Claims conference, York University, Toronto, 2009.

Finn, Janet L. *Tracing the Veins: Of Copper, Culture, and Community from Butte to Chuquicamata*. Berkeley: University of California Press, 1998.

Flanagan, Tom. *First Nations? Second Thoughts*. Montreal: McGill-Queens University Press, 2000.

Flanagan, T., and C. Alcantara. "Individual Property Rights on Canadian Indian Reserves." Fraser Institute Occasional Paper 60, 2002.

Fong, Linda. "An Overview of Some of the Issues, Attitudes, and Events Concerning the Proposed Uranium Development in the Kitts-Michelin Area of Labrador." Happy Valley, NL: Labrador Resources Advisory Council, 1977.

Fort McKay Group of Companies LP. "Corporate Information." Accessed June 15, 2013. www.fortmckaygroup.com.

Fort McKay Indian Band. "From Where We Stand." Fort McMurray, AB: Fort McKay Indian Band, 1983.

———. "An Issues Assessment for Concerns Regarding Ongoing Oil Sands Developments and the Community of Fort McKay." Fort McKay, AB: Fort McKay Indian Band, 1986.

Foster, Terry, and Ronne Heming, eds. *Yellowknife Tales: Sixty Years of Stories from Yellowknife*. Yellowknife, NWT: Outcrop, 2003.

Foucault, Michel, and Michel Senellart. *The Birth of Biopolitics: Lectures at the College de France, 1978–79*. New York: Palgrave Macmillan, 2008.

Fox, Bonnie. *Hidden in the Household: Women's Domestic Labour under Capitalism*. Toronto: Women's Educational Press, 1980.

Fox, Michael G. "The Impact of Oil Sands Development on Trapping with Management Implications." MA thesis, University of Calgary, 1977.

Francaviglia, Richard V. *Hard Places: Reading the Landscape of America's Historic Mining Districts*. Iowa City: University of Iowa Press, 1991.

Fraser, Nancy. *Justice Interruptus: Critical Reflections on the "Postsocialist" Condition*. New York: Routledge Press, 1997.

Frickel, Scott, and William R. Freudenburg. "Mining the Past: Historical Context and the Changing Implications of Natural Resource Extraction." *Social Problems* 43, no. 4 (1996): 444–66.

Fumoleau, René. *As Long as This Land Shall Last*. Toronto: McClelland and Stewart, 1975.

Galbraith, Lindsay. "Understanding the Need for Supraregulatory Agreements in Environmental Assessment: An Evaluation from the Northwest Territories, Canada." MA thesis, Simon Fraser University, 2005.

Galbraith, Lindsay, Ben Bradshaw, and Murray Rutherford. "Towards a New Supraregulatory Approach to Environmental Assessment in Northern Canada." *Impact Assessment and Project Appraisal* 25, no. 1 (2007): 27–41.

Gallagher, Nuala, ed. "New Millennium: Advancing the New Millennium Iron Range." *International Resource Journal* 6, no. 3 (June 2011): 116–25.

Gallup, D. N. "Impact Assessment of Discharge." In *Great Canadian Oil Sands Dyke Discharge Water*. Edmonton: Alberta Department of the Environment, August 1977.

Garbutt, Gordon C. *Uranium in Canada*. Ottawa: Eldorado Mining and Refining Ltd., 1964.

Gedicks, Al. *Resource Rebels: Native Challenges to Mining and Oil Companies*. Cambridge, MA: South End Press, 2001.

Geren, Richard, Blake McCullogh, and Iron Ore Company of Canada. *Cain's Legacy: The Building of Iron Ore Company of Canada*. Sept-Îles, QC: Iron Ore Company of Canada, 1990.

Gibson, Ginger, and Ciaran O'Faircheallaigh. *IBA Community Toolkit: Negotiation and Implementation of Impact and Benefit Agreements*. Toronto: Walter and Duncan Gordon Foundation, 2010.

Gibson, Ginger, and Jason Klinck. "Resilient North: The Impact of Mining on Aboriginal Communities." *Pimatisiwin: A Journal of Aboriginal and Indigenous Community Health* 3, no. 1 (2005): 115–39.

Gibson, Robert. "The Strathcona Sound Mining Project: A Case Study in Decision Making." Science Council of Canada, No. 42, 1978.

Gibson, Virginia Valerie. "Negotiated Spaces: Work, Home, and Relationships in the Dene Diamond Economy." PhD diss., University of British Columbia, 2008.

Gogal, Sandra, Richard Reigert, and JoAnn Jamieson. "Aboriginal Impact and Benefit Agreements: Practical Considerations." *Alberta Law Review* 43, no. 1 (2006): 129–58.

Goin, Peter, and C. Elizabeth Raymond. *Changing Mines in America*. Santa Fe: Center for American Places, 2004.

———. "Living in Anthracite: Mining Landscape and Sense of Place in Wyoming Valley, Pennsylvania." *The Public Historian* 23, no. 2 (2001): 29–45.

Goodall, Heather. "Indigenous Peoples, Colonialism, and Memories of Environmental Injustice." In *Echoes from the Poisoned Well: Global Memories of Environmental Injustice*, edited by Sylvia Hood Washington, Paul C. Rosier, Heather Goodall, 74–75. Oxford: Lexington Books, 2006.

Gordon, Todd, and Jeffery R. Webber. "Imperialism and Resistance: Canadian Mining Companies in Latin America." *Third World Quarterly* 29, no. 1 (2008): 63–87.

Gosselin, Pierre, Steve E. Hrudey, M. Anne Naeth, Andre Plourde, Rene Therrien, Glen Van Der Kraak, and Zhenghe Xu. "Environmental and Health Impacts of Canada's Oil Sands Industry: The Royal Society of Canada Expert Panel Report." Ottawa: The Royal Society of Canada, 2010.

Government of Alberta. "Regulatory Enhancement Project." Accessed October 10, 2012. http://www.energy.alberta.ca/initiatives/regulatoryenhancement.asp.

Government of Newfoundland and Labrador. "Land Claims Policy." St. John's, NL: Intergovernmental Affairs Secretariat, Native Policy Unit, 1987.

———. "Statement by Premier A. Brian Peckford on the Question of Native Land Claims in the Province." Telecommunication to Toby Andersen, President of LRAC, Happy Valley, Labrador, October 10, 1980.

Government of the Northwest Territories. Approach to Regulatory Improvement. March 2009.

———. "An Investigation of Atmospheric Emissions from the Royal Oak Giant Yellowknife Mine." Department of Renewable Resources, June 1993.

Grace, Sherrill. *Canada and the Idea of North*. Montreal and Kingston: McGill-Queen's University Press, 2001.

Graham, Katherine. *The Development of the Polaris Mine*. Eastern Arctic Study 3. Kingston, ON: Institute of Local Government, 1982.

Grant, Shelagh. *Sovereignty Or Security?: Government Policy in the Canadian North, 1936–1950*. Vancouver: UBC Press, 1988.

Grygier, Pat Sandiford. *A Long Way from Home: The Tuberculosis Epidemic among the Inuit*. Montreal: McGill-Queen's University Press, 1994.

Haida Nation v. British Columbia (Minister of Forests), 2004. S.C.J. No. 70.

Hale, Charles. "Neoliberal Multiculturalism: The Remaking of Cultural Rights and Racial Dominance in Central America." *PoLAR* 28, no. 1 (2005): 10–28.

Hall, Rebecca. "Diamond Mining in Canada's Northwest Territories: A Colonial Continuity." *Antipode* 45, no. 2 (2013): 376–93.

Hall, Valerie. "Differing Gender Roles: Women in Mining and Fishing Communities in Northumberland, England, 1880–1914." *Women's Studies International Forum* 27, no. 5/6 (November 2004): 521–30.

Halvaksz, Jamon Alex, II. "Whose Closure? Appearances, Temporality, and Mineral Extraction in Papua New Guinea." *Journal of the Royal Anthropological Institute* 14 (2008): 21–37.

Hamilton, Ed, Buchans Historic Research Group, and Red Indian Lake Development Association. *Khaki Dodgers: The History of Mining and the People of the Buchans Area*. Grand Falls-Windsor, NL: Red Indian Lake Development Association, 1992.

Hamilton, Paula, and Linda Shopes. "Introduction: Building Partnerships Between Oral History and Memory Studies." In *Oral History and Public Memories*, edited by Paula Hamilton, vii–xvii. Philadelphia, PA: Temple University Press, 2008.

Hamlet of Arctic Bay Working Group. "Submission to the Nunavut Water Board." 2002. Accessed April 2011. ftp://nunavutwaterboard.org/1%20PRUC/1%20INDUSTRIAL/1A/1AR%20-%20Remediation/1AR-NAN0914/2%20ADMIN/4%20HEARINGS/2%20HEARING/2002%20Renewal/Exhibits%20Undertakings/020726NWB1NAN9702%20Exhibit11-Submission%20of%20Hamlet%20AB%20Wrkg%20Group-English.pdf.

Hare, F. Kenneth. "The Labrador Frontier." *Geographical Review* 42, no. 3 (July 1952): 405–24.

Harner, John. "Place Identity and Copper Mining in Sonora, Mexico." *Annals of the Association of American Geographers* 91, no. 4 (2001): 660–80.

Harvey, David. *A Brief History of Neoliberalism*. New York: Oxford, 2005.

Hatfield, C. T. and G. L. Williams. "A Summary of Possible Environmental Effects of Disposing Mine Tailings into Strathcona Sound, Baffin Island." Hatfield Consultants Limited Arctic Land Use Research Program, 1976.

Haysom, V. "Labrador Inuit Land Claims: Aboriginal Rights and Interests vs. Federal and Provincial Responsibilities and Authorities." *Northern Perspectives* 18, no. 2 (1990).

Hayter, Roger, and Trever Barnes. "Neoliberalization and Its Geographic Limits: Comparative Reflections from Forest Peripheries in the Global North." *Economic Geography* 88, no. 2 (2012): 197–221.

Heath, Bill. "Exhibit 1: Opening Remarks by Bill Heath, Vice-President of CanZinco." Public Meeting at Arctic Bay, 2002. Accessed April 2011. ftp://nunavutwaterboard.org/1%20PRUC/1%20INDUSTRIAL/1A/1AR%20-%20Remediation/1AR-NAN0914/2%20

ADMIN/4%20HEARINGS/2%20HEARING/2002%20Renewal/Exhibits%20Undertakings/.

Heber, Robert Wesley. "Indigenous Knowledge, Resources Use, and the Dene of Northern Saskatchewan." *Canadian Journal of Development Studies* 26 (2005): 247–56.

Hefferton, S. J. (Minister of Public Welfare, Government of Newfoundland). "Letter to Minister of Citizenship and Immigration, Ottawa, 26 March 1959." In Government of Newfoundland (Department of Public Welfare), 1964, The Administration of Northern Labrador Affairs, Appendix 1: 97–98.

Helin, Calvin. *Dances with Dependency: Indigenous Success through Self-Reliance*. Vancouver: Orca Spirit Publishing and Communications, 2006.

Henningson, David. "Somba Ke: The Money Place." 45 min. 2006.

Hickling-Partners Inc. "Evaluation of the Nanisivik Project." Canada Department of Indian Affairs and Northern Development, 1981.

High, Steven. *Industrial Sunset: The Making of North America's Rust Belt, 1969–1984*. Toronto: University of Toronto Press, 2003.

High, Steven, and David Lewis. *Corporate Wasteland: The Landscape and Memory of Deindustrialization*. Toronto: Between the Lines, 2007.

Hipwell, William, Katy Mamen, Viviane Weitzner, and Gail Whiteman. "Aboriginal People and Mining in Canada: Consultation, Participation and Prospects for Change: Working Discussion Paper." Ottawa: North-South Institute, 2002.

Hitch, M. "Impact and Benefit Agreements and the Political Ecology of Mineral Development in Nunavut." PhD diss., University of Waterloo, 2005.

Hiyate, Alisha. "The New Normal." *Mining Markets* 4, no. 3 (September 2011): 5.

Holden, William, Kathleen Nadeau, and R. Daniel Jacobson. "Exemplifying Accumulation by Dispossession: Mining and Indigenous Peoples in the Philippines." *Geografiska Annaler: Series B, Human Geography* 93, no. 2 (2011): 141–61.

House of Commons Standing Committee on Environment and Sustainable Development. "It's about Our Health! Towards Pollution Prevention, CEPA Revisited." Report, June 1995.

Howitt, Richard. *Rethinking Resource Management: Justice, Sustainability, and Indigenous Peoples*. London: Routledge, 2001.

Howlett, Cathy, Monica Seini, Diana Mcallum, and Natalie Osborne. "Neoliberalism, Mineral Development and Indigenous People: A

Framework for Analysis." *Australian Geographer* 42, no. 3 (2011): 309–23.

Hrudey, S. E. "Characterization of Wastewaters from the Great Canadian Oil Sands Bitumen Extraction and Upgrading Plant." Ottawa: Water Pollution Control Section, Environmental Protection Service, Northwest Region, Environment Canada, 1975.

Humpage, Louise. "Tackling Indigenous Disadvantage in the Twenty-First Century: 'Social Inclusion' and Maori in New Zealand." In *Indigenous Peoples and Poverty: An International Perspective*, edited by Robyn Eversole, John-Andrew McNeish, and Alberto Cimadamore, 158–84. London: Zed Books, 2005.

"IBA Research Network." Accessed November 30, 2012. http://www.impactandbenefit.com/IBA_Database_List/.

Indian and Northern Affairs Canada (INAC). "Labrador Inuit Land Claims Agreement." Ottawa: INAC, 2005.

———. "Mine Site Reclamation Policy for Nunavut." Minister of Indian Affairs and Northern Development. Ottawa: INAC, 2002.

Indian and Northern Affairs Canada, and Government of the Northwest Territories. "Giant Mine Remediation Project Developers' Assessment Report." October 2010. Accessed February 11, 2013. http://www.reviewboard.ca/upload/project_document/EA0809-001_Giant_DAR_1328896950.PDF.

Innis, Harold Adams. *The Fur Trade in Canada: An Introduction to Canadian Economic History*. Rev. ed. Toronto: University of Toronto Press, 1956.

———. *The Problems of Staple Production in Canada*. Toronto: Ryerson Press, 1933.

———. *Settlement and the Mining Frontier*. Toronto: Macmillan, 1936.

Irlbacher-Fox, Stephanie. *Finding Dashaa: Self-Government, Social Suffering, and Aboriginal Policy in Canada*. Vancouver: UBC Press, 2009.

Isaac, Thomas, and Anthony Knox. "Canadian Aboriginal Law: Creating Certainty in Resource Development." *University of New Brunswick Law Journal* 53, no. 3 (2004): 3–42.

Isenberg, Andrew C. *Mining California: An Ecological History*. New York: Hill and Wang, 2005.

Jackson, Jason Baird. "The Story of Colonialism, or Rethinking the Ox-Hide Purchase in Native North America and Beyond." *Journal of American Folklore* 126, no. 499 (2013): 31–54.

Jackson, Susan, ed. *Yellowknife, NWT: An Illustrated History*. Yellowknife, NWT: Nor'West Publishing, 1990.

Jameson, Elizabeth. *All That Glitters: Class, Conflict, and Community in Cripple Creek*. Urbana: University of Illinois Press, 1998.

Jenish, D'Arcy. "Destined for Grandeur: Chinese Steel Giant Invests Heavily in Quebec Ore Property." *Canadian Mining Journal* 133, no. 3 (April 2012): 16–18.

John, Angela. *By the Sweat of Their Brow: Women Workers at Victorian Coal Mines*. London: Croom Helm, 1980.

Jorgenson, J. D. "Challenges Facing the North American Iron Ore Industry." Reston, VA: US Geological Survey, 2006. Open-File Report 2006-1061.

Journeaux, Dean. "Breaking New Ground in the Labrador Trough." Presentation at Northern Exposure, St. John's, NL, January 22–24, 2013.

Jules, Manny. "Foreword." In *Beyond the Indian Act: Restoring Aboriginal Property Rights*, edited by Tom Flanagan, Chris Alcantara, and Andre Le Dressay. Montreal and Kingston: McGill-Queen's University Press, 2012.

Justus, Roger, and Joanne Simonetta. "Major Resource Impact Evaluation, Prepared for the Cold Lake Band and the Indian and Inuit Affairs Program." Vancouver: Justus-Simonetta Development Consultants Limited, December 1979.

Kalin, M. "Ecological Engineering for Gold and Base Metal Mining Operations in the Northwest Territories." *Environmental Studies*, no. 59 (1987).

Kealey, Linda, ed. *Pursuing Equality: Historical Perspectives on Women in Newfoundland and Labrador*. St. John's, NL: Institute of Social and Economic Research, Memorial University of Newfoundland, 1993.

Keeling, Arn. "'Born in an Atomic Test Tube': Landscapes of Cyclonic Development at Uranium City, Saskatchewan." *Canadian Geographer* 54, no. 2 (2010): 228–52.

———. "The Rancher and the Regulator: Public Challenges to Sour-Gas Industry Regulation in Alberta 1970–1994." In *Writing Off the Rural West: Globalization, Governments, and the Transformation of Rural Communities*, edited by Roger Epp and Dave Whitson, 279–300. Edmonton: University of Alberta Press, 2001.

Keeling, Arn, and John Sandlos. "Environmental Justice Goes Underground? Historical Notes from Canada's Mining Frontier." *Environmental Justice* 2 (2009): 117–25.

———. "Shooting the Archives: Document Digitization for Historical-Geographical Collaboration." *History Compass* 9, no. 5 (May 2011): 423–32.

Keeping, Janet. "Local Benefits and Mineral Rights Disposition in the Northwest Territories: Law and Policy." In *Disposition of Natural Resources: Options and Issues for Northern Lands*, edited by Monique Ross and John Owen Saunders. Calgary: Canadian Institute of Resources Law, 1997.

———. "Thinking about Benefits Agreements: An Analytical Framework." Prepared for Canadian Arctic Resource Committee, 1998.

Kelly, Erin N., David W. Schindler, Peter V. Hodson, Jeffrey W. Short, Roseanna Radmanovich, and Charlene C. Nielsen. "Oil Sands Development Contributes Elements Toxic at Low Concentrations to the Athabasca River and Its Tributaries." *PNAS Environmental Sciences* 107, no. 37 (2010): 16178–83.

Kennedy, John. "Aboriginal Organizations and Their Claims: The Case of Newfoundland and Labrador." *Canadian Ethnic Studies* 19, no. 2 (1987): 13–25.

———. "The Changing Significance of Labrador Settler Ethnicity." *Canadian Ethnic Studies* 20, no. 3 (1988): 94–111.

———. *Holding the Line: Ethnic Boundaries in a Northern Labrador Community*. St. John's, NL: ISER, 1982.

———. "Local Government and Ethnic Boundaries in Makkovik, 1972." In *The White Arctic: Anthropological Essays on Tutelage and Ethnicity*, edited by Robert Paine, 359–75. St. John's, NL: ISER, 1977.

———. *People of the Bays and Headlands: Anthropological History and the Fate of Communities in the Unknown Labrador*. Toronto: University of Toronto Press, 1995.

Kennett, Steven. *Issues and Options for a Policy on Impact and Benefit Agreements for the Northern Territories*. Calgary: Canadian Institute of Resource Law, 1999.

King, Thomas. *The Truth about Stories: A Native Narrative*. Minneapolis: University of Minnesota Press, 2008.

Kirsch, Stuart. *Reverse Anthropology: Indigenous Analysis of Social and Environmental Relations in New Guinea*. Palo Alto, CA: Stanford University Press, 2006.

Klubock, Thomas Miller. *Contested Communities: Class, Gender, and Politics in Chile's El Teniente Copper Mine, 1904–1951*. Durham, NC: Duke University Press, 1998.

Kovach, Margaret. "Emerging from the Margins: Indigenous Methodologies." In *Research as Resistance: Critical, Indigenous, and Anti-oppressive Approaches*, edited by Leslie A. Brown and Susan Strega, 19–36. Toronto: Canadian Scholars' Press/Women's Press, 2005.

———. *Indigenous Methodologies – Characteristics, Conversations, and Contexts.* Toronto: University of Toronto Press, 2009.

Kulchyski, Peter K., and Frank Tester. *Kiumajut (Talking Back): Game Management and Inuit Rights, 1900–1970.* Vancouver: UBC Press, 2007.

Kuokkanen, Rauna. "From Indigenous Economies to Market-Based Self-Governance: A Feminist Political Economy Analysis." *Canadian Journal of Political Science* 44, no. 2 (2011): 275–97.

Kupsch, W. O. "From Erzgebirge to Cluff Lake – A Scientific Journey through Time." *The Musk-Ox* 23 (1978): 7–87.

Kurek, Joshua, Jane L. Kirk, Derek C. G. Muir, Xiaowa Wang, Marlene S. Evans, and John P. Smol. "Legacy of a Half Century of Athabasca Oil Sands Development Recorded by Lake Ecosystems." Proceedings of the National Academy of Sciences (2013).

Kyhn, Curt, and Bo Elberling. "Frozen Cover Actions Limiting AMD from Mine Waste Deposited on Land in Arctic Canada." *Cold Regions Science and Technology* 32 (2001): 133–42.

———. "Port Radium: The Story of Its Beginnings." *The Refiner*, 1960.

Labrador Institute of Northern Studies. *Labrador in the 90s.* Happy Valley–Goose Bay, Labrador, NL: October 29–November 1, 1990, 56.

Labrador West Status of Women Council. "Submission to the Employment Practices Commission," 1989.

Lackenbauer, P. Whitney, and Matthew Farish, "The Cold War on Canadian Soil: Militarizing a Northern Environment." *Environmental History* 12, no. 4 (2007): 920–50.

Laforest, Richard, Jacques Frenette, Robert Comtois, and Michel Mongeon. "Occupation et utilisation du territoire par les Montagnais de Schefferville." Rapport au Conseil Attikamek-Montagnais, Village des Hurons, 1983.

Lalonde, Marc. "The National Energy Program," edited by Department of Energy and Natural Resources. Ottawa: Minister of Supply and Services, 1980.

Lane, Marcus B., and E. Rickson Roy. "Resource Development and Resource Dependency of Indigenous Communities: Australia's Jawoyn Aborigines and Mining at Coronation Hill." *Society and Natural Resources* 10 (1997): 121–42.

Lapointe, Ugo. "L'héritage du principe de *free mining* au Québec et au Canada." *Recherches amérindiennes au Québec* 40, no. 3 (2010): 9–25.

Larner, Wendy. "Neo-liberalism: Policy, Ideology, Governmentality." *Studies in Political Economy* 63, no. 1 (2000): 5–26.

Lawrence, Daina. "Canadian Government Fostering Arctic Exploration." *Resource World* 11 (December 2011/January 2012): 82–83.

LeCain, Timothy J. "The Limits of 'Eco-Efficiency': Arsenic Pollution and the Cottrell Electric Precipitator in the U.S. Copper Smelter Industry." *Environmental History* 5 (2000): 336–51.

———. *Mass Destruction: The Men and Giant Mines That Wired America and Scarred the Planet.* New Brunswick, NJ: Rutgers University Press, 2009.

Leddy, Lianne. "Cold War Colonialism: The Serpent River First Nation and Uranium Mining, 1953–1988." PhD diss., Wilfrid Laurier University, 2011.

———. "Interviewing Nookomis and Other Reflections: The Promise of Community Collaboration." *Oral History Forum d'histoire orale* 30 (2010): 1–18.

Legros, Dominique. "Oral History as History: Tutchone Athapaskan in the Period 1840–1920." Occasional Papers in Yukon History No. 3 (2). Whitehorse: Yukon Cultural Services Branch, 2007.

Levitan, Tyler. "Impact and Benefit Agreements in Relation to the Neoliberal State: The Case of Diamond Mines in the Northwest Territories." MA thesis, Carleton University, 2012.

Li, Tania. "Governmentality." *Anthropologica* 49, no. 2 (2007): 275–81.

Liddington, Jill. "Gender Authority and Mining in Industrial Landscape: Anne Lister 1791–1840." *History Workshop Journal* 42 (1996): 58–86.

Lim, Tee W. "Inuit Encounters with Colonial Capital: Nanisivik – Canada's First High Arctic Mine." MA thesis, University of British Columbia, 2013.

Loder, Richard L. "Changing Familial Patterns in a Newfoundland Mining Town: A Town of Widows and Orphans." Paper presented to Professor Louise Charimonte. Centre for Newfoundland Studies, Memorial University of Newfoundland, 1972.

Loo, Tina. "Disturbing the Peace: Environmental Change and the Scales of Justice on a Northern River." *Environmental History* 12 (October 2007): 895–919.

Low, Nicholas, and Brendan Gleeson. *Justice, Society, and Nature: An Exploration of Political Ecology.* London: Routledge, 1998.

———. "Situating Justice in the Environment: The Case of BHP at the Ok Tedi Copper Mine." *Antipode* 30 (1998): 201–26.

Lowenthal, David. *The Past Is a Foreign Country.* Cambridge: Cambridge University Press, 1985.

Luxton, Meg. *More Than a Labour of Love: Three Generations of Women's Work in the Home*. Toronto: Women's Educational Press, 1980.

Lynch, Martin. *Mining in World History*. London: Reaktion, 2002.

Maaka, R., and A. Fleras. *The Politics of Indigeneity: Challenging the State in Canada and Aotearoa New Zealand*. Dunedin, NZ: University of Otago Press, 2005.

MacDonald, Fiona. "Indigenous Peoples and Neoliberal 'Privatization' in Canada: Opportunities, Cautions, and Constraints." *Canadian Journal of Political Science* 44, no. 2 (2011): 257–73.

Mackay, W. C. "Toxicity of GCOS Tailings Pond Dyke Discharge." In *Great Canadian Oil Sands Dyke Discharge Water*. Edmonton: Alberta Department of the Environment, August 1977.

Mackenzie Valley Environmental Impact Review Board. "Report of Environmental Assessment and Reasons for Decision on Giant Mine Remediation Project EA0809-0001." June 20, 2013. Accessed July 1, 2013. http://www.reviewboard.ca/upload/project_document/EA0809-001_Giant_Report_of_Environmental_Assessment_June_20_2013.PDF.

Macpherson, Janet E. "The Pine Point Mine." In *Northern Transitions, Volume I: Northern Resource Use and Land Use Policy Study*, edited by Everett B. Peterson and Janet B. Wright, 65–110. Ottawa: Canadian Arctic Resources Committee, 1978.

Mahon, Rianne. "Canadian Public Policy: The Unequal Structure of Representation." In *The Canadian State: Political Economy and Political Power*, edited by L. Panitch, 165–98. Toronto: University of Toronto Press, 1977.

Markey, Sean, Greg Halseth, and Don Manson. "Challenging the Inevitability of Rural Decline: Advancing the Policy of Place in Northern British Columbia." *Journal of Rural Studies* 24 (2008): 409–21.

Marsh, Ben. "Continuity and Decline in the Anthracite Towns of Pennsylvania." *Annals of the Association of American Geographers* 77, no. 3 (1987): 337–52.

Martinez-Alier, Joan. "Mining Conflicts, Environmental Justice, and Valuation." *Journal of Hazardous Materials* 86, no. 1–3 (2001): 153–70.

McAllister, Mary Louise. "Shifting Foundations in a Mature Staples Industry: A Political Economic History of Canadian Mineral Policy." *Canadian Political Science Review* 1 (June 2007): 73–90.

McCarthy, James. "First World Political Ecology: Lessons from the Wise Use Movement." *Environment and Planning* A 34 (2002): 1281–1302.

McCormack, Patricia A. *Fort Chipewyan and the Shaping of Canadian History, 1788–1920s: "We Like to Be Free in This Country."* Vancouver: UBC Press, 2010.

McCrank, Neil. *Road to Improvement: The Review of the Regulatory Systems across the North.* Ottawa: Canada, Minster of Public Works and Government Services Canada, 2008.

McKenzie, Réal. "En finir avec la discrimination." *Recherches amérindiennes au Québec* 41, no. 1 (2011): 71–76.

McMahon, Fred, and Miguel Cervantes. "Fraser Institute Annual Survey of Mining Companies, 2011/2012." Vancouver: Fraser Institute, February 2012.

McPherson, Kathryn, Cecilia Louise Morgan, and Nancy Forestell. *Gendered Pasts: Historical Essays in Femininity and Masculinity in Canada.* Canadian Social History Series. Toronto: University of Toronto Press, 2003.

McPherson, Robert. *New Owners in Their Own Land: Minerals and Inuit Land Claims.* Calgary: University of Calgary Press, 2003.

Metheny, Karen Bescherer. *From the Miner's Doublehouse: Archaeology and Landscape in a Pennsylvania Coal Company Town.* Knoxville: University of Tennessee Press, 2007.

Mochoruk, Jim. *Formidable Heritage: Manitoba's North and the Cost of Development, 1870 to 1930.* Winnipeg: University of Manitoba Press, 2004.

Moehring, Eugene. *Urbanism and Empire in the Far West, 1840–1890.* Reno: University of Nevada Press, 2004.

Momaday, N. Scott. *The Man Made of Words.* New York: St. Martin's Press, 1997.

Montrie, Chad. *To Save the Land and People: A History of Opposition to Surface Coal Mining in Appalachia.* Chapel Hill: University of North Carolina Press, 2002.

Morse, Kathryn. *The Nature of Gold: An Environmental History of the Klondike Gold Rush.* Seattle: University of Washington Press, 2003.

Murphy, Mary. *Mining Cultures: Men, Women, and Leisure in Butte, 1914–41.* Women in American History series, edited by Anne Firor Scott, Nancy A. Hewitt, and Stephanie Shaw. Urbana: University of Illinois Press, 1997.

Myers, Heather. *Uranium Mining in Port Radium, N.W.T: Old Wastes, New Concerns.* Ottawa: Canadian Arctic Resources Committee, August 1982.

Nadasdy, Paul. *Hunters and Bureaucrats: Power, Knowledge, and Aboriginal-State Relations in the Southwest Yukon*. Vancouver: UBC Press, 2003.

Nanisivik Mine. "Letter from Nanisivik mine to Thomas Kudloo of NWB regarding final submissions." 2002. Accessed April 2011. ftp://nunavutwaterboard.org/1%20PRUC/1%20INDUSTRIAL/1A/1AR%20-%20Remediation/1AR-NAN0914/2%20ADMIN/4%20HEARINGS/2%20HEARING/2002%20Renewal/Post%20Hearing%20Submissions/BWR/English/.

———. "Letter from Nanisivik mine to Thomas Kudloo of NWB regarding public hearing outcomes." 2002. Accessed April 2011. ftp://nunavutwaterboard.org/1%20PRUC/1%20INDUSTRIAL/1A/1AR%20-%20Remediation/1AR-NAN0914\Admin#\hearings\hearings\2002 hearing\post hearing submissions\BWR.

NASA Earth Observatory. "World of Change: Athabasca Oil Sands." NASA Goddard Space Flight Centre. Accessed April 5, 2013. http://earthobservatory.nasa.gov/Features/WorldOfChange/athabasca.php?all=y.

Nash, June. *We Eat the Mines and the Mines Eat Us: Dependency and Exploitation in Bolivian Tin Mines*. 2nd ed. New York: Columbia University Press, 1993.

Natural Resources Canada. "Canada: A Diamond-Producing Nation." Accessed January 31, 2014. http://www.nrcan.gc.ca/minerals-metals/business-market/3630.

Natural Resources Canada, Legal Surveys Division. Historical Review – Matimekosh (undated). Accessed August 10, 2013. http://clss-satc.nrcan-rncan.gc.ca/data-donnees/publications/indlanhisque-hisfonterindque/matimekosh_ang.pdf.

Neale, Stacy. "The Rankin Inlet Ceramics Project: A Study in Development and Influence." MA thesis, Concordia University, 1997.

Neil, Cecily, Markku Tykklainen, and John Bradbury, eds. *Coping with Closure: An International Comparison of Mine Town Experiences*. New York: Routledge, 1992.

Neufeld, David. "Parks Canada, The Commemoration of Canada, and Northern Aboriginal Oral History." In *Oral History and Public Memories*, edited by Paula Hamilton, 26–48. Philadelphia, PA: Temple University Press, 2008.

New Millennium Iron. "On the Path to Production." Corporate presentation, September 2011.

Newman, Peter C. "Gilbert Labine: Adventurous Bushwhacker." *The Beaver* (1959): 48–53.

Nikiforuk, Andrew. *Saboteurs: Wiebo Ludwig's War against Big Oil.* Toronto: Macfarlane Walter and Ross, 2001.

———. *Tar Sands: Dirty Oil and the Future of a Continent.* Vancouver: Greystone Books, 2010.

Noble, Russell B. "La hommage [sic] au 'Plan Nord.'" *Canadian Mining Journal* 133, no. 3 (April 2012): 5.

———, ed. "Ahead of the Curve: Working with Aboriginal Partners in the Race for Canada's Iron Ore." *Canadian Mining Journal* 133, no. 3 (April 2012): 26–27.

Northern Regulatory Improvement Initiative. Submission by the Northwest Territories and Nunavut Chamber of Mines, the Prospectors and Developers Association of Canada, and the Mining Association of Canada to Neil McCrank, 2008.

Northwest Territories Mining Heritage Society. "The Gold Mines Built Yellowknife." Accessed February 28, 2012. http://www.nwtminingheritage.com/files/frontend-static mininghistory/The%20Gold%20Mines%20Built%20Yellowknife.pdf.

Notzke, Claudia. *Aboriginal Peoples and Natural Resources in Canada.* North York, ON: Captus University Publications, 1994.

NRCan. "Agreements between Mining Companies and Aboriginal Communities or Governments." Accessed February 4, 2014. https://www.nrcan.gc.ca/mining-materials/aboriginal/14694.

Nunavummiut Makitagunarningit. "Discussion Paper – Kiggavik Draft Socioeconomic Impact Statement." June 2012. Accessed November 10, 2012. http://makitanunavut.files.wordpress.com/2012/06/makita-socioeconomic-discussion-paper.pdf.

———. "Submission to the Study of the United Nations Special Rapporteur on the Rights of Indigenous Peoples on Extractive and Energy Industries in and near Indigenous Territories." April 1, 2013. Accessed May 9, 2013. http://makitanunavut.files.wordpress.com/2013/04/2013-04-01-makita-to-james-anaya1.pdf.

Nunavut Water Board. "Final Closure and Reclamation of the Nanisivik Mine." Nunavut Water Board Official Public Hearing Transcripts, June 2004. Accessed April 2011. ftp://nunavutwaterboard.org/1%20PRUC/1%20INDUSTRIAL/1A/1AR%20-%20Remediation/1AR-NAN0914/2%20ADMIN/4%20HEARINGS/2%20HEARING/2004/.

———. "Nanisivik Mine Type A Water License Renewal and Amendment Application." Nunavut Water Board Official Public Hearing Transcripts, February 2009. Accessed April 2011. ftp://nunavutwaterboard.org/1%20PRUC/1%20INDUSTRIAL/1A/1AR%20

-%20Remediation/1AR-NAN0914/2%20ADMIN/4%20
HEARINGS/2%20HEARING/2008%20Hearing/.

———. "Public Hearing Meeting Notes." 2004. Accessed April 2011. ftp://
nunavutwaterboard.org/1%20PRUC/1%20INDUSTRIAL/1A/1AR%20
-%20Remediation/1AR-NAN0914/2%20ADMIN/4%20
HEARINGS/0%20GENERAL/2004/.

———. "Public Hearing Minutes." Nunavut Water Board Official Public Hearing Transcripts, July 2002. Accessed April 2011. ftp://nunavutwaterboard.org/1%20PRUC/1%20INDUSTRIAL/1A/1AR%20-%20 Remediation/1AR-NAN0914/2%20ADMIN/4%20HEARINGS/2%20 HEARING/2002%20Renewal/PH%20July%2022-24,%202002/.

NWT and Nunavut Chamber of Mines. *Sustainable Economies: Aboriginal Participation in the NWT Mining Industry, 1990–2004.* Yellowknife, NWT, 2004.

Obed, Enoch. *Kinatuinamut Ilingajuk.* Nain, August 10, 1979.

Ocean Equities. "Iron Ore: The Labrador Trough." London: Ocean Equities, January 18, 2013.

O'Faircheallaigh, Ciaran. "Aboriginal – Mining Company Contractual Agreements in Australia and Canada: Implications for Political Autonomy and Community Development." *Canadian Journal of Development Studies* 3, nos. 1–2 (2010): 69–86.

O'Faircheallaigh, Ciaran, and Ginger Gibson. "Economic Risk and Mineral Taxation on Indigenous Lands." *Resources Policy* 37, no. 1 (2012): 10–18.

Offen, Karl. "Historical Political Ecology: An Introduction." *Historical Geography* 32 (2004): 7–18.

Office of the Auditor General. "2002 Report of the Commissioner of the Environment and Sustainable Development," Chapter 3, *Abandoned Mines in the North.* Ottawa: Minister of Public Works and Services, 2002. Accessed March 10, 2013. http://www.oag-bvg.gc.ca/internet/English/parl_cesd_200210_03_e_12409.html.

———. "Spring 2012 Report of the Commissioner of the Environment and Sustainable Development," Chapter 3, *Federal Contaminated Sites and Their Impact.* Accessed March 10, 2013. http://www.oag-bvg.gc.ca/internet/docs/parl_cesd_201205_03_e.pdf.

———. "Fall 2012 Report of the Commissioner of the Environment and Sustainable Development," Chapter 2, *Financial Assurances for Environmental Risk.* Accessed February 11, 2013. http://www.oag-bvg.gc.ca/internet/docs/parl_cesd_201212_02_e.pdf.

O'Hara, Paul. "Model Cities, Mill Towns, and Industrial Peripheries: Small Industrial Cities in Twentieth-Century America." In *After The Factory:*

Reinventing America's Industrial Small Cities, edited by James Connolly, 19–48. Toronto: Lexington Books, 2010.

O'Reilly, Kevin. "Giant Mine, Giant Legacy." *Northern Public Affairs* 1, no. 2 (2012): 50–53.

O'Reilly, Kevin, and Erin Eacott. "Aboriginal Peoples and Impact and Benefit Agreements: Summary of a National Workshop." *Northern Perspectives* 25, no. 1 (1999).

O'Rourke, J. C. "Comments on Public Hearings." In *Summary of Kitts-Michelin Hearings*, 1–4. St. John's, NL: Brinex Ltd., 1979.

Overton, James. "Progressive Conservatism? A Critical Look at Politics, Culture, and Development in Newfoundland." In *Ethnicity in Atlantic Canada*. Social Science Monograph Series 5. Saint John: University of New Brunswick, 1985.

Pamack-Jeddore, Rosina. "What Confederation Has Meant to the Labrador Eskimo." *Decks Awash* 3, no. 5 (1974): 6–7.

Panitch, Leo, and Sam Gindin. *The Making of Global Capitalism: The Political Economy of American Empire*. London: Verso, 2012.

Papillon, Martin. "Les peuples autochtones et la citoyenneté: quelques effets contradictoires de la gouvernance néolibérale." éthique publique 14, no. 1 (2012).

Paquette, Pierre. *Les mines du Québec, 1867–1975: une évaluation critique d'un mode historique d'industrialisation nationale*. Outremont, QC: Carte blanche, 2000.

Parsons, Linda Ann. "Labrador City by Design: Corporate Visions of Women." In *Their Lives and Times: Women of Newfoundland and Labrador: A Collage*, edited by Carmelita McGrath, Barbara Neis, and Marilyn Porter. St. John's, NL: Killick Press, 1995.

———. "Passing the Time: The Lives of Women in a Northern Industrial Town." MA thesis, Memorial University of Newfoundland, 1987.

Pasternak, Shiri. "How Capitalism Will Save Colonialism: Hernando De Soto, the Settler Colony of Canada, and the Privatization of Reserve Lands." Presentation at the Association of American Geographers Conference, New York, NY, February 25, 2012.

Peck, Jamie. "Neoliberalizing States: Thin Policies/Hard Outcomes." *Progress in Human Geography* 25, no. 3 (2001): 445–55.

Peck, Jamie, and Adam Tickell. "Neoliberalizing Space." *Antipode* 34, no. 3 (2002): 380–404.

Pini, Barbara, Robyn Mayes, and Paula McDonald. "The Emotional Geography of a Mine Closure: A Study of the Ravensthorpe Nickel Mine in

Western Australia." *Social & Cultural Geography* 11, no. 6 (September 2010): 559–74.

Piper, Liza. *The Industrial Transformation of Subarctic Canada*. Vancouver: UBC Press, 2009.

———. "Innis, Biss, and Industrial Circuitry in the Canadian North, 1921–1965." In *Harold Innis and the North: Appraisals and Contestations*, edited by William J. Buxton, 127–48. Montreal and Kingston: McGill-Queen's University Press, 2013.

———. "Subterranean Bodies: Mining the Large Lakes of North-West Canada, 1921–1960." *Environment and History* 13, no. 2 (2007): 155–86.

Plaice, Evelyn. "'Making Indians': Debating Indigeneity in Canada and South Africa." In *Culture Wars: Context, Models, and Anthropologists' Accounts*, edited by Deborah James, Evelyn Plaice, and Christina Toren. New York: Berghahn Books, 2010.

———. "Response to Kuper's Return of the Native." *Current Anthropology* 44, no. 3 (2003): 396–97.

Poloski, Murray. "Minutes of Public Meeting." Makkovik, NL: Labrador Resources Advisory Council, July 27–28, 1976.

Pope, Sharon Gray, and Jane Burnham. "Change Within and Without: The Modern Women's Movement in Newfoundland." In *Pursuing Equality: Historical Perspectives on Women in Newfoundland and Labrador*, edited by Linda Kealey. St. John's, NL: Institute of Social and Economic Research, Memorial University of Newfoundland, 1993.

Povinelli, Elizabeth. *The Cunning of Recognition: Indigenous Alterities and the Making of Australian Multiculturalism*. Durham, NC: Duke University Press, 2002.

Power, Rosemary. "'After the Black Gold': A View of Mining Heritage from Coalfield Areas of Britain." *Folklore* 119 (2008): 160–81.

Pratt, Larry. *The Tar Sands: Syncrude and the Politics of Oil*. Edmonton: Hurtig Publishers, 1976.

Price, Richard, ed. *The Spirit of the Alberta Indian Treaties*. Montreal: Institute for Research on Public Policy, 1979.

Prior, Timothy, Damien Giurco, Gavin Mudd, Leah Mason, and Johannes Behrisch. "Resource Depletion, Peak Minerals, and the Implications for Sustainable Resource Management." *Global Environmental Change* 22, no. 3 (2012): 577–87.

Prno, Jason. "Assessing the Effectiveness of Impact and Benefit Agreements from the Perspective of Their Aboriginal Signatories." MA thesis, University of Guelph, 2007.

Prno, J., B. Bradshaw, and D. Lapierre. "Impact and Benefit Agreements: Are They Working?" Paper presented at the Canadian Institute of Mining, Metallurgy, and Petroleum Annual Conference, Vancouver, BC, May 11, 2010.

Prno, Jason, and Scott D. Slocombe. "Exploring the Origins of 'Social License to Operate' in the Mining Sector: Perspectives from Governance and Sustainability Theories." *Resources Policy* 37, no. 3 (2012): 346–57.

Purcell, M. *Recapturing Democracy: Neoliberalization and the Struggle for Alternative Urban Futures.* New York: Routledge, 2008.

Québec. "Plan Nord: Building Northern Québec Together, The Project of a Generation." Ministère des Ressources naturelles et de la Faune, 2011.

Quiring, David. "CCF Colonialism." In *Northern Saskatchewan: Battling Parish Priests, Bootleggers, and Fur Sharks.* Vancouver: UBC Press, 2004.

R. v. Suncor Inc. 1983. Alberta Court of Appeal, 219. Appeal #16352. September 15, 1983.

Rea, Kenneth J. *The Political Economy of the Canadian North.* Toronto: University of Toronto Press, 1968.

Reedyk, S. "Profiles of Baffin Communities Affected by Closure of the Nanisivik Mine and a Summary of the Impact of Closure on the Communities." Indian and Northern Affairs Canada, Mining Management and Infrastructure Division, 1987.

Regional Aquatics Monitoring Program. "Joint Community Update 2008 Reporting Our Environmental Activities to the Community." Fort McMurray, AB: Regional Aquatics Monitoring Program (RAMP), Wood Buffalo Environmental Association (WBEA), Cumulative Environmental Management Association (CEMA), 2008.

Rennie, Rick. *The Dirt: Industrial Disease and Conflict at St. Lawrence, Newfoundland.* Halifax, NS: Fernwood Publishing, 2008.

Resource Futures International. "Socio-Economic Analysis of Three Management Options to Reduce Atmospheric Emissions of Arsenic from Gold Roasting." September 9, 1996.

Richards, John, and Larry Pratt. *Prairie Capitalism: Power and Influence in the New West.* Toronto: McClelland and Stewart Limited, 1979.

Ritchie, Donald A. *Doing Oral History: A Practical Guide.* 2nd ed. Oxford: Oxford University Press, 2003.

Robbins, Paul. *Political Ecology: A Critical Introduction.* Oxford: Blackwell, 2004.

Roberts, Leslie. "Living on Radium." *Collier's Weekly,* June 5, 1937.

Robertson, David. *Hard as the Rock Itself: Place and Identity in the American Mining Town.* Boulder: University Press of Colorado, 2006.

Rodon, Thierry, ed. *Teach an Eskimo How to Read . . .: Conversations with Peter Freuchen Ittinuar*. Iqaluit, NU: Nunavut Arctic College, 2008.

Rollwagen, Katharine. "When Ghosts Hovered: Community and Crisis in a Company Town, Brittania Beach, British Columbia, 1957–1965." *Urban History Review/Revue d'histoire urbaine* 35, no. 2 (2007): 25–36.

Rompkey, William. *The Story of Labrador*. Montreal: McGill-Queen's Press, 2003.

Rowbotham, Sheila. "More Than Just a Memory: Some Political Implications of Women's Involvement in the Miners' Strike, 1984–85." *Feminist Review* 23 (1986): 407–21.

Royal Oak Mines and EBA Engineering Limited. "Arsenic Trioxide Surface Storage and Handling Project Scoping Document." December 1997.

Sabin, Jerald, and Frances Abele. *State and Society in a Northern Capital: Yellowknife's Social Economy in Hard Times*. Ottawa: Carleton Centre for Community Innovation, 2010.

Salverson, Julie. "They Never Told Us These Things." *Maisonneuve*, August 12, 2011.

Sandlos, John. *Hunters at the Margin: Native People and Wildlife Conservation in the Northwest Territories*. Vancouver: UBC Press, 2007.

Sandlos, John, and Arn Keeling. "Claiming the New North: Development and Colonialism at the Pine Point Mine, Northwest Territories, Canada." *Environment and History* 18, no. 1 (2012): 5–34.

———. "Giant Mine: Historical Summary, Report Submitted to the Mackenzie Valley Environmental Impact Review Board," August 12, 2012. Accessed February 11, 2013. http://www.reviewboard.ca/upload/project_document/EA0809-001_Giant_Mine__History_Summary.PDF.

———. "Zombie Mines and the (Over)burden of History." *Solutions Journal* 4, no. 3 (June 2013): 80–83.

Satzewich, Vic, and Terry Wotherspoon. *First Nations: Race, Class, and Gender Relations*. Regina: Canadian Plains Research Centre, 2000.

Scales, Marylin. "Big Money: Mont-Wright Expansion to Pump $2.1 Billion into Quebec." *Canadian Mining Journal* 133, no. 3 (April 2012): 13–15.

Schroeder, Richard A., Kevin St. Martin, and E. Albert Katherine. "Political Ecology in North America: Discovering the Third World Within?" *Geoforum* 37 (2006): 163–68.

Searle, John R. *Expression and Meaning*. Cambridge: Cambridge University Press, 1979.

Searles, Edmund. "Anthropology in an Era of Inuit Empowerment." In *Critical Inuit Studies: An Anthology of Contemporary Arctic Ethnography*, edited by Pamela Stern and Lisa Stevenson, 89–101. Lincoln: University of Nebraska Press, 2006.

Seddon, Vicky. *The Cutting Edge: Women and the Pit Strike*. London: Lawrence and Wishart, 1986.

Sénéchal, Christian, and Hélène Brown (dir.). "La ville de Gagnon." 19 min. 11 s. Plate-Formes-Prods, 2008. Accessed August 10, 2013. http://parolecitoyenne.org/gagnon-le-film-documentaire?dossier_nid=20476.

Shoebridge, Paul, and Michael Simons. *Welcome to Pine Point*. Montreal: National Film Board, 2011.

Shopes, Linda. "Oral History and the Study of Communities: Problems, Paradoxes, and Possibilities." In *The Oral History Reading*, edited by Robert Preks. 2nd ed. New York: Routledge Taylor and Francis Group, 2006.

Shore, Cris, and Susan Wright. "Policy: A New Field of Anthropology." In *Anthropology of Policy: Critical Perspectives on Governance and Power*, edited by Cris Shore and Susan Wright, 3–42. London: Routledge, 1997.

Siebenmorgen, P. "Developing an Ideal Mining Agenda: Impact and Benefit Agreements as Instruments of Community Development in Northern Ontario." MA thesis, Guelph University, 2009.

Silke, Ryan. *High-Grade Tales: Stories from the Mining Camps of the Northwest Territories*. Yellowknife, NWT: Ryan Silke, 2012.

Slaney, Rennie. *More Incredible Than Fiction: The True Story of the Indomitable Men and Women of St. Lawrence, Newfoundland from the Time of Settlement to 1965: History of Fluorspar Mining at St. Lawrence Newfoundland*. St. John's, NL: Confederation of National Trade Unions, 1975.

Slowey, Gabrielle. *Navigating Neoliberalism: Self-Determination and the Mikisew Cree First Nation*. Vancouver: UBC Press, 2008.

Smith, Allan. "The Myth of the Self-Made Man in English Canada, 1850–1914." *Canadian Historical Review* 59, no. 2 (1978): 189–219.

Smith, Dale, ed. *Keewatin Journal*. Rankin Inlet, NU: Dale Smith, 1979.

Smith, Duane A. *Mining America: The Industry and the Environment, 1800–1980*. Lawrence: University Press of Kansas, 1987.

Smith, Linda Tuhiwai. "The Native and the Neoliberal Down Under: Neoliberalism and 'Endangered Authenticities.'" In *Indigenous Experience Today*, edited by Marisol de la Cadena and Orin Starn, 333–52. Oxford: Berg, 2007.

Smith, Philip. *Brinco: The Story of Churchill Falls.* Toronto: McClelland and Stewart Ltd., 1975.

Solodzuk, W., N. R. Morgenstern, N. L. Iverson, E. J. Klohn, M. A. J. Matich, B. D. Prasad, and I. H. Anderson. "Report on Great Canadian Oil Sands Tar Island Tailings Dyke." Design Review Panel, Alberta Environment, February 1977.

Sosa, Irene, and Karyn Keenan. "Impact Benefit Agreements between Aboriginal Communities and Mining Companies: Their Use in Canada." Calgary: Canadian Environmental Law Association, 2001.

Spracklin, Andrea. "No Business like Sew Business (re: Cassell's Sewing Business in Labrador City)." *Downhomer* 14, no. 10 (March 2002): 35.

Standing Committee on Environment and Sustainable Development. "Evidence of Meeting #122." May 11, 1995. Accessed January 27, 2013. http://www.parl.gc.ca/content/hoc/archives/committee/351/sust/evidence/122_95-05-11/sust-122-cover-e.html.

———. "Evidence of Meeting #123." May 11, 1995. Accessed January 27, 2013. http://www.parl.gc.ca/content/hoc/archives/committee/351/sust/evidence/123_95-05-11/sust-123-cover-e.html.

Stein, J. N., and M. R. Miller. "An Investigation into the Effects of a Lead-Zinc Mine on the Aquatic Environment of Great Slave Lake." Unpublished report. Winnipeg: Resource Development Branch, Fisheries Service, Department of Environment, 1972.

Steinbock, Bernd. *Social Memory in Athenian Public Discourse: Uses and Meanings of the Past.* Ann Arbor: University of Michigan Press, 2013.

Stern, Pamela. "Upside-Down and Backwards: Time Discipline in a Canadian Inuit Town." *Anthropologica* 45, no. 1 (2003): 147–61.

Stevenson, D. S. "Problems of Eskimo Relocation for Industrial Employment: A Preliminary Study." Northern Science Research Centre report, Department of Indian Affairs and Northern Development, May 1968.

Stiller, David. *Wounding the West: Montana, Mining, and the Environment.* Lincoln: University of Nebraska Press, 2000.

Sukagawa, Paul. "Is Iron Ore Priced as a Commodity? Past and Current Practice." *Resources Policy* 35, no. 1 (March 2010): 54–63.

Sumi, Lisa, and Sandra Thomsen. *Mining in Remote Areas: Issues and Impacts.* Ottawa: MiningWatch Canada, 2001.

Summerby-Murray, Robert. "Interpreting Personalized Industrial Heritage in the Mining Towns of Cumberland County, Nova Scotia: Landscape Examples From Springhill and River Hebert." *Urban History Review* 35, no. 3 (2007): 51–59.

Syncrude. "Biophysical Impact Assessment for the New Facilities at the Syncrude Canada Ltd. Mildred Lake Plant." Calgary: Syncrude Canada Ltd., 1984.

Tanner, Adrian, John C. Kennedy, Susan McCorquodale, and Gordon Inglis. "Aboriginal Peoples and Governance in Newfoundland and Labrador." A Report for the Governance Project, Royal Commission on Aboriginal Peoples, St. John's, NL, 1994.

Tata Steel Minerals Canada. "Welcome to Tata Steel Minerals Canada Limited." Accessed August 10, 2013. http://www.tatasteelcanada.com.

Taussig, Michael T. *The Devil and Commodity Fetishism in South America*. Chapel Hill: University of North Carolina Press, 1980.

Taylor, Mary Josephine. "The Development of Mineral Policy for the Eastern Arctic, 1953–1985." MA thesis, Carleton University, 1985.

Terriplan Consultants. "Giant Mine Underground Arsenic Trioxide Management Alternatives. Draft Workshop Summary Report, January 14–15, 2003." Prepared for the Department of Indian Affairs and Northern Development, March 2003.

———. "Giant Mine Underground Arsenic Trioxide Management Alternatives, Moving Forward: Selecting a Management Alternative, Draft Summary Workshop Report." May 26–27, 2003. Prepared for the Department of Indian Affairs and Northern Development.

Tester, Frank, and Chris Flannelly. "The Rankin File: Public Responsibility, Private Provision, and Health Care at the Edge of the Canadian Liberal Welfare State." Paper presented at the ArcticNet Annual Scientific Meeting, Ottawa, ON, December 14–17, 2010.

Tester, Frank, and Peter Kulchyski. *Tammarniit (Mistakes): Inuit Relocation in the Eastern Arctic, 1939–63*. Vancouver: UBC Press, 1994.

Thomson, Alistair. "Four Paradigm Transformations in Oral History." *Oral History Review* 34, no. 1: 49–70.

Tompkins, Edward. "Pencilled Out: Newfoundland and Labrador's Native People and Canadian Confederation, 1947–1954." Report written for Jack Harris, MP. Ottawa: House of Commons, 1988.

Tonts, Matthew, Kirsten Martinus, and Paul Plummer. "Regional Development, Redistribution, and the Extraction of Mineral: The Western Australian Goldfields as a Resource Bank." *Applied Geography* 45 (2013): 365–74.

Tough, Frank. *As Their Natural Resources Fail: Native People and the Economic History of Northern Manitoba, 1870–1930*. Vancouver: UBC Press, 1997.

Trigger, David S. "Mining, Landscape and the Culture of Developmental Ideology in Australia." *Cultural Geographies* 4, no. 2 (1997): 161–80.

Vachon, Daniel. *L'histoire montagnaise de Sept-Îles*. Québec: Éditions Innu, 1985.

Valdivia, Gabriela. "On Indigeneity, Change, and Representation in the Northeastern Ecuadorian Amazon." *Environment and Planning* A 37 (2005): 285–303.

Vallee, F.G. *Kabloona and the Eskimo in the Central Keewatin*. Ottawa: Northern Co-ordination and Research Centre, 1962.

van Wyck, Peter C. *The Highway of the Atom*. Montreal: McGill-Queen's University Press, 2010.

———. "The Highway of the Atom: Recollections Along a Route." *Topia* 7 (2002): 99–115.

Vérificateur général du Québec. "Report of the Auditor General of Québec to the National Assembly for 2008–2009." Vol. 2. April 1, 2009.

Walker, Brett. *Toxic Archipelago: A History of Industrial Disease in Japan*. Seattle: University of Washington Press, 2010.

Walker, Peter. "Politics of Nature: An Overview of Political Ecology." *Capitalism, Nature, Socialism* 9 (March 1998): 131–44.

———. "Reconsidering 'Regional' Political Ecologies: Toward a Political Ecology of the Rural American West." *Progress in Human Geography* 27 (2003): 7–24.

Ward, Ashley. "Reclaiming Place through Remembrance: Using Oral Histories in Geographic Research." *Historical Geography* 40 (2012): 133–45.

Watkins, Mel, ed. *Dene Nation: The Colony Within*. Toronto: University of Toronto Press, 1977.

Watts, Michael. "Development and Governmentality." *Singapore Journal of Tropical Geography* 24, no. 1 (2003): 6–34.

Wenig, Michael, and Kevin O'Reilly. "The Mining Reclamation Regime in the Northwest Territories: A Comparison with Selected Canadian and U.S. Jurisdictions." Canadian Institute of Resources Law and Canadian Arctic Resources Committee, January 2005. Accessed February 11, 2013. http://carc.org/pdfs/mining49_nwtminingreclam_final_21jan05.pdf.

Williamson, Robert G. *Eskimo Underground: Socio-Cultural Change in the Canadian Central Arctic*. Uppsala, Sweden: Institutionen För Allmän och Jämförande Etnografi, 1974.

Williamson, Robert G., and Terrence W. Foster. "Eskimo Relocation in Canada." Ottawa: Department of Indian and Northern Affairs, 1975.

Wilson, Jeffrey D. "Chinese Resource Security Policies and the Restructuring of the Asia-Pacific Iron Ore Market." *Resources Policy* 37, no. 3 (September 2012): 331–39.

Wilson, Lisa J. "Riding the Resource Roller Coaster: Understanding Socioeconomic Differences between Mining Communities." *Rural Sociology* 69, no. 2 (2009): 261–81.

Wilson, Shawn. *Research Is Ceremony: Indigenous Research Methods*. Winnipeg: Fernwood Press, 2008.

Wirth, John D. *Smelter Smoke in North America: The Politics of Transborder Pollution*. Lawrence: University of Press of Kansas, 2000.

"Wood Buffalo Environmental Association Human Exposure Monitoring Program (HEMP) Methods Report and 2005 Monitoring Year Results." Fort McMurray, AB: Wood Buffalo Environmental Monitoring Association, 2007.

World Steel Association. "Crude Steel Production 2012." Accessed August 10, 2013. http://worldsteel.org/statistics/crude-steel-production.html.

Wyckoff, William. "Postindustrial Butte." *Geographical Review* 85, no. 4 (1995): 478–96.

Wyllie, Irvin G. *The Self-Made Man in America: The Myth of Rags to Riches*. New York: The Free Press, 1966 [1954].

Yazzie-Lewis, Esther, and Jim Zion. "Leetso, the Powerful Yellow Monster." In *The Navajo People and Uranium Mining*, edited by Doug Brugge, Timothy Benally, and Esther Yazzie-Lewis, 1–10. Albuquerque: University of New Mexico Press, 2006.

Yellowknives Dene First Nation. *Weledeh Yellowknives Dene: A History*. Dettah, NWT: Yellowknives Dene First Nation Council, 1997.

Yergin, Daniel. *The Prize: The Epic Quest for Oil, Money, and Power*. New York: Free Press, 1991.

Yoshimatsu, Donna. "The Legacy of the Rail Lives On, But Could It Be Built Today?" *Canadian Mining Journal* 130, no. 5 (June/July 2009): 8.

Young, B. "Memo: Wolfden Resources: Massive Sulphide Discovery at High Lake." 2003. Accessed April 2011. ftp://nunavutwaterboard.org/1%20 PRUC/1%20INDUSTRIAL/1A/1AR%20-%20Remediation/1AR-NAN0914/Tech/10 A and R/2003.

Yukon Territory. *From the Trenches*. Volume 3, Issue 1. Whitehorse: Energy Mines and Resources, Spring 2010.

———. *Silver Trail Tourism Development Plan*. Whitehorse: Yukon Department of Tourism, 1989.

———. *United Keno Hill Mine: Closure (1989) Analysis*. Whitehorse: Yukon, Economic Development, 1995.

Zaslow, Morris. *The Northward Expansion of Canada, 1914–1967.* Toronto: McClelland and Stewart, 1988.

———. *The Opening of the Canadian North, 1870–1914.* Toronto: McClelland and Stewart, 1971.

Zimmerer, Karl S., and Thomas J. Bassett. "Approaching Political Ecology: Society, Nature, and Scale in Human-Environment Studies." In *Political Ecology: An Integrative Approach to Geography and Environment-Development Studies,* edited by Karl S. Zimmerer and Thomas J. Bassett. New York: Guilford Press, 2003.

Index

Note: Page numbers in bold refer to photographs.

A

AANDC. *See* Aboriginal Affairs and Northern Development Canada (AANDC)
abandoned mines
  in Auditor General's report, 11–12
  definition, xi
  Giant Mine, **12**
  Pine Point, 142, **145**
  Schefferville, 182, **182**, **183**, **184**, **186**
"Abandoned Mines in Northern Canada" (research project), 5, 140–41, 378
Aboriginal Affairs and Northern Development Canada (AANDC). *See also* Canadian government
  comments on IBAs, 273, 275
  remediation projects, 12, 358–65
Aboriginal rights
  Crown's duty to protect, 272
  IBAs as infringement of, 270–71
  Labrador Inuit movement, 240–41
  northern and southern activists on, 326
Aboriginal self-determination
  and IBAs, 279, 281
  and neoliberal practices, 233, 235, 252, 266–67, 268, 279, 280

acculturation, Rankin Inuit, 45–46, 52
acid mine drainage
  defined, xi
  in Nanisivik tailings, 303–4
Adriana Resources, 194
affirmative action programs, 126, 218–19, 222
Agnico-Eagle Meadowbank Mine project, 1
agreements. *See also* comprehensive land claim agreements (CLCAs); impact and benefit agreements (IBAs)
  Alsands, 222
  connected to hegemonic neoliberal values, 233, 235
  IOC's gender equality programs, 126
  Labrador Inuit land claim, 240, 243–44, 245, 248
  Nanisivik employment agreement, 326
  Strathcona Sound, 296–97
  Syncrude and community, 223
Agricola, Georgius, 337
air quality
  Fort McKay area, 211–12
  Yellowknife, 347
Akumalik, Joanasie, 305
Akumalik, Moses, 298
Akumalik, Mucktar, 299

425

Alberta
  GCOS expansion approval, 214
  on GCOS tailings ponds seepage, 211
  health warnings to Fort McKay, 212
  on indigenous employment in oil sands, 218–19
  "one window" review policy, 217
  priority of oil development, 208, 210
  on Suncor pollution spill, 215–16
Alberta Clean Water Act regulations, 211, 216
Alberta Conservation and Utilization Committee, 218
Alberta oil sands. *See* oil sands industry
Albo, Greg, 265–66
Alexco Resource Corporation, 104–5, 113
Alsands project, 216–17, 221–22
Alternatives North, 365, 385
Andersen, William, III, 243–44
Andre, Leroy, 72
ArcelorMittal (Luxembourg), 192
Arctic Bay, Ikpiarjuk Inuit
  concerns about tailings, 303
  mine impacts on, 297–99
  on participating in monitoring, 305–6
arsenic trioxide at Giant Mine
  environmental assessments, 347
  management and remediation plan, 355–57, **360**
  in mining processes, 343–44
  roaster complex, 362
  storage remediation options, 350–51
assimilation. *See also* modernization of the North
  Alberta position on, 218
  Rankin Inuit, 45–46, 52
  with resource development, 7
ATC. *See* Athabasca Tribal Council (ATC)
Athabasca Chipewyan First Nation, 221, 225
Athabasca River, 208, 210–11, 213–14, 215–16
Athabasca Tribal Council (ATC), 209, 221–22
Atkinson, Marion, 126
Atlantic Gateway and Trade Corridor Strategy, 193
Aurora Energy Resources Ltd., 245, 249, 251
Austin, John L., 60
Ayha (Dene Prophet), 72–74

B

Bacon, Joséphine, 169
Balsillie, George, 144
Bankeno Mines, 322
Barnes, Trevor, 6
Barnett, Clive, 279
Barrett, J. E. and Associates, 322, 323
BC Research, 325
Beaulieu, Angus, 144, 149, 157
Beaulieu, Gord, 148, 152, 156, 157
Beaulieu, Leander, 149
Beaulieu, Leonard, 146–48, 153
Beaulieu, Ronald, 152
Beaulieu, Tommy, 152
Beckwith, Karen, 119
Belgium global radium supply, 63–64
Bell, J. Macintosh report, 66–67
Bellekeno Mine, 88, 99, 104
Berger, Thomas, 139, 144
Beyonnie (Sahtúot'įnę elder), 70–71
Blondin, George, 72
Blondin, Joe, 72
Bolivian indigenous mining experiences, 38, 61
boom and bust cycles
  brief history in northern mining, 1–2
  gold mining, 342
  iron ore, 178
  Keno Hill mining district, 88–90
  studies on community impacts, 5–6
Boucher, Catherine, 154, 157
Boucher, Jim, 221
Boulter, Patricia J.
  biography, 383
  chapter by, 18, 35–58
Boutet, Jean-Sébastien, 53
  biography, 383
  chapter by, 19–20, 53, 169–206
  comments, 331
Bowes-Lyon, Léa-Marie, 329
Bradbury, John H., 6, 181
Bridge, Gavin, 37–38
Brinco. *See* British Investment Company (Brinco)
Brinex. *See also* British Investment Company (Brinco)
  Kitts-Michelin project, 239, 242–43

response to Aboriginal rights recognition, 242
British Investment Company (Brinco), 236, 237, 239
Brodie Consulting, 306
Bugghins, Sam, 145
Buyck, Debbie, 97–98
Buyck, Helen, 112

C

Caine, Ken, 261, 270, 282
Cameron, Emilie
  biography, 383
  chapter by, 259–90
  comments, 141
Cameron Bay, 63
Canada–Délįne Uranium Table (CDUT), 64–65, 79
Canadian Arctic Resources Committee (CARC), 7, 139, 324–26
Canadian Environmental Protection Act (CEPA), 348
Canadian government. *See also* Aboriginal Affairs and Northern Development Canada (AANDC); Department of Indian Affairs and Northern Development (DIAND); Department of Northern Affairs and National Resources; Indian and Northern Affairs Canada (INAC)
  1969 White Paper, 240
  affirmative action programs, 126, 222
  on Alberta's indigenous employment policy, 219
  archives on northern mining, 14
  assistance to industry, 1, 7–8, 63–64, 138, 295
  authority over resources, 342
  clawbacks, 49, 278
  colonial legacy in the North, 280
  defense of wage gap, 44
  employment programs, 49
  environmental reports, 148
  Giant Mine maintenance, 346, 353–54
  health reports, 347
  and IBAs, 262, 271–78
  land and asset purchases, 48, 178
  and oil development, 210
  voiced criticisms of, 71, 81, 139, 266, 267–68
Canadian Public Health Association, 346
CanZinco Ltd.
  on Nanisivik as pioneer project, 295–96
  Nanisivik closure, 297, 308
  on soundness of scientific methods, 306–7
  tailings cover engineering, 304
  valuation of reclamation costs, 300–303
CARC. *See* Canadian Arctic Resources Committee (CARC)
Cardinal, Lloyd, 145–46, 157
Careen, Noreen, 127
Carol Lake Mine. *See* Iron Ore Company of Canada (IOC)
Carreau, Bob, 297
Cassell, Linda, 124
Cater, Tara, 317
CDUT. *See* Canada–Délįne Uranium Table (CDUT)
Chesterfield Inlet, 41, 46, 49
childcare facilities, 124
children, recruited for secondary industry jobs, 128
Chinese steel industry, 170, 191, 192, 193
Chipewyan Prairie First Nation, 221, 225
climate panels, warming estimates, 304
Cloutier, Richard, 139
collective memory of mining
  formation and value of, 316, 317
  Resolute experience, 320–21, 331, 332–33
College of the North Atlantic, 127
colonialism
  and legacy of the state, 266–67
  mineral development as, 7, 8, 17–18, 37–38, 196–97
  neoliberalization as new form of, 268
  Port Radium as symbol of, 61, 70, 74–75, 80–82
  redressing through IBAs, 276, 280–82
Cominco
  criticisms of, 145, 146, 330
  development of Polaris Mine, 321, 322
  hiring plans for Polaris Mine, 326–27
  meetings with Resolute Inuit, 322–24
  Pine Point Mine, 138, 142–43

Polaris Mine closure, 329
town facilities built by, 151
commemorations of mining life and landscape. *See also* collective memory of mining
   instances of, 9–10, 35–36, 100–101, 139–40
   oral history as correction to, 138
Committee for Original People's Entitlement (COPE), 322
communities. *See also* specific groups and towns
   company influence on, 46–47, 119, 120, 123, 151
   connectedness to, 49, 50–51, 120, 140
   debates on environment vs. economy, 80–81, 109–10, 111–13, 245–51, 252
   recording of experiences, 13–17
compensation expectations
   Arctic Bay Inuit, 298–99
   by CARC on behalf of Resolute, 325
   Pine Point, 143–44, 145–46, 157
   Port Radium, 71
   Schefferville, 179, 180–81, 189, 198
comprehensive land claim agreements (CLCAs), 10, 103–4, 263–64, 271
Con Mine, 17, 50, 343, 344, 353
consent. *See* consultation and consent; impact and benefit agreements (IBAs)
Consolidated Mining and Smelting Company. *See* Cominco
consultation and consent. *See also* impact and benefit agreements (IBAs)
   absent in *Plan Nord*, 170
   absent in Schefferville development, 175, 198–99
   in Alberta oil sands industry, 222, 223
   Arctic Bay Inuit, 305–6
   Fort McKay excluded from, 216–17
   Labrador calls for, 239
   Pine Point expectations, 142, 145–46, 157
   Resolute Inuit experiences, 321, 323–24, 330
   Yellowknife exclusions and concerns, 350, 356–57, 358–63
Contact Lake, 63
Courteorielle, Lawrence, 222
Crown. *See* Canadian government
Cruikshank, Julie
   on narratives, 60, 62, 68, 75, 81, 94, 113
   on oral history approach, 93
culture
   diversity at Pine Point, 151–52
   and language barriers, 41, 44–45
   in oral history research, 15
   protection through resource development, 247
cyclonics of hinterland resource development, 6
Cyprus-Anvil Mine, 2, 8, 17

D

Dailey, Robert C. and Lois, 46
Davidson, Jacob A., 88
Délįnę, Northwest Territories
   land and resource ethics, 77–79
   mine impacts on, 59–60, 64
   narratives about Port Radium, 62, 69–74
   on remediation efforts, 65, 79–80
demolition of Schefferville infrastructure, 181, 188–89
Dempsey, Jessica, 268
Dene Nation. *See* Sahtúot'įnę
Deninu K'ue First Nation, 143–44, 264
Department of Indian Affairs and Northern Development (DIAND). *See also* Aboriginal Affairs and Northern Development Canada (AANDC); Canadian government
   on Alsands project assessment, 217
   Giant Mine management and remediation planning, 350–51, 355–57
   and IBAs, 262–63, 274–75
   Polaris development, 326
   sale of Giant Mine, 353
   Strathcona Sound Agreement, 296–97
   subsidies to Miramar, 354
   working with Royal Oak Mines, 352
Department of Northern Affairs and National Resources. *See also* Canadian government
   advocacy for Inuit miners, 45
   monitoring of relocations, 50

promotion of wage labour for Inuit, 39–40, 45–46
responses to mine closure, 47–48
settlement policies, 42–44
Deprez, Paul, 139
DEW Line, 39, 46
DIAND. *See* Department of Indian Affairs and Northern Development (DIAND)
Dickins, Clennell Haggerston "Punch," 76
Dion, Joe, 222
Distant Early Warning (DEW) Line, 39, 46
distrust. *See* trust and distrust
Doran, Barbara, 126
Dragon, Larry, 151
Dunn, Lawrence, 241
Durham Miners' Gala, 317

E

Easton, J. Andrew, 41, 45, 49
EBA Engineering, 302
economies. *See also* land-based living
combining wage labour with subsistence, 10, 22, 45, 98, 101–2, 150, 153, 158, 190–91, 378
Fort McKay, 208–9, 218–23
Labrador Inuit, 236, 242
limitations of mining to improve, 252
modernization, 7–8, 138
Nunatsiavut, 246–49, 252
Resolute, 321, 328–29, 330–31
Sahtúot'įnę, 80–81
shaped by hiring practices, 120
Yellowknives Dene, 342
education
"Employee of the Future" program, 127
integration with industry needs, 194, 276–77
as taking control of one's future, 247
EIAs. *See* environmental impact review processes
Ekati diamond mine, IBAs, 261, 278
Eldorado Gold Mines Ltd. (later Eldorado Mining and Refining Ltd.), 63–64, 66–68
"Elizabeth Andrews," 120–21, 123, 125
Elsa, Yukon
hub of Keno Hill district, 90

life in, 97–98, 106
mine closure, 102
townsite remediation, 104
employment
affirmative action programs, 126, 221–22
after mine closures, 49–50, 102
Alberta oil sands, 218–20
Keno Hill mining district, 95–96, 97, 98–99, 101
Labrador City gendered hiring practices, 119–22, 126–27, 129
Labrador City secondary industries, 124, 127–28
Nanisivik, 296–97
North Rankin Nickel Mine, 38, 40–41, 44–46
Pine Point Mine, 144–45
Polaris Mine, 321, 326–27, 327–28
Port Radium, 64, 79–80
preference for local workers, 41, 194
Rankin Inlet industries, 49
Schefferville, 175
employment agreements, 276, 296–97, 326
employment promises, 142, 144–45, 175, 190, 194, 237–38, 251, 323
energy crisis, 209–10
Energy Resources Conservation Board (ERCB)
on affirmative action hiring, 221
Alsands hearings, 216–17
environmental reviews commissioned by, 213
GCOS expansion hearings, 213–14
entertainment. *See* recreation and entertainment
Environmental Assessment and Review Process (EARP), 325
environmental impact reviews
to avoid long-term liabilities, 296, 380–81
concerns about process, 157, 210, 211–13, 216–17, 325, 358–63
model for future, 366
environmental impacts
Alberta oil sands, 210–16
anticipated in Labrador, 246
Giant Mine, 344, 347, 349
of industrial mining, 4, 8–9, 225
Keno Hill mining district, 89, 90, 103

Nanisivik, 297–99
Pine Point Mine, 141–44, 146–48
Polaris Mine, 321, 323–24, 327, 329–30
Port Radium, 63, 64–65
understanding through oral histories, 13
environmental legislation
    Alberta Clean Water Act, 211, 216
    apparent weaknesses, 224
    attempts to create, 346–48
    Giant Mine negotiations, 349, 351, 358
    in Nanisivik reclamation, 300
    Polaris development, 324–25
environmental monitoring
    advice ignored, 344, 346
    demands for participation in, 305–6, 332
    Joint Oil Sands Monitoring Program, 225
Environmental Rights Act (NWT), 347
ERCB. *See* Energy Resources Conservation Board (ERCB)
exploration
    Keno Hill, 88, 104
    Labrador Trough, 173, 196
    Pine Point seismic, 144–45
    Port Radium area, 66–67

F

Fabian, Roy, 146
Fabien, Kevin, 144
Fafard (Father), 44
family lifestyles and company policies, 120, 123–24
Faro (Yukon), 2, 8
Feit, Harvey, 269, 279
ferroducts, 193
Fidler, Courtney, 272
financial independence for women, 120, 121, 126, 127, 128–29
First Air, 329
fish and fishing
    impacts on, 64, 103, 146, 211, 213, 215, 216, 349
    overfishing, 187
food. *See also* land-based living
    importance of country food, 298
    purchasing, 152–53
foreign control of resources, 173–74, 197
foreign workers, 128

forests
    deforestation in Pine Point, 142, 143
    health of in Fort McKay area, 213
    used in mining, 8, 60, 64, 89, 96
Fort Franklin. *See* Déline
Fort McKay, Alberta
    environmental impacts on, 210–13, 214–16
    excluded from Alsands project hearings, 216–17
    at GCOS expansion hearings, 213–14
    and oil sands development, 208–9
    in the oil sands workforce, 219–22
    partnership with Syncrude, 223
Fort McKay First Nation, 221, 225
Fort McMurray, Alberta, 209, 211, 212
Fort McMurray First Nation, 221, 225
Fort Resolution, Northwest Territories
    compared to Pine Point, 150–51
    concerns about Pine Point Mine, 141–48
    excluded from economic benefits, 139, 145–46, 154
    mine reopening assessment hearings, 157
    mixed feelings about Pine Point Mine, 140–41, 156
    quality of life changes, 152–53
    social changes, 149
Foster, Terrence W., 50
Fox, Bonnie, 118
Francaviglia, Richard V., 9
Frederiksen, Tomas, 37–38

G

Gagnon, Quebec, 178
Galbraith, Lindsay, 263
Garrow Lake tailings disposal, 324, 325, 332
GCOS. *See* Great Canadian Oil Sands Ltd. (GCOS); Suncor
generational differences in views, 80–81, 112–13
Geological Survey of Canada, 66
Giant Mine, **12**
    arsenic byproducts, 342–44
    arsenic storage proposals, 350–51, 352, **360**
    environmental assessments, 347
    environmental issues, 8

heritage preservation, 10
maintenance and remediation costs, 346, 353–54
Oversight Body, 365, 367
remediation proposals, 12, 355–65
roaster demolition, 362
water pollution concerns, 349
Gibson, Virginia Valerie, 19, 22–23
Gilchrist, W. M., 344
GN. *See* Nunavut
GNWT. *See* Northwest Territories
Gogal, Sandra, 264, 273, 274, 275, 277
Gold Roast Discharge Control Regulations, 348
Gordon, Sarah M.
biography, 383–84
chapter, 18, 59–85
Grace, Sherrill, 68
Great Bear Lake. *See also* Sahtúot'ı̨nę
Dene settlements, 59
Gilbert Labine at, 66–67, 70
as "uninhabited," 75
Great Canadian Oil Sands Ltd. (GCOS). *See also* Suncor
development, 208, 209
ERCB hearings, 213–14
tailings pond seepage, 211
Great Slave Lake, 138, 341
Green, Heather
biography, 384
chapter by, 21, 315–39

H

Hager, Dave, 88, 92, 95–96, **95**, 107
Haida Nation vs. British Columbia, 272
Hale, Charles, 235
Halvaksz, Jamon Alex, 11, 190, 198
Hamilton, Paula, 320
Hammond, Jane
biography, 384
chapter by, 19, 117–35
Harper, Stephen, on vision for the North, 1
harvesting. *See also* hunting and trapping
Fort McKay, 213
Na-Cho Nyäk Dun, 89, 90, 97–98
Pine Point area, 143
Schefferville Innu, 175
Yellowknives Dene, 344

Hay River Reserve. *See* K'atl'odeeche First Nation
health impacts
Alberta oil sands pollution, 211–13, 214–15, 215–17, 225
demolition of hospitals, 181, 188–89
from mine development, 8–9
from mining work, 99–100, 110–11
Pine Point Mine, 146–48
Port Radium, 64–65, 71–72
Yellowknife gold mines, 344, 347
Hebron, Newfoundland and Labrador, 238, 250
Hickling-Partners Inc., 297
Hills of Silver (Aho), 96
Hiroshima in Délı̨nę narrative, 72–74
historical political ecology framework, 7
historical records
Délı̨nę community, 60
exclusions of Aboriginal peoples, 14, 74–76
history societies, 9–10, 101, 138
Hitch, Michael, 272
HNSE. *See* Hollinger North Shore and Exploration Company (HNSE)
Hogan, Joella
biography, 384
chapter by, 18–19, 87–116
Hollinger North Shore and Exploration Company (HNSE), 173, 176
Horatio Alger myth, 68, 75
housing
Labrador City, 123–24
Rankin Inlet, 42–44, **43**
Howlett, Cathy, 268–69
Human Health and Ecological Risk Assessment (HHERA), 301–2
Hummel, Kaylie-Ann, 94, 100, 102
hunting and trapping. *See also* land-based living
activities maintained, 10, 101–2, 107–8, 153, 158
retaining connections to, 42, 45–46, 220, 323–24
territories damaged, 8, 139, 142–44, 175, 183, 210, 214, 298–99
Hydro-Quebec, 191
hydroelectric dams, 8, 90, 175

I

IAA. *See* Indian Association of Alberta (IAA)
IBAs. *See* impact and benefit agreements (IBAs)
identity
    with mining, 9, 11, 36, 51–52, 139–40
    and solidarity challenges, 240, 241
    with territory, 184–87
Idle No More movement, 267
Iglukak, David, 48
impact and benefit agreements (IBAs)
    and government clawbacks, 278
    negotiations, 262–64
    private nature of, 260–62
    as private resource deals, 272–75
    role for, 279, 281–82
    as solution to social problems, 275–77
    undermining Aboriginal rights, 270–71
    and unplanned events, 277
    use of, 10, 145
INAC. *See* Indian and Northern Affairs Canada (INAC)
incomes
    clawbacks from, 49, 278
    from mining work, 152
    pensions, 180–81
    wage gaps, 44
    workers compensation, 100
Indian Act (1876) on enfranchisement, 97
Indian and Northern Affairs Canada (INAC), 261, 300. *See also* Aboriginal Affairs and Northern Development Canada (AANDC); Canadian government
Indian Association of Alberta (IAA), 217, 218, 222
Indian Eskimo Association (IEA), 322
Indian steel industry, 170, 192–93
indigenous peoples
    in colonial and neocolonial processes, 81
    experiences with mining, 6–7, 37–38, 61
    experiences with the state, 266, 280
    relationship with environment, 61–62
infrastructure. *See also* hydroelectric dams; roads and railways
    access to markets, 174–75, 193
    access to resources, 191
    in boom and bust cycles, 1, 4
    demolitions of, 181, 188–89
    linking communities, 4, 104, 149
    negotiations in IBAs, 275
    subsidization of, 7–8, 138, 295, 296
    value after mine closure, 301
Innis, Harold Adams, 5–6, 139
Innu
    combining mine work with subsistence, 190–91
    feelings after mine closure, 179–81, 186–87, 188–89, 197–98
    identification with territory, 184–85
    on landscape damages, 181–84
    lawsuit against IOC/Rio Tinto, 198
    on life in Schefferville, 187–88
    relocation to Schefferville, 175
    on reopening of mines, 199
Inuit and qallunaat (non-Inuit) relations, 45, 46–47
Inuit Tapirisat of Canada (ITC), 322, 324–26
IOC. *See* Iron Ore Company of Canada (IOC)
Ipkarnerk, Peter, 44, 45, 50, 51
Irniq, Piita, 52
iron and steel industry, 170, 177–78, 191–94
Iron Ore Company of Canada (IOC)
    founding of, 174, 177
    gender equality actions, 127, 128
    Innu and Naskapi criticisms of, 180, 184, 188–89
    Innu lawsuit against, 198
    labour division by gender, 119–22, 126–27, 129
    Labrador City housing policies, 123–24
    Schefferville mine closure, 178, 180–82
    training and recruitment, 127–28
Isaac, Thomas, 262
ITC. *See* Inuit Tapirisat of Canada (ITC)
Ittinuar, Ollie, 41, 50
Ittinuar, Peter, 47, 48

J

John, Angela, 118
Joint Oil Sands Monitoring Program, 225
justice

law courts on affirmative action hiring, 221–22
lawsuits, 198
pollution charges, 216
in reciprocity and exchange values, 71–72, 78–79
Justus, Roger, 219–20

K

Kablunângajuit (people of mixed Inuit-settler ancestry), 240, 241, 251
Kabvitok, Jack, 45, 49
Kangiqiniq. *See* Rankin Inlet (Kangiqiniq), Nunavut
Kapuk, Francis, 41, 42, 49
K'atl'odeeche First Nation, on mine impacts, 141, 145, 146, 148, 156, 157
Kavik, Joachim, 41, 42, 50
Keeling, Arn
  biography, 384
  chapter by, 18, 35–58, 377–81
  comments, 317
Keewatin (Kivalliq) region, 41
Keewatin Journal (Smith, ed.), 42
Kennett, Steven, 262–63, 274–75
Kenny, Andrew John "AJ" and Dennis, 70–72, 81
Keno City, Yukon, 10, 88, 104
Keno Hill mining district
  closure, 90, 102
  development, 2, 88–89
  environmental impacts, 103
  history preservation, 10
  imported workforce, 107
  redevelopment, 104–5
  as zombie mine, 12
Kitts-Michelin project, 239, 242–43, 245
Kivalliq (Keewatin) region, 41
Klondike gold rush, 62, 88, 342
Knox, Anthony, 262
Kovach, Margaret, 93
Krogman, Naomi, 261, 270, 282
Kuokkanen, Rauna, 268, 281

L

Labine, Gilbert
  Eldorado Gold Mines Ltd., 63
  narratives about, 66–69, 70–71, 76
Labrador, governance by Newfoundland, 236–37
Labrador City, Newfoundland and Labrador
  development of, 117, 120
  employment of women in, 119–21
  housing, 123–24
  iron mining, 4
  recruiting challenges in, 127–28
Labrador Inuit
  before resettlement, 236
  Brinex response to, 242
  debate on resource development, 245–49
  founding of LIA, 240–41
  lingering social issues, 250–51
  relocations and resettlement, 237–39
Labrador Inuit Association (LIA), 240–41, 243–44
Labrador Trough
  development potential, 170, 177, 191–92, 196–97
  exploration and leases, 173–74
Labrador West Status of Women Committee, 126
Lafferty, Cecil, 142–43
Lafferty, Priscilla, 148
Lake Athabasca, 209, 215
land and resources. *See also* mineral resource development
  Canadian authority over, 342
  in colonialism, 74–75
  indigenous relationship to, 61–62
  Sahtúot'įnę ethics on using, 78–79
  self-sufficiency with, 252
land-based living. *See also* fish and fishing; harvesting; hunting and trapping
  policies to phase out, 7, 175, 190–91, 218, 238
  precariousness of, 38, 39, 40, 42
  resumption after mine closure, 48–49, 51–52, 90, 298–99
  way of life protection, 240–42, 246–47
land claims. *See also* comprehensive land claim agreements (CLCAs)

IBAs overcoming unresolved, 263–64, 274
Labrador Inuit, 240, 243–44, 245, 248
Na-Cho Nyäk Dun, 103–4
Nunavut, 322
landscapes of mining
   creation of, 8, 142–44, 181–85
   identification with, 9, 36, 185
   out of sight and mind, 320, 331–32
   as site of scientific knowledge, 308–9
language and cultural barriers
   in oral history research, 15
   overcome at work, 41, 44–45
Laurier, Sir Wilfred, on northern expansion, 69
lawsuit against IOC/Rio Tinto, 198
Leddy, Lianne, 93
Leetso (Navajo monster), 61
Levitan, Tyler
   biography, 385
   chapter by, 20, 259–90
LIA. *See* Labrador Inuit Association (LIA)
Little Cornwallis Island, 322, 330, 331
Lizotte, Garvin, 150, 151–52, 157
Longley, Hereward
   biography, 385
   chapter by, 207–32
Lorax Environmental Services, 302
Luxton, Meg, 118
Lynn Lake, Manitoba, 8, 50

M

M. A. Hanna Company, 173, 176
MacDonald, Dorothy, 215, 216–17
MacDonald, Fiona, 266, 268
McKay (Quebec) Explorers (MQEC), 173
Mckay, Darrin, 153, 155
McKay, Denise, 148, 154
McKay, Eddy, 150–51, 154
McKay, Henry, 146
McKay, Linda, 150
Mckay, Lorraine, 150, 151, 157
McKay, Ron, 149, 151
McKenzie, Réal, 189, 198
Mackenzie Valley Environmental Impact Review Board, 360–62, 363–64, 366
Mackenzie Valley Pipeline Inquiry

on Pine Point Mine, 139, 141–42, 144
   and rise of Aboriginal rights movement, 240
Mackenzie Valley Resource Management Act, 358
McPherson, Robert, 322
McQuesten area, 88, 109
Makkovik, Newfoundland and Labrador, 237, 238–39
Mandeville, Melvin, 148, 154
Manhattan project uranium supply, 63–64
Manilak, Veronica, 41, 42, 48, 49
Manitoba mine development, 4
marriages
   breakdowns, 126
   to non-Natives, 92, 96, 97
   women's dependence on, 123–24, 127
Marsh, Ben, 36, 294, 317
Matimekush-Lac John (Schefferville), 178, 198
Mayo, Yukon
   after mine closure, 102, 104
   development of, 88–89
   hydroelectric dam, 90
   Na-Cho Nyäk Dun citizens in, 97
Meadowbank Gold Mine development, 1
Melancon, Bobbie-Lee, 94, 96, 109
Melancon, Herman
   about, 87, 92–93, **92**, 96–97
   on hunting and trapping, 101–2, 108
   interview approach, 93–94
   on living in Elsa, 106
   on mining work, 98–101, 102, 105, 107–8, 110–11
   views on mining, 106–7, 109, 111–12
Melancon, Maurice, 92, 96–97
Mellor, M., 317
memories negative and positive juxtaposed, 141, 149–50, 154–56, 187–88
Metheny, Karen, 16
Midgley, Scott
   biography, 385
   chapter by, 20–21, 293–314
migration experiences
   Labrador Inuit, 237–39
   Rankin Inuit, 41
Mikisew Cree First Nation, 221, 225, 268
Millennium Iron Range, 193

mine closure impacts
    Keno Hill, 90, 102, 104
    Nanisivik, 293–94
    Pine Point, 139, 142, 154
    Rankin Inuit, 47–51
    Schefferville, 179–84, 186–89
mine closures. *See also* CanZinco Ltd.; Nanisivik
    nature of, 52, 315–18, 320–21
    process, 294
Mine Site Reclamation Policy for Nunavut, 300
mineral resource development
    Canadian visions and policies, 1, 7–8, 69, 138, 173–76, 197, 321–22
    as colonialism, 7, 8, 17–18, 37–38, 81
    community debates on, 80–81, 109–10, 111–13, 245–51, 252
    and IBAs, 260, 262, 270–71, 272, 273–74
    Nanisivik as test in High Arctic, 295–96
    need for reflective examination, 377–81
    Plan Nord, 169–72, 189–92, 194
    policy alternatives, 194–96
    studies on community impacts, 5–6
Mineral Resources International (MRI), 295, 296
mining heritage, 316, 317
"mining imaginary," 2, 294, 303, 308, 309
mining industries. *See also* boom and bust cycles; redevelopment of mines
    access to markets, 174–75, 193
    gold markets, 342, 351
    history in northern Canada, 1–4
    iron and steel, 177–78, 191–94
    oil sands in Alberta, 208, 209–10, 224
    sustainability questions, 379–80
    uranium, 63–64, 239, 243, 244, 249, 251
mining work
    barriers to getting, 107, 220–23, 327
    experiences, 44–45, 99–100, 107–8
    gender relations in, 122–23, 129, 130
    health impacts of, 110–11
    as lifelong vocation, 49–50, 97
Miramar Mining Corporation, 353, 354
modernization of the North
    Labrador Inuit, 238, 247, 251
    Pine Point Mine, 138
    Rankin Inuit, 38, 39–40, 47, 50

Resolute Bay, 322
Schefferville, 175, 190
Momaday, N. Scott, 60–61
money
    Dave Hager's remark on, 96
    Sahtúot'įnę ethic on, 78
Moore, Harold, 146
Moses, David, 89–90
motherhood and employment, 124–25
MQEC. *See* McKay (Quebec) Explorers (MQEC)
MRI. *See* Mineral Resources International (MRI)
Mulroney, Brian (president of IOC), 178

N

Na-Cho Nyäk Dun
    adaptation to mine development, 87, 89–95, 97–98, 101
    after mine closure, 102
    mining development debates, 111–13
    self-governance, 103–4, 105
    territory, 88
Na-Cho Nyäk Dun Development Corporation, 104, 105
Nanisivik
    baseball games with Polaris team, 330
    community concerns about reclamation, 305–6
    community responses to mine closure, 293–94, 297–99
    development of, 295–97
    field monitoring program, 304
    Inuit employment agreements, 326
    Inuit employment at, 50
    reclamation cost valuation, 300–303
    as scientifically valued minescape, 303, 308–9
    as zombie mine, 12
narratives
    as bridges to understanding, 93–95
    as corrective to boosterism, 137–38
    as reflection of values, 60–62
Nash, June, 61
Naskapi
    feelings after mine closure, 179–81, 186–87, 188–89, 197–98
    relocation to Schefferville, 175

National Energy Program conditions on hiring, 222
National Film Board, 10, 35, 139–40
Navajo mines, 61
neoliberalism
   against self-determination, 268
   features of IBAs as, 269–70
   and Labrador Inuit goals, 233, 235, 247–48, 251, 252
   and self-determination goals, 266–67, 279
   and the state, 265–66, 280–81
New Millennium Iron (NML), 193
newcomers influx, 89, 92, 96, 97, 107
Newell, Eric, 223
Newfoundland
   governance of Labrador, 236–37
   land claims negotiations, 243–44
   on protecting traditional way of life, 241
   women in the workforce, 119, 121, 127
Newfoundlanders, in Pine Point, 148, 151
Neyelle, Morris, 77–78, 79
NG. *See* Nunatsiavut (NG)
"Nichole Churchill," 120
non-Inuit (qallunaat) and Inuit relations, 45, 46–47
North Rankin Nickel Mine (NRNM), **36, 37**
   closure, 47–48
   history, 2, 35, 38, 39
   housing and settlement policies, 42–44
   post-closure influence, 317
   recruitment of Inuit, 40–41
   segregation policies, 46–47
   workplace, 44–46
   as zombie mine, 12
North Slave Métis Alliance, 141
Northern Affairs. *See* Department of Northern Affairs and National Resources
northern development. *See also* mineral resource development; modernization of the North
   alternative policies, 194–96
   Canadian policies, 7–8, 69, 138, 176, 321–22
   Plan Nord, 169–72, 189–92, 194
   visions and realities, 1, 2, 4
"northern narrative," 68–69

Northern Transportation Company Ltd. (NTCL), 64
Northwest Territories
   environmental legislation attempts, 347–48
   Giant Mine remediation plans, 357
   and IBAs, 275, 278
   liability for Giant Mine, 353–54
   Socioeconomic Action Plan, 323, 325
   subsidies to Miramar, 354
   subsidies to Royal Oak Mines, 351–52
Northwest Territories Water Board
   Giant Mine negotiations, 349, 351
   Pine Point reclamation plan hearings, 142–43
   Polaris development, 324–25
Notley, Grant, 215, 219
NRNM. *See* North Rankin Nickel Mine (NRNM)
Nunatsiavut (NG), 244–50, 251
Nunavut
   concerns about abandoned projects, 299–300
   creation of, 322
   Nanisivik closure and reclamation plan, 300, 302
Nunavut Tunngavik Incorporated, 271
Nunavut Water Board (NWB), 300
Nutak, Newfoundland and Labrador, 238, 250
NWB. *See* Nunavut Water Board (NWB)
NWT Mining Heritage Society, 9–10

O

O'Faircheallaigh, Ciaran, 264, 278
oil sands industry in Alberta
   Alsands project hearings, 216–17
   and communities, 221–22, 223, 225
   development of, 207–8, 209–10, 224
   GCOS expansion hearings, 213–14
Ontario mine development, 4
OPEC. *See* Organization of the Petroleum Exporting Countries (OPEC)
open pit mining, xi, 109, 117
Oqalluk, Leah, 303
oral history research methods
   Délı̨nę community, 62

Kangiqiniq, 38–39
Labrador City, 118, 131n5
Na-Cho Nyäk Dun, 93–95
objectives and experiences, 13–17, 138
Pine Point, 140–41
Resolute community, 320
ore concentrates, xii, 327
O'Reilly, Kevin, chapter by, 21, 341–76
biography, 385–86
Organization of the Petroleum Exporting Countries (OPEC), 209
Oyukuluk, Kunuk, 298

P

Padlirmiut starvation crisis, 44
Pamack-Jeddore, Rosina, 238–39
Papillon, Martin, 268
Paquette, Pierre, 176
Pearson, Lester B., on northern development, 69
Peck, Jamie, 265
Peet, Fred J. "Tiny," 76
pensions, 180–81
"People of the Rock" (National Film Board documentary), 35
permafrost as pollution containment
Garrow Lake, 304
Yellowknife, 344
Pine Point, Northwest Territories
after mine closure, 142, 154
commemoration of memories, 10, 139–40
concerns about Pine Point Mine, 141–48
hydroelectric dams, 8
quality of life in, 150–53, 155–56
racism in, 148
Pine Point Mine, **140, 145, 147**
community feelings about, 8, 140–50, 154–56
development of, 2, 138
employment at, 50
mine closure, 142
reopening project, 156–57
as zombie mine, 12
"Pine Point Revisited," 139–40
pitchblende discovery, 63, 66–68, 70, 71
Plaice, Evelyn, 241

Plan Nord, 169–72, 189–92, 194
Pochon, Marcel, 75
Polaris Mine, **328**
closure, 329, 331
development, 322–27
operations and employment, 318, 327–28
Resolute community and, 316
"policy of dispersal," 39
political activism
CARC and ITC on Polaris Mine, 324–26
movements in Northern Canada, 322
pollutants. *See* toxins and pollutants
Poloski, Murray, 242
Port Radium
in historical records, 75–76
history preservation, 10
mine history, 59–60, 63–64
narratives on, 66–69, 70, 72, 74
pollution and radiation, 8, 64–65
remediation, 79–80
as symbol of colonialism, 61, 80–82
as zombie mine, 12
Power, Rosemary, 316
Prno, Jason, 262
Procter, Andrea
biography, 386
chapter by, 232–58

Q

qallunaat (non-Inuit) and Inuit relations, 45, 46–47
QNS&L. *See* Quebec North Shore and Labrador Railway Company (QNS&L)
Quebec
demolition of Schefferville, 181, 188–89
infrastructure investments, 191
mineral development policy, 173–76, 197
Plan Nord, 169–72, 189–92, 194
resource economies, 176, 177–78
Quebec–Labrador iron ore region, 6, 10. *See also* Labrador Trough; Millennium Iron Range; Ungava region
Quebec North Shore and Labrador Railway Company (QNS&L), 174–75, 198

R

racism memories, 148, 151
Rankin Inlet (Kangiqiniq), Nunavut
    government-sponsored enterprises, 49
    memories of mining, 35–36, 38–39
    mine closure, 47–48
    resilience and identity, 10, 50–51, 58n50, 317–18
    segregation policies, 46–47
    townsite design, 42–44
Rankin Inuit, 35–36
    adjustment to mine closure, 47–51
    government policies towards, 39–40
    identity and resilience, 51–53
    migration and recruitment, 41–42
    in the mining workforce, 38, 40–41
    settlement experiences, 42–44, **43**
    social life, 46–47
    workplace relations, 44–46
RCMP. *See* Royal Canadian Mounted Police (RCMP)
Rea, Kenneth J., 139
Reclamation Security Trust, 353
reclamations
    definition, xii
    as mining project milestone, 297
    Nanisivik cost valuations, 300–308
    in Strathcona Agreement, 296
recreation and entertainment
    Pine Point, 140, 151
    Polaris Mine, 330
    Rankin Inlet, 46–47
    Schefferville, 187–88
redevelopment of mines
    implications to communities, 11–12
    Keno Hill, 104–5
    Labrador Trough, 189–94
relocation programs
    Labrador Inuit, 237–39, 250–51
    Na-Cho Nyäk Dun, 90
    Naskapi, 175
    Padlirmiut, 44
    Rankin Inuit, 50–51
remediation
    costs, 11–12
    definition, xii
    Giant Mine, 346, 350–51, 353–54, 355–65
    implications to communities, 11–12
    Keno Hill, 104
    political action suggestions, 365–66
    Port Radium, 65, 79–80
    Schefferville, 182
Rennie, Rick, 119
research framework, 6–7, 10
residential schooling, 80, 90, 221, 266
resilience of mining communities, 9–11, 158–59
Resolute Bay (Qausuittuq), Nunavut
    collective mining memory, 316, 320–21
    criticisms of Cominco, 330–31
    employment expectations, 326–27
    experiences at Polaris Mine, 321, 330
    meetings with Cominco, 322–24
resources. *See* land and resources; mineral resource development
Rice, Terry, 246
roads and railways
    Fort Resolution, 149, 153
    Pine Point Mine, 138, 142, 143
    Schefferville Mine, 174–75
    Silver Trail, 104
    Yukon, 90
Robertson, David, 294
Robertson, R. Gordon, 40, 47
Rollwagen, Katharine, 52
Royal Canadian Mounted Police (RCMP) at Rankin Inlet, 39, 40
Royal Commission on Aboriginal Peoples hearings, 143–44
Royal Commission on the Status of Women, 120
Royal Oak Mines
    arsenic removal and reprocessing proposal, 349–50
    failing finances, 351–52
    labour tragedy, 347
    receivership, 343, 352–53
    resistance to air quality regulations, 348
    water license negotiations, 349
Rumily, Robert, 173

S

Sahtú landscape, 59, 61, 75–76
Sahtúot'įnę
  exclusions from historical accounts, 74–76
  mine impacts on, 60, 64–65
  narratives of Port Radium origin, 69–74
  reciprocity and exchange ethic, 77–79
  on remediation efforts, 65, 79–81
St. Paul, Charles, 67–68
Sandlos, John
  biography, 386
  chapter by, 19, 137–65, 377–81
Saskatchewan mines, 4, 17
Sawmill Bay, Northwest Territories, 59, 79
Sayine, Robert, 142
Schefferville, Quebec
  after mine closure, 181, 188–89, 197–98
  barricades at, 198
  community resilience, 10
  quality of life in, 187–88
  relocations to, 175
Schefferville Mine, 180–81, **182**, **183**, **184**, **186**
  closure, 178, 179–81
  development, 4, 174–75
  economic value, 177–78
  landscape impacts, 181–84
  as zombie mine, 12, 198
scientific expertise, as authoritative and reliable, 307–8
scientific methods in reclamations
  modelling for tailings cover, 306
  using evidence for cost valuation, 301–3
Searle, John R., 60
security bonds
  adequacy of, 380–81
  Giant Mine remediation, 351, 353
  Nanisivik reclamation project, 300–303
Seddon, Vicky, 118
segregation policies, 42–44, 46–47, 98
Sept-Îles, Quebec, 175, 193, 198
settlement
  Makkovik, 238–39
  Na-Cho Nyäk Dun, 89
  Rankin Inuit experiences, 42–44, **43**
  Yellowknives Dene First Nation, 342

Shoebridge, Paul, 139
Shopes, Linda, 15, 320
Shore, Cris, 250
Sidbec-Normines (mine), 178
Silver Trail (Yukon), 104
Simonetta, Joanne, 219–20
Simons, Michael, 139
Singiituk (Inuk foreman), 40, **40**, 48, 55n16
Slocombe, Scott D., 262
Slowey, Gabrielle, 235, 266, 267
Smallwood, Joey
  at Labrador Conference, 237
  Terms of Union agreement, 236
Smith, Kitty, 87
Smith, Linda Tuhiwai, 252
snow blindness, 67–69
Sǫ́bak'e (the money place), 59
social impacts
  Arctic Bay Inuit, 298–99
  Fort Resolution and Pine Point, 148, 149, 150–53, 154, 156
  Keno Hill, 107, 111–13
  Labrador Inuit relocations, 238–39, 241, 250–51
  of mine development, 4, 8
social license to operate, 194, 260
social life
  Labrador City, 122–23, 128
  Pine Point, 150–52, 155–56
  Rankin Inlet, 46–47
  Resolute, 330
  Schefferville, 187–88
Socioeconomic Action Plan, 323, 325
Sonnefrere, Daniel, 156
South Slave region. *See* Pine Point Mine
southern Canadians
  comments on fate of Rankin Inuit, 47
  cultural views of achievement, 75, 76–77, 81
  disconnected from realities of mining, 380
  and "northern narrative," 68–69
  perceptions of northern Canada as underdeveloped, 7–8
southern influences on communities, 149, 152, 156
Ste Marguerite River, 175
steelmaking industry, 177–78, 191–94

Stephenson, C., 317
Stern, Pamela, 51
Stewart River, 89, 96
stories. *See* narratives
Strathcona Agreement, 296–97
Strathcona Sound dock
    contaminants at, 305, 327
    investment in, 295, 296
subsidies
    infrastructure investments, 7–8, 138, 295, 296
    Northwest Territories to Miramar, 353
    Northwest Territories to Royal Oak Mines, 351–52
subsistence living. *See* land-based living
sulphur dioxide emissions at Giant Mine, 347
Suncor. *See also* Great Canadian Oil Sands Ltd. (GCOS)
    pollution spill (1981), 214–16
Syncrude
    employment of indigenous people, 218, 219–20, 223
    environmental impact assessments, 210
    founding of, 209
    particulate emissions, 212
    partnership with communities, 223

T

tailings and tailings ponds
    cover for Nanisivik, 303–8
    definitions, xiii
    GCOS, 211
    Giant Mine, 344, 349
    Pine Point Mine, 142, 143, 148
    Polaris Mine (Garrow Lake), 324, 325, 331–32
    uranium in, 63
Tamerlane Ventures Pine Point Mine project, 156–57
Tata Steel (Mumbai), 192–93
temporary foreign workers, 128
Thiobacillus bacteria in tailings, 303, 304
Timmins, Jules, 173, 175
tourism development, 104
Towtoongie, John, 41, 46, 49

toxins and pollutants. *See also* tailings and tailings ponds
    Fort McKay area, 210–12
    Giant Mine, 342–44, 347, 348, 362
    as loaded terminology, 302
    mentions in oral history accounts, 138
    permafrost as containment of, 304, 344
    Pine Point Mine, 146–48
    Port Radium, 64–65
traditional lifeways. *See* land-based living
Trigger, David S., 322
trust and distrust
    on protection of environment, 305–6, 348
    in remediation process, 80, 362, 366
Tudlik, Thomas, 41, 42, 45
Twardy, Stanley, 47

U

Uashat mak Mani-utenam (Sept-Îles), 198
UKHM. *See* United Keno Hill Mines Ltd. (UKHM)
unemployment insurance, 49
Ungava region, 172, 173–78. *See also* Labrador Trough
Union Miniére (Belgium), 63
union participation in education programs, 127
United Keno Hill Mines Ltd. (UKHM), 90, 98, 102
United Nations Declaration on the Rights of Indigenous Peoples, 267
United States
    iron and steel industry, 177, 196–97
    uranium oxide demand, 63–64
Unka, Bernadette, 143–44
Unka, Tommy, 149, 156
uranium
    Brinex mining proposal, 240
    discovery in Labrador, 236–37
    industry and markets, 63–64, 239, 243, 244, 249, 251
Uranium City, Saskatchewan, 8, 10, 209

V

values
  conveyed in community debates, 80–81, 109–10, 111–13, 245–51, 252
  conveyed in narratives, 60–61, 81
  of minerals versus development costs, 377–81
  on reciprocity and exchange, 78–79
  regarding hard work, 51
  regarding land protection, 111–13
Villeneuve, Greg, 145
Voisey's Bay nickel deposit, 244
volunteer work, 128

W

wage labour
  attitudes towards, 51
  as economy modernization, 39–40, 138, 175, 237–38, 276–77, 296
  indigenous experiences with, 10, 44, 45, 50
waste rock, **140**
  definition, xiii
water quality
  Athabasca River, 213–14
  Fort McKay area, 211, 212–13, 214–16
  Keno Hill, 103
  Pine Point Mine, 142–43, 146–48
  Port Radium, 63, 64
  Yellowknife, 344, 349
Weber, W.W., 40–41
welfare payments, 49, 278
Wernecke Mountains, 110
Whatmough, Ken, 40
Williamson, Robert G., 46, 50
Wilson, Shawn, 93
Winton, Alexandra
  biography, 386
  chapter by, 18–19, 87–116
WISCO International Resources Development & Investment (WISCO), 193–94
Wiseman, Alison, 124
Wolfden Resources, Nanisivik, 301, 308
women in mining
  labour divisions, 119–21
  mine worker relations, 122–23
  models for studying, 118
  push for opportunities, 126–27, 128–29, 130
  social attitudes about, 121–22, 124–26
Women's Centre (Labrador City), 126
workers compensation, 100
working mothers, 124–25
workplace culture
  Carol Lake Mine, 129, 130
  North Rankin Nickel Mine, **36**, **37**, 45
  Schefferville Mine, 187
Wright, Susan, 250
Wyckoff, William, 294

Y

Yellowknife, Northwest Territories
  air quality concerns, 347
  area history, 2, 341–42
  community today, 9–10, 59
  concessions and subsidies, 351–52, 353, 354–55
  disappointment with remediation planning, 357–58
  principles for remediation plan, 356
  push for pollution regulations, 348
  push for review process, 358–59
  recommendations accepted, 363
  water pollution concerns, 349
Yellowknives Dene First Nation
  excluded from workshops, 350
  government briefing of, 357–58
  health impacts from gold mines, 344
  lands and settlement, 341–42
  loss of trust in government, 348
  push for review process, 358–59
  recommendations accepted, 363
youth
  disconnected from the land, 78
  employment in mining towns, 127, 128
Yukon
  agreements with First Nations, 103
  economic development, 104
  employment, 96

Z

"zombie" mines, 11, 12, 198

www.ingramcontent.com/pod-product-compliance
Lightning Source LLC
Chambersburg PA
CBHW042117300426
44117CB00021B/2975